T0253048

Mechanical Vibrations

Tony L. Schmitz • K. Scott Smith

Mechanical Vibrations

Modeling and Measurement

 Springer

Tony L. Schmitz
Department of Mechanical Engineering
and Engineering Science
University of North Carolina at Charlotte
Charlotte, NC, USA
tony.schmitz@uncc.edu

K. Scott Smith
Department of Mechanical Engineering
and Engineering Science
University of North Carolina at Charlotte
Charlotte, NC, USA
kssmith@uncc.edu

Please note that additional material for this book can be downloaded from
http://extras.springer.com

ISBN 978-1-4614-0459-0   e-ISBN 978-1-4614-0460-6
DOI 10.1007/978-1-4614-0460-6
Springer New York Dordrecht Heidelberg London

Library of Congress Control Number: 2011934974

Printed on acid-free paper

Springer is part of Springer Science+Business Media (www.springer.com)

To our children, Jake, BK, Kellye, and Kyle.

Preface

In this textbook, we describe essential concepts in the vibration analysis of mechanical systems. The book incorporates the required mathematics, experimental techniques, fundamentals of modal analysis, and beam theory into a unified framework and is written to be accessible to undergraduate students, researchers, and practicing engineers alike. We based the book on undergraduate courses in mechanical vibrations that we have previously offered and developed the text to be applied in a traditional 15-week course format. It is appropriate for undergraduate engineering students who have completed the basic courses in mathematics (through differential equations) and physics and the introductory mechanical engineering courses including statics, dynamics, and mechanics of materials.

We organized the book into nine chapters. The chapter topics are summarized here.

- Chapter 1 – We introduce the types of mechanical vibrations, damping, and periodic motion.
- Chapter 2 – We explore topics in single degree of freedom free vibration, including the equation of motion, the damped harmonic oscillator, and unstable behavior.
- Chapter 3 – We introduce single degree of freedom forced vibration and discuss the frequency response function, rotating unbalance, base motion, and the impulse response.
- Chapter 4 – We extend the Chap. 2 analysis to consider two degree of freedom free vibration. This includes the eigensolution for the equations of motion and modal analysis.
- Chapter 5 – We extend the Chap. 3 analysis to consider two degree of freedom forced vibration. We describe complex matrix inversion, modal analysis, and the dynamic absorber.
- Chapter 6 – In this chapter we analyze model development by modal analysis. This incorporates the peak picking approach for identifying modal parameters from a system frequency response measurement and mode shape measurement.

- Chapter 7 – We describe frequency response function measurement techniques in this chapter. Impact testing is highlighted.
- Chapter 8 – The topic of this chapter is continuous beam modeling. Closed-form frequency response function expressions are developed for transverse beam vibration, torsion vibration, and axial vibration of beams.
- Chapter 9 – This chapter introduces the concept of receptance coupling, where frequency response functions (receptances) are coupled to predict assembly dynamics.

To demonstrate and unify the various concepts, the Beam Experimental Platform (BEP) is used throughout the text. Engineering drawings for the BEP are included in Appendix A so that instructors can provide their own demonstrations in the classroom. Additionally, MATLAB® programming solutions are integrated into the text through many numerical examples.

Special features of the book include: (1) MATLAB® MOJO code examples; (2) *By the Numbers* numerical solutions; (3) chapter problems and solutions, including MATLAB® code; (4) non-mathematical *In a Nutshell* explanations that summarize selected concepts in layman's terms; and (5) discussions and numerical examples of model uncertainty.

We conclude by acknowledging the many contributors to this text. These naturally include our instructors, colleagues, collaborators, and students. Among these, we'd like to particularly recognize the contributions of J. Tlusty, M. Davies, T. Burns, J. Pratt, and H.S. Kim. We also thank the reviewers of this book, including D. Blood, A. Burdge, M. Mitchell, and M. Pope, for their helpful suggestions and proofreading.

Charlotte, NC
Charlotte, NC

Tony L. Schmitz
K. Scott Smith

Contents

Chapter 1
Introduction

The last thing one discovers in composing a work is what to put first.

– Blaise Pascal

1.1 Mechanical Vibrations

The subject of mechanical vibrations deals with the oscillating response of elastic bodies to disturbances, such as an external force or other perturbation of the system from its equilibrium position. All bodies that possess mass and have finite stiffness are capable of vibrations. While some vibrations are desirable, such as the "silent ring" mode for cell phones or the simulation of belly-shaking laughter by the Tickle Me Elmo doll,[1] it is often the engineer's objective to reduce or eliminate vibrations. Examples include:

- automobile vibrations, which can lead to passenger discomfort
- building vibrations during earthquakes
- bridge vibrations due to high winds
- cutting tool vibrations during machining operations.

 IN A NUTSHELL   Vibratiorn can be thought of as a periodic exchange of potential and kinetic energy (stored energy and the energy of motion). All mechanical systems that vibrate have mass and stiffness. The mass is the part of the system that relates force and acceleration. When the mass is in motion, the system has kinetic energy. The stiffness is the part of the system that relates force and displacement. When the stiffness element is displaced, the system has potential energy. All real physical systems also possess damping, which is the part of the system that dissipates energy. Damping may be caused by friction between moving elements, flow of a fluid through a

---

[1] Both of these "desirable" vibrations are a product of the same phenomenon. See Sect. 3.5.

T.L. Schmitz and K.S. Smith, *Mechanical Vibrations: Modeling and Measurement*, DOI 10.1007/978-1-4614-0460-6_1, © Springer Science+Business Media, LLC 2012

restriction, or other means, but whatever the source, damping converts kinetic and potential energy into heat, which is lost. During vibration, energy is periodically transformed back and forth between kinetic and potential until all the energy is lost through damping.

1.2 Types of Vibrations

Mechanical vibrations can be classified into three general categories. These are: *free vibration*, *forced vibration*, and *self-excited vibration* (or *flutter*). In the following paragraphs, the three vibration types are described in more detail.

1.2.1 Free Vibration

Free vibration is encountered when a body is disturbed from its equilibrium position and a corresponding vibration occurs. However, there is no long-term external force acting on the system after the initial disturbance. When describing the motion of a vibrating body that can be modeled as a simple spring-mass-damper system, for example, free vibration results when some initial conditions are applied, such as an initial displacement or velocity, to obtain the solution to its homogeneous second-order differential equation of motion (Kreyszig 1983).

Free vibration is observed as an exponentially decaying, periodic response to the initial conditions as shown in Fig. 1.1. This periodic motion occurs at the system's (damped) natural frequency. We will discuss these concepts in more detail in Sect. 2.4. A good example of free vibration is the motion and resulting sound of a

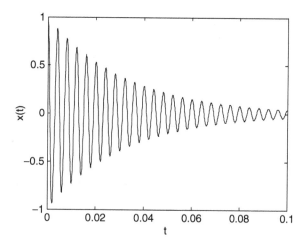

Fig. 1.1 Example of free vibration. The magnitude of the oscillating motion decays over time and the periodic vibration occurs at the natural frequency

Fig. 1.2 Forced vibration is
often described in the
frequency domain rather than
the time domain. Resonance
is identified where the forcing
frequency is equal to the
natural frequency

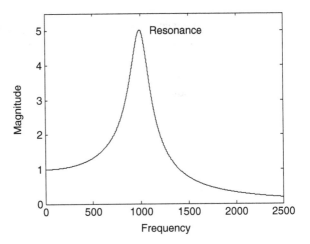

guitar string after it is plucked. The pitch of the sound (the natural frequency of
vibration) depends on the string's length and diameter; a shorter string produces a
higher natural frequency for a selected diameter, while a larger diameter string
produces a lower natural frequency for a given length. The pitch also depends on
the tension in the string; tighter strings produce higher frequencies.

1.2.2   Forced Vibration

In this case, a continuing periodic excitation is applied to the system. After some
initial transients (i.e., the homogeneous solution to the differential equation), the
system reaches steady state behavior (i.e., the particular solution). At steady state,
the system response resembles the forcing function and the vibrating frequency
matches the forcing frequency.

A special situation arises when the forcing frequency is equal to the system's
natural frequency. This results in the largest vibration magnitude (for the selected
force magnitude) and is referred to as *resonance*. Unlike free vibration, where the
response of the system to the initial conditions is typically plotted as a function of
time, forced vibration is most often described as a function of the forcing frequency.
See Fig. 1.2, where the peak corresponds to resonance.

Rotating unbalance represents a common type of forced vibration. Consider a
wheel/tire assembly on an automobile, for example. If the mass of the wheel/tire
is not distributed evenly around the circumference, then a once-per-revolution
forcing function is produced by the unbalanced mass. This periodic forcing function
(whose frequency depends on the rotating speed of the wheel/tire) can serve to
excite one of the car frame or drive train natural frequencies and can lead to
significant vibration magnitude. For this reason, it is common practice to balance
wheel/tire assemblies before installing them on a vehicle. We will discuss rotating
unbalance more in Sect. 3.5.

IN A NUTSHELL   Almost all forced vibrations are man-made. The vibration persists as long as the excitation is present. When the excitation stops, the vibration becomes a free vibration and dies away due to damping. Many readers will be intuitively familiar with the frequency-dependent nature of forced vibrations and the concept of resonance. If the automobile with the unbalanced tire is driven very slowly, then vibration is not felt. As the car speeds up, the amplitude of the vibration increases, but if the driver continues to increase the speed, then after a certain speed, the vibration amplitude diminishes. The largest vibration amplitude occurs at resonance. As another example, shower singers may have noticed that there is a particular note that is very loud when singing in a shower stall. That frequency corresponds to a resonance of the enclosure.

1.2.3   Self-Excited Vibration

Self-excited vibration, or flutter, occurs when a steady input force is modulated into vibration near the system's natural frequency. An intuitive example is whistling. Here, the steady blowing of air across your lips produces sound (vibration) at a frequency which depends on the tension in your lips (which governs the natural frequency). The diaphragm does not move at the high frequency of the sound, but rather the steady push of air is converted into a vibration of the "structure." Similarly, the steady pull of the bow across a violin string causes sound near the string's natural frequency (which, like the guitar example, depends on the string's length, diameter, and tension).

 IN A NUTSHELL   The rosin on a violin bow has an interesting characteristic – the coefficient of friction between the bow and the string changes depending on the relative velocity between them. When the string and the bow move in the same direction, the relative velocity is low and the coefficient of friction is high. Energy is put into the string motion during this phase. When the string and the bow move in opposite directions, the relative velocity is high and the coefficient of friction is low. Less energy is lost here than was input during the first part of the motion, and the difference is enough to sustain the vibration over the damping losses.

This behavior differentiates self-excited vibrations from both free and forced vibrations. Unlike free vibration, a long-term external force is present. Contrary to forced vibration, the excitation in a self-excited vibration is steady rather than periodic and the vibration occurs near the natural frequency. An example time-domain response for self-excited vibration is provided in Fig. 1.3.

Fig. 1.3 The magnitude of self-excited vibration can increase over time until limited in some way (typically by a nonlinear effect). The oscillating frequency is close to the system's natural frequency

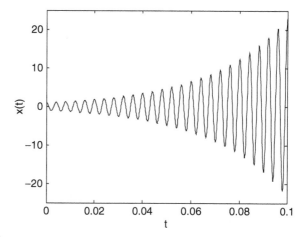

The term flutter is based on the common presence of self-excited vibrations in aeroelastic applications. It occurs due to the air (fluid) flow over wings during flight. This steady flow is modulated into vibration near the wings' natural frequency. This vibration can be minor (a small magnitude vibration manifested as in-flight "buzz") or can be catastrophic in extreme cases. The phenomenon is not limited to aircraft structures, however. A well-known example is the destruction of the Tacoma Narrows Bridge on November 7, 1940, in Washington state (http://en.wikipedia.org/wiki/Tacoma_Narrows_Bridge_(1940)).

Another common example of self-excited vibration is *chatter* in machining processes. Chatter in milling occurs when the steady forced excitation caused by the teeth impacting the workpiece (and removing material in the form of small chips) is modulated into vibration near the system's natural frequency. This modulation can occur due to the inherent feedback mechanism which is present in milling. The feedback is the result of the dependence of the thickness of the chip being removed on not only the tool's current vibration state but also on the vibration state of the previous tooth when removing material at the same angular location. Because the cutting force is proportional to the chip thickness, the variable chip thickness causes variation in the cutting force. This force variation, in turn, also affects the corresponding tool vibration. This phenomenon is referred to as "regeneration of waviness" and results in a time-delayed differential equation of motion for milling. If chatter occurs, the resulting large forces and vibrations can lead to poor surface finish and damage to the tool, workpiece, and/or spindle (Schmitz and Smith 2009).

IN A NUTSHELL   The three classes of vibration are distinguished not by the physical characteristics of the vibrating components, but rather by the nature of the excitation and the character of the resulting motion. Vibration problems have different solutions depending on the class. For example, free vibration problems are often solved by modifying the initial conditions or by increasing the damping so that the transient vibration attenuates more quickly. Forced vibration problems are

Table 1.1 Vibration classification

Vibration class	Excitation	Frequency of resulting motion	Characteristic
Free vibration	Initial conditions	(Damped) natural frequency	
Forced vibration	Persistent, periodic external source	Frequency of the external source	
Self-excited vibration	Persistent, steady external source	Self-selected and usually close to the natural frequency	

often solved by reducing the external excitation or changing the system so that the excitation is not close to resonance. Self-excited vibration problems are often solved by disturbing or eliminating the self-excitation mechanism. Table 1.1 summarizes the three vibration types.

1.3 Damping

All vibrating systems are subject to *damping*, or energy dissipation due to fluid motion, friction at contacting surfaces, or other mechanisms. This causes the response to decay over time for free vibration. In forced vibration, the input force must overcome the damping in order to sustain the constant magnitude response. We will discuss types of damping and their mathematical models in more detail in Sect. 2.4.

1.4 Modeling

As noted previously, it is often the responsibility of the engineer to model vibrating systems to determine their response to arbitrary inputs or decide how to modify a structure to mitigate the effects of a particular forcing function. Regardless, each system requires a certain number of independent coordinates to adequately describe its motion. These coordinates are the *degrees of freedom*.

For a particle in three-dimensional space, three coordinates are necessary – one for each of the three translation directions. A rigid body, on the other hand, requires six degrees of freedom to describe its motion: three translations and three rotations. See Fig. 1.4.

For an elastic body, or one that possesses mass and has finite stiffness, an infinite number of coordinates is required to fully describe its motion. To demonstrate this, let's imagine a ruler overhung from the edge of a table and clamped flat; see Fig. 1.5. We can consider each tick mark as a point on the beam. Naturally, the motion of each of these points differs as a function of time in response to the end of the ruler being displaced and released. Therefore, we require a coordinate at each of the tick marks. If we switched that ruler with another ruler of identical geometry and overhang length, but with twice as many tick marks, we could repeat the experiment and see that a coordinate was again required at each tick mark. At the

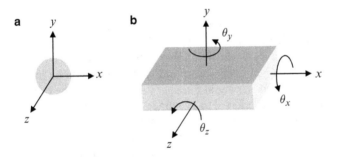

Fig. 1.4 (a) Three degrees of freedom for a particle and (b) six degrees of freedom for a rigid body

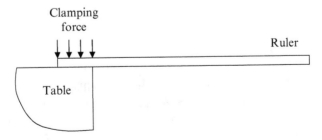

Fig. 1.5 A ruler clamped against the edge of a table

Fig. 1.6 Fundamental mode shape for a cantilever beam

Fig. 1.7 Single degree of freedom spring-mass-damper model used to approximate the cantilever beam behavior

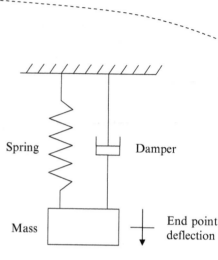

limit, we could assign an infinite number of coordinates to an infinite number of tick marks. An infinite number of coordinates, however, means we would have an infinite number of degrees of freedom. We could make the same argument about any other structure we might consider.

This poses a problem for our modeling efforts. Must all our models possess an infinite number of degrees of freedom? Luckily, it is often acceptable to use only a few degrees of freedom when describing a body's vibration behavior. For the cantilever beam (clamped ruler) shown in Fig. 1.5, we might be interested in only the lowest, or fundamental, natural frequency. The shape of the beam as it vibrates at this natural frequency, referred to as a mode shape, is shown in Fig. 1.6. If we are only interested in motion that occurs in this single mode shape (and corresponding natural frequency), we could pick a single point on the beam (at its end, for example) and describe the motion using this single degree of freedom. We could then approximate the continuous beam with a (discrete) single degree of freedom model as shown in Fig. 1.7. We will discuss how to determine the parameters for this spring-mass-damper model in Sects. 3.4 and 6.2.

1.5 Periodic Motion

To this point, we have mentioned that vibrations are oscillatory in nature. Let's now add some definitions to this discussion. A signal that repeats at regular intervals in time is referred to as *periodic*. An example is shown in Fig. 1.8. The simplest form of periodic motion is *harmonic motion*. Common examples of harmonic motion are the sine and cosine functions.

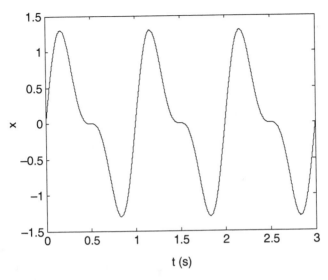

Fig. 1.8 Periodic motion example, $x(t)$. The signal repeats over an interval of 1 s. Three full periods are shown

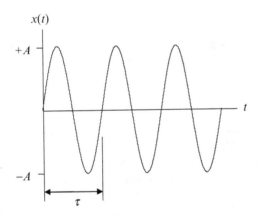

Fig. 1.9 The sine function is an example of harmonic motion

Figure 1.9 shows the sine function $x(t) = A \sin(\omega t)$, where A is the magnitude with the units of x (if x is a displacement, then units of millimeters or micrometers may be appropriate, for example) and ω is the circular frequency expressed in radians/second (or rad/s). As shown in the figure, the function repeats every τ seconds and we can write that $x(t + \tau) = x(t)$. The variable τ is referred to as the *time period* (or just the period) of the signal $x(t)$. It is related to ω (rad/s) as shown in Eq. 1.1. There is

Fig. 1.10 Argand diagram
for the function $x(t) = A \sin(\omega t)$

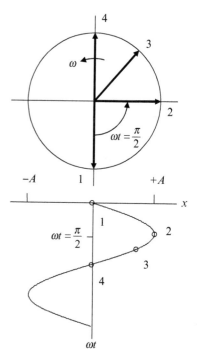

really no mystery to this relationship. It is simply based on two facts: (1) there are 2π rad per cycle of vibration; and (2) τ gives the number of seconds per cycle.

$$\tau = \frac{2\pi}{\omega} \tag{1.1}$$

IN A NUTSHELL   Many readers may be more familiar with expressing vibration frequencies in cycles/second, or Hertz (Hz). The vibration motion repeats each cycle, or whenever the sine or cosine argument changes by 2π. For that reason, there are 2π radians per cycle. We will use the Greek letter ω to represent a vibration frequency in radians/second and the letter f to indicate a vibration frequency in cycles/second. The conversion between the two is provided in Eq. 1.4.

The reason we refer to ω as the circular frequency is that we can represent $x(t)$ as a vector rotating in the complex (real–imaginary) plane with the frequency ω. The *Argand* diagram[2] in Fig. 1.10 shows that the sine function is the projection of the counterclockwise rotating vector with a length (magnitude) A on the horizontal (or real) axis. The angle of the vector at any instant is the product ωt.

---

[2] A plot of complex numbers as points in the complex plane is referred to as an Argand diagram (Weisstein 2010); we will use this description throughout the text.

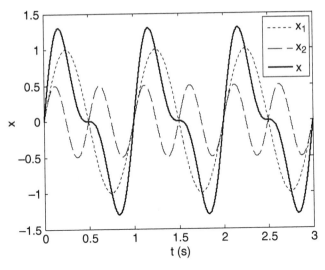

Fig. 1.11 The periodic motion in Fig. 1.8 is the sum of two sine functions, x_1 and x_2

The French physicist and mathematician Jean Baptiste Joseph Fourier made an important observation about periodic functions. He found that a periodic function $x(t)$ can be represented by an infinite sum of sine and cosine terms as shown in Eq. 1.2, where $\omega_2 = 2\omega_1$, $\omega_3 = 3\omega_1$, etc. This is referred to as the *Fourier series* and the coefficients a_0, a_1, b_1, a_2, b_2, etc., are called the Fourier coefficients (Kreyszig 1983).

$$x(t) = a_0 + a_1 \cos(\omega_1 t) + b_1 \sin(\omega_1 t) + a_2 \cos(\omega_2 t) + b_2 \sin(\omega_2 t)$$
$$+ a_3 \cos(\omega_3 t) + b_3 \sin(\omega_3 t) + \cdots \tag{1.2}$$

Let's take another look at Fig. 1.8. This signal is actually the sum of two sine functions, $x(t) = x_1 + x_2 = \sin(2\pi t) + 0.5 \sin(2 \cdot 2\pi t)$. All three signals, x_1, x_2, and $x(t)$, are shown in Fig. 1.11. Comparing this function to Eq. 1.2, we see that $\omega_1 = 2\pi$ rad/s, $b_1 = 1$, $b_2 = 0.5$, and the remaining coefficients are zero. Also, from the figure we see that the period of $x(t)$ is 1 s. Let's use MATLAB®[3] to see how we can define and plot this function. See MATLAB® MOJO 1.1.

[3] MATLAB® and Simulink® are registered trademarks of The MathWorks, Inc. For MATLAB® and Simulink® product information, please contact: The MathWorks, Inc., 3 Apple Hill Drive, Natick, MA, 01760–2098 USA, Tel: (508) 647–7000, Fax: (508) 647–7001, E-mail: info@mathworks.com, Web: www.mathworks.com.

MATLAB® MOJO 1.1

```
% matlab_mojo_1_1.m

clc
clear all
close all

% define variables
t = 0:0.001:3;                    % s
omega1 = 2*pi;                    % rad/s
omega2 = 2*omega1;                % rad/s
b1 = 1;
b2 = 0.5;

% define functions
x1 = b1*sin(omega1*t);
x2 = b2*sin(omega2*t);
x = x1 + x2;

figure(1)
plot(t, x1, 'k:', t, x2, 'k--', t, x, 'k-')
set(gca,'FontSize', 14)
xlabel('t (s)')
ylabel('x')
legend('x_1', 'x_2', 'x')
```

Next, consider the square wave shown in Fig. 1.12. We see that this function is also periodic with a period of 1 s. Can it be represented using Eq. 1.2? The answer is yes – its Fourier series is the sum of the odd sine harmonics[4]: $x(t) = b_1 \sin(\omega_1 t) + b_3 \sin(\omega_3 t) + \cdots$. This makes sense when we observe that the square wave is an *odd function*, which means that $x(t) = -x(-t)$. In other words, the function does not simply mirror about the $t = 0$ point; it must also be inverted. Based on the period of 1 s and using Eq. 1.1, we can find that $\omega_1 = \frac{2\pi}{\tau} = \frac{2\pi}{1} = 2\pi$ rad/s. Furthermore, the Fourier coefficients are given by: $b_n = \frac{4}{n\pi}$, where n represents the odd integers. These odd integers can be represented by the expression $n = 2k - 1$, where $k = 1, 2, 3, \ldots$.

The top panel of Fig. 1.13 shows the square wave and Fourier series approximation using the first three terms ($k = 1, 2,$ and 3, so $n = 1, 3,$ and 5). As with any Fourier series, increasing the number of terms tends to increase its ability to reproduce the signal in question. The bottom panel of Fig. 1.13 shows the result for 20 terms. In this case, the square wave is reproduced quite accurately. However, if we look closely, we see oscillations near the discontinuities in the piecewise continuous square wave. This reflects the inherent challenge in representing a discontinuous function by a finite series of continuous sine and cosine functions and is referred to as *Gibbs' phenomenon* (http://en.wikipedia.org/wiki/Gibbs_phenomenon). Figure 1.13 was produced using the m-file provided in MATLAB® MOJO 1.2.

---

[4] "Harmonics" is used to indicate the terms in the Fourier series.

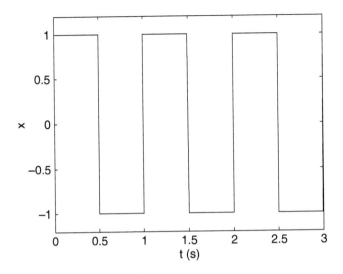

Fig. 1.12 Square wave for $t \geq 0$

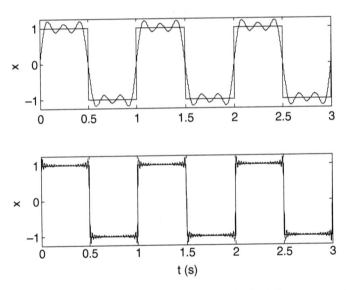

Fig. 1.13 Square wave and Fourier series approximations for (*top*) three terms and (*bottom*) 20 terms

MATLAB® MOJO 1.1
```
% matlab_mojo_1_2.m

clc
clear all
close all

% define variables
t = 0:0.001:3;                  % s
omega1 = 2*pi;                  % rad/s

% define square wave as a piecewise function
for cnt = 1:length(t)
    if t(cnt) < 0.5
        x_ref(cnt) = 1;
    elseif t(cnt) >= 0.5 & t(cnt) < 1
        x_ref(cnt) = -1;
    elseif t(cnt) >= 1 & t(cnt) < 1.5
        x_ref(cnt) = 1;
    elseif t(cnt) >= 1.5 & t(cnt) < 2
        x_ref(cnt) = -1;
    elseif t(cnt) >= 2 & t(cnt) < 2.5
        x_ref(cnt) = 1;
    elseif t(cnt) >= 2.5 & t(cnt) < 3
        x_ref(cnt) = -1;
    elseif t(cnt) >= 3
        x_ref(cnt) = 1;
    end
end

% define square wave using its Fourier series
terms = 3;
x = 0;

for cnt = 1:terms
    b = 4/((2*cnt - 1)*pi);
    omega = (2*cnt - 1)*omega1;
    x = x + b*sin(omega*t);
end

figure(1)
hold on
subplot(211)
plot(t, x_ref, 'k', t, x, 'k')
set(gca,'FontSize', 14)
ylabel('x')
axis([0 3 -1.25 1.25])

% define square wave using its Fourier series
terms = 20;
x = 0;

for cnt = 1:terms
    b = 4/((2*cnt - 1)*pi);
    omega = (2*cnt - 1)*omega1;
    x = x + b*sin(omega*t);
end

subplot(212)
plot(t, x_ref, 'k', t, x, 'k')
set(gca,'FontSize', 14)
xlabel('t (s)')
ylabel('x')
axis([0 3 -1.25 1.25])
```

Fig. 1.14 Argand diagram showing the derivation of Eq. 1.3

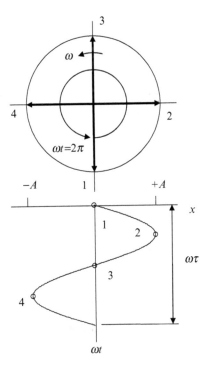

IN A NUTSHELL Fourier's genius was to recognize that any time-varying signal can be represented as a sum of sine and cosine functions of different frequencies and amplitudes. This means that if we determine how to treat these sine and cosine terms, then we can extend the analysis to any time-varying signal. This will be very useful as we proceed with our study of mechanical vibrations.

Returning to our sine wave in Figs. 1.9 and 1.10, we see in Fig. 1.14 that the path of the rotating vector is repeated every 2π rad in the Argand diagram. Because the vector angle at any instant in time is ωt rad and one full cycle of vibration (i.e., one rotation of the vector) takes τ seconds, we can write $2\pi = \omega\tau$. This equation can be rearranged to give Eq. 1.1, or we can solve for the circular frequency.

$$\omega = \frac{2\pi}{\tau} \ \text{rad/s.} \tag{1.3}$$

Alternately, we can define the frequency in cycles/s, or *Hertz* (Hz). By convention in this text, we will use the frequency variable f when applying units of Hz and ω when using rad/s. Because there are 2π rad per cycle of vibration, the conversion from ω to f is:

$$f = \frac{\omega}{2\pi}. \tag{1.4}$$

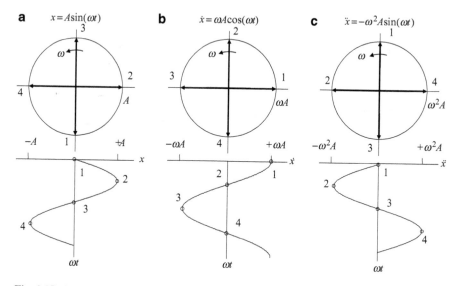

Fig. 1.15 Argand diagram showing (**a**) position, (**b**) velocity, and (**c**) acceleration

Combining Eqs. 1.3 and 1.4, we can relate the frequency in Hz directly to the vibration period in seconds.

$$f = \frac{1}{\tau}. \tag{1.5}$$

If $x(t) = A \sin(\omega t)$ in Fig. 1.14 represents position, what can we say about the corresponding velocity, $v(t)$, and acceleration, $a(t)$? For velocity, we calculate the first time derivative of position to obtain:

$$v(t) = \frac{dx}{dt} = \omega A \cos(\omega t). \tag{1.6}$$

Acceleration is the time derivative of velocity (or second derivative of position). We can therefore write acceleration as:

$$a(t) = \frac{dv}{dt} = -\omega^2 A \sin(\omega t). \tag{1.7}$$

Using the Argand diagrams in Fig. 1.15a–c, we can see that velocity leads position by $\frac{\pi}{2}$ rad (90°) and acceleration leads position by π rad (180°). Although all three vectors are rotating at the circular frequency ω, they have different *phase* values. That is, they reach their maximum values at different instants in time. This is demonstrated in Fig. 1.16, where the three vectors are shown together for an arbitrary instant in time. We can see from this figure that position and acceleration always point in opposite directions. The position reaches its maximum at the moment that the acceleration reaches its minimum. Therefore, with proper scaling

Fig. 1.16 Argand diagram
showing position, velocity,
and acceleration vectors
together

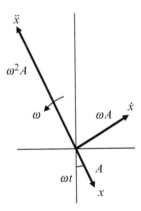

the two vectors could be summed to zero for all time, t. Using $x(t) = A \sin(\omega t)$ and Eq. 1.7, we see that the required scaling factor is $-\omega^2$. Rewriting Eq. 1.7, we have:

$$\ddot{x} = \frac{d^2x}{dt^2} = -\omega^2 A \sin(\omega t) = \omega^2 A \sin(\omega t + \pi) \tag{1.8}$$

where the second equality emphasizes the π rad phase shift between acceleration and position. Combining Eq. 1.8 and $x(t) = A \sin(\omega t)$ with the $-\omega^2$ scaling factor, we can write:

$$\ddot{x} + \omega^2 x = 0 \text{ or} \tag{1.9}$$

$$\frac{d^2x}{dt^2} + \omega^2 x = 0. \tag{1.10}$$

Equation 1.10 is referred to as the *differential equation of harmonic motion*. We will revisit this equation in Sect. 2.1.

As we mentioned previously, the horizontal projection axis we used for our Argand diagram is the *real axis*. The vertical axis is the *imaginary axis*. These names do not identify the "existence" of the axes. Rather, given that we can describe an arbitrary vibration as a sum of cosine and sine components, the real part is the cosine component and the imaginary part is the sine component of the signal. We can therefore represent a unit magnitude vector in the complex plane as $x(t) = \cos(\omega t) + i \sin(\omega t)$, where i is the imaginary variable and $i = \sqrt{-1}$. The projection of this unit vector on the real and imaginary axes is demonstrated in Fig. 1.17.

IN A NUTSHELL   The idea of a real and imaginary axis leads to a mathematical convenience as discussed here. It is sometimes difficult for students to see the reason why this notation would be useful, but it eliminates many trigonometric manipulations that would otherwise be required. Bear with us for a little while and trust that this will be important (and useful) as we proceed.

Fig. 1.17 The unit vector in the complex plane with its projections on the real and imaginary axes

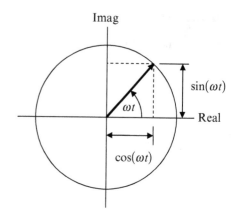

Exponential notation is often used to describe vectors in the complex plane. In this case, we write the unit vector as $x(t) = e^{i\omega t} = \cos(\omega t) + i\sin(\omega t)$. This is referred to as an *Euler's formula*. We can show it is true using three Maclaurin series:[5]

1. $e^y = 1 + y + \frac{y^2}{2!} + \frac{y^3}{3!} + \frac{y^4}{4!} + \cdots$

2. $\cos(y) = 1 - \frac{y^2}{2!} + \frac{y^4}{4!} - \frac{y^6}{6!} + \frac{y^8}{8!} - \cdots$ Note that cosine is an *even function*, where $x(t) = x(-t)$, so we use the even harmonics in its series expansion.

3. $\sin(y) = y - \frac{y^3}{3!} + \frac{y^5}{5!} - \frac{y^7}{7!} + \frac{y^9}{9!} - \cdots$.

If we equate y in these series with $i\omega t$, then we find that $y^2 = i^2(\omega t)^2 = -(\omega t)^2$, $y^3 = -i(\omega t)^3$, $y^4 = (\omega t)^4$, $y^5 = i(\omega t)^5$, and so on. Substituting into the individual series, we have:

1. $e^{i\omega t} = 1 + i\omega t - \frac{(\omega t)^2}{2!} - \frac{i(\omega t)^3}{3!} + \frac{(\omega t)^4}{4!} + \cdots$

2. $\cos(i\omega t) = 1 - \frac{(\omega t)^2}{2!} + \frac{(\omega t)^4}{4!} - \frac{(\omega t)^6}{6!} + \frac{(\omega t)^8}{8!} - \cdots$

3. $\sin(i\omega t) = \omega t - \frac{(\omega t)^3}{3!} + \frac{(\omega t)^5}{5!} - \frac{(\omega t)^7}{7!} + \frac{(\omega t)^9}{9!} - \cdots$.

By inspection, we see that the sum $\cos(\omega t) + i\sin(\omega t)$ is equal to $e^{i\omega t}$. The conclusion is that we can express harmonic motion using exponential notation and still satisfy the differential equation of harmonic motion.

Let's now look at two additional relationships derived from Euler's formula. Consider the two counter-rotating unit vectors, x_1 and x_2, shown in Fig. 1.18. Both are rotating at the circular frequency ω but in opposite directions. At $t = 0$, both are horizontal and pointing to the right. We can write these two vectors as: $x_1 = \cos(\omega t) + i\sin(\omega t)$ and $x_2 = \cos(\omega t) - i\sin(\omega t)$. Alternately, we can use

---

[5] A Maclaurin series is a Taylor series expansion of a function evaluated about zero.

Fig. 1.18 Two counter-
rotating unit vectors used to
derive additional Euler's
formula relationships

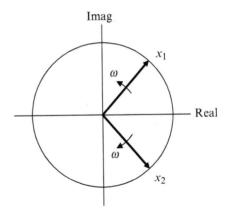

exponential notation to write $x_1 = e^{i\omega t}$ and $x_2 = e^{-i\omega t}$. When the two vectors are
added, the imaginary parts always cancel so that we are left with:

$$x_1 + x_2 = (\cos(\omega t) + i\sin(\omega t)) + (\cos(\omega t) - i\sin(\omega t))$$
$$x_1 + x_2 = 2\cos(\omega t). \tag{1.11}$$

Alternately, the sum can be expressed as: $x_1 + x_2 = e^{i\omega t} + e^{-i\omega t}$. The right-hand
side of this equation can then be equated to the right-hand side of the second line in
Eq. 1.11 to arrive at the relationship:

$$\cos(\omega t) = \frac{e^{i\omega t} + e^{-i\omega t}}{2}. \tag{1.12}$$

Similarly, taking the difference between x_1 and x_2 gives the relationship:

$$\sin(\omega t) = i\left(\frac{e^{i\omega t} - e^{-i\omega t}}{2}\right). \tag{1.13}$$

By the Numbers 1.1

Let's complete an example to quantitatively explore the vector representation of
vibration using the Argand diagram. Consider the signal $x(t) = 5e^{i2,700t} =$
$5(\cos(2,700t) + i\sin(2,700t))$. The circular frequency is 2,700 rad/s or $\frac{2,700}{2\pi} =$
429.7 Hz (according to Eq. 1.4).[6] The angle of the vector in the complex plane is
defined by $\omega t = 2,700t$ rad, where t is expressed in seconds. For $t = 0.001$ s, the
angle is 2.7 rad $= 2.7\frac{180}{\pi} = 154.7°$. The vector at this time is displayed in Fig. 1.19.
The projections on the real (cosine component) and imaginary (sine component)

---

[6] To give this frequency some frame of reference, middle C on the musical scale has a frequency of
261.63 Hz. The frequency for this example would therefore be well within the audible range if its
amplitude was large enough.

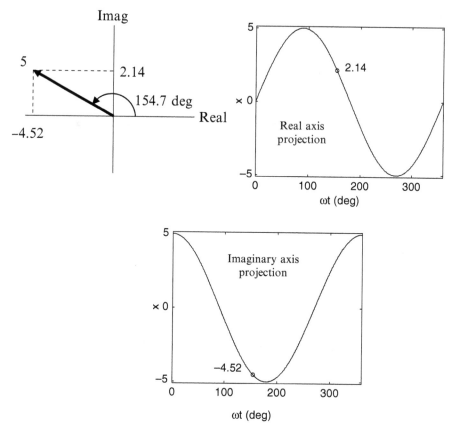

Fig. 1.19 *By the Numbers 1.1* – the vector $x(t) = 5e^{i2,700t}$ is shown in the complex plane for $t = 0.001$ s

axes are also shown. The real axis projection is: $5\cos(154.7) = -4.52$, while the imaginary axis projection is: $5\sin(154.7) = 2.14$. Therefore, for $t = 0.001$ s, we can write: $x(0.001) = 5e^{i2.7} = -4.52 + i2.14$.

Chapter Summary

- There are three primary categories of vibration: free, forced, and self-excited. Free vibration occurs at the system's natural frequency when there is no long-term forcing function. Forced vibration is the response to a periodic forcing function; the vibrating frequency matches the forcing frequency under steady-state conditions. Forced vibration is generally described in the frequency domain. Self-excited vibration exists when a steady input force is modulated into vibration at a system's natural frequency.

- All physical systems are subject to some form of damping or energy dissipation.
- The number of independent coordinates required to describe a body's motion is the number of degrees of freedom. While mechanical systems possess an infinite number of degrees of freedom in general, it is typically possible to adequately describe a system's behavior with a limited set of coordinates. The activity of describing a continuous system by a discrete number of coordinates is referred to as "modeling" the vibratory system.
- Periodic motion repeats at regular intervals; the time for each interval is called the time constant. The vibrating frequency is related to the time constant.
- Examples of harmonic motion are the sine and cosine functions.
- An Argand diagram is the representation of complex numbers as points in the complex plane. The complex plane axes are labeled as "Real" and "Imaginary" to correspond to the real and imaginary parts of complex numbers.
- The Fourier series represents periodic motion as an infinite sum of sine and cosine terms.
- Typical frequency units are rad/s and Hz.
- The exponential function can be used to represent harmonic motion.

Exercises

1. Answer the following questions.
   (a) All bodies which possess ——————— and ——————— are capable of vibrations.
   (b) Name the three fundamental categories of vibration.
   (c) How many degrees of freedom are necessary to fully describe the vibratory motion of an elastic body?

2. For the waveform shown below, answer the following questions.

   (a) Is this motion periodic?
   (b) If the motion is periodic, what is the period of the motion (in seconds)?

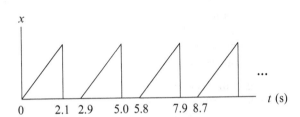

Fig. P1.2 Example waveform

    0   2.1 2.9   5.0 5.8   7.9 8.7

   (c) If the motion is periodic, what is the frequency of its motion (express your answer in both rad/s and Hz)?

3. To explore the Fourier series representation of signals, complete the following.

(a) Approximate a square wave by plotting the following function in MATLAB®. Begin a new m-file by selecting File/New/M-File.

$$x = \frac{4}{\pi} \sin(\omega t) + \frac{4}{3\pi} \sin(3\omega t) + \frac{4}{5\pi} \sin(5\omega t) + \frac{4}{7\pi} \sin(7\omega t)$$

Use $\omega = 2\pi$ rad/s and plot your results from $t = 0$ to 5 s in steps of 0.001 s. You can define your time vector using the following statement:

```
t = 0:0.001:5;
```

Also, π can be defined in MATLAB® using pi. Finally, the argument for sine (or any harmonic function) must be in rad/s in MATLAB® (not degrees). For example, you can define the displacement using the following statement:

```
x = 4/pi*sin(omega*t) + 4/(3*pi)*sin(3*omega*t) +
4/(5*pi)*sin(5*omega*t) + 4/(7*pi)*sin(7*omega*t);
```

To plot the result, you can use the following statements:

```
figure(1)
plot(t, x)
xlabel('t (s)')
ylabel('x(t)')
```

(b) What is the period (in seconds) of the waveform?
(c) What is the frequency in Hz?
(d) Replot the function using 50 terms (following the pattern of odd multiples of ω with the $\frac{4}{n\pi}$ coefficients where $n = 1, 3, 5, \ldots$). What is the effect of including additional terms?

4. If displacement can be described as $x = 5\cos(\omega t)$ mm, where $\omega = 6\pi$ rad/s, complete the following.

(a) Plot the displacement over the time interval from $t = 0$ to 3 s in steps of 0.02 s. What is the period (in seconds) of the harmonic displacement?
(b) Plot the velocity (in mm/s) over the same time interval.
(c) Plot the acceleration (in mm/s^2) over the same time interval.
(d) Calculate the maximum velocity (i.e., calculate the time derivatives and find the maximum values) and acceleration and verify your results using your plots.

5. The complex exponential function, $x = e^{i\omega t}$, can be used to describe harmonic motion (the function can be defined in MATLAB® using x = exp (1i*omega*t);). Complete the following to explore this function.

(a) Plot the real part of the function for $\omega = \pi$ rad/s over a time interval of $t = 0$ to 10 s using time steps of 0.05 s. Use the command plot (t, real (x)) to complete this task.

(b) Plot the imaginary part of the function. Use the command `plot(t, imag (x))`.

(c) Describe your results from parts (a) and (b) in terms of sine and cosine functions.

(d) Sketch the Argand diagram for x at $t = 0.25$ s and show its projections on the real and imaginary axes. What is the numerical value of these projections? How do these results relate to parts (a) and (b)?

6. A harmonic motion has an amplitude of 0.2 cm and a period of 15 s.

(a) Determine the maximum velocity (m/s) and maximum acceleration (m/s^2) of the periodic motion.

(b) Assume that the motion expresses the free vibration of an undamped single degree of freedom system and that the motion was initiated with an initial displacement and no initial velocity. Express the motion (in units of meters) in each of the following four forms:

- $A\cos(\omega_n t + \Phi_c)$
- $A\sin(\omega_n t + \Phi_s)$
- $B\cos(\omega_n t) + C\sin(\omega_n t)$
- $De^{i(\omega_n t)} + Ee^{-i(\omega_n t)}$.

7. Determine the sum of the two vectors $x_1 = 6e^{i\frac{\pi}{6}}$ and $x_2 = -1e^{i\frac{\pi}{3}}$.

8. If the velocity at a particular point on a body is $v(t) = 250\sin(100t)$, complete the following.

(a) Plot the velocity in the complex plane at $t = 0.1$ s.

(b) Using the velocity equation, determine the corresponding expression for displacement.

9. In bungee jumping, a person leaps from a tall structure while attached to a long elastic cord. Would the resulting oscillation be best described as free, forced, or self-excited vibration?

10. The sine function can be represented as $\sin(\theta) = \theta - \frac{\theta^3}{3!} + \frac{\theta^5}{5!} - \frac{\theta^7}{7!} + \frac{\theta^9}{9!} \cdots$. Plot the percent error between $\sin(\theta)$ and:

- θ (rad)
- $\theta - \frac{\theta^3}{3!}$ (rad)
- $\theta - \frac{\theta^3}{3!} + \frac{\theta^5}{5!}$ (rad)

for a range of θ values from 0.001 rad to $\frac{\pi}{2}$ rad in steps of 0.001 rad.

Calculate the percent error using $\left(\frac{\theta - \sin(\theta)}{\sin(\theta)}\right) \cdot 100$ for the θ approximation, $\left(\frac{\left(\theta - \frac{\theta^3}{3!}\right) - \sin(\theta)}{\sin(\theta)}\right) \cdot 100$ for the $\theta - \frac{\theta^3}{3!}$ approximation, and so on. How do these results relate to the small angle approximation?

References

http://en.wikipedia.org/wiki/Gibbs_phenomenon

http://en.wikipedia.org/wiki/Tacoma_Narrows_Bridge_(1940)

Kreyszig E (1983) Advanced engineering mathematics, 5th edn. Wiley, New York

Schmitz T, Smith KS (2009) Machining dynamics: Frequency response to improved productivity. Springer, New York

Weisstein EW (2010) "Argand diagram" from MathWorld – a Wolfram web resource, http://mathworld.wolfram.com/ArgandDiagram.html

Chapter 2
Single Degree of Freedom Free Vibration

*The least movement is of importance to all nature. The entire
ocean is affected by a pebble.*

– Blaise Pascal

2.1 Equation of Motion

For the discussions in this chapter, we will use what is referred to as a *lumped
parameter model* to describe free vibration. The "lumped" designation means that
the mass is concentrated at a single coordinate (degree of freedom) and it is
supported by a massless spring and damper. Recall from Sect. 1.2.1 that free
vibration means that the mass is disturbed from its equilibrium position and
vibration occurs at the natural frequency, but a long-term external force is not
present. The lumped parameter model is typically depicted as shown in Fig. 2.1.
Here, the linear spring, k, exerts a force, f, proportional to displacement, x. See
Fig. 2.2, where the slope of the line represents the spring constant, k. This linear
relationship is referred to as *Hooke's law*. Typical SI units for k are N/m.

$$f = kx \qquad (2.1)$$

 IN A NUTSHELL   Lumped parameters do not exist in the real
world. All springs have some mass. All physical masses deform in
the presence of a force and, therefore, have stiffness. However, it is
often possible to identify system components where mass, stiffness,
or damping is the dominant feature. A lumped parameter model is
dramatically simpler and often sufficiently accurate for our purposes. It is the
essence of engineering to make the problem simple enough to be tractable, yet
sophisticated enough to reasonably represent reality.

T.L. Schmitz and K.S. Smith, *Mechanical Vibrations: Modeling and Measurement*,
DOI 10.1007/978-1-4614-0460-6_2, © Springer Science+Business Media, LLC 2012

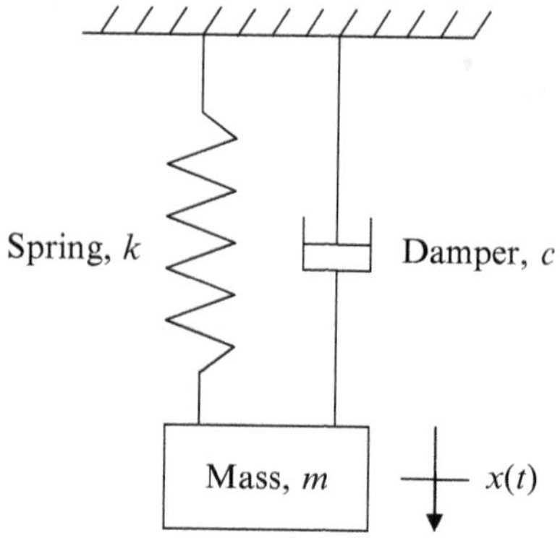

Fig. 2.1 Lumped parameter
spring–mass–damper
model for single degree of
freedom free vibration

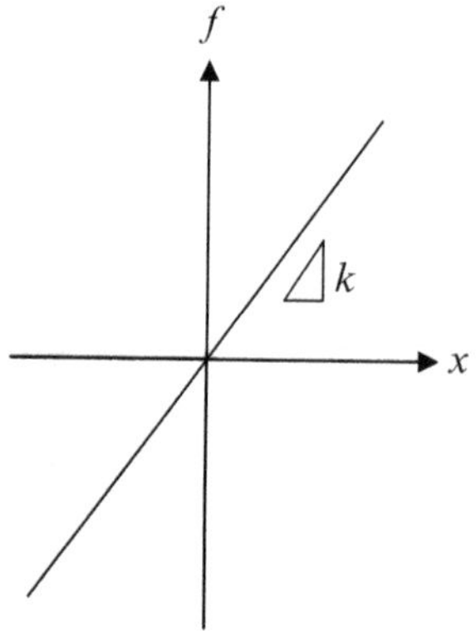

Spring, k Damper, c

Mass, m $x(t)$

Fig. 2.2 Hooke's law for
a linear spring; the spring
constant, k, is the slope
of the line

f

k

x

A viscous damping model is assumed in Fig. 2.1. For this type of damping, the force is proportional to velocity; see Eq. 2.2, where c is the viscous damping coefficient with units of force per velocity (in SI, the units are N-s/m). Physically, viscous damping is observed by forcing a body through a fluid. For example, if you pull your hand through the water in a swimming pool, you will find that the force increases with your hand's speed. Another example is a shock absorber, or dashpot, where fluid is forced though one or more small holes. If you attempt to collapse or expand the shock absorber more rapidly, the force increases proportionally. Damping is discussed in more detail in Sect. 2.4.

$$f = c\dot{x} \tag{2.2}$$

IN A NUTSHELL   If we assume lumped parameters and viscous damping, then the equations of motion will be ordinary linear differential equations with constant coefficients – one of the few classes of differential equations that we can easily solve. This advantage is so powerful that we often assume "equivalent" viscous damping even when we are certain that the damping is not actually viscous.

Fig. 2.3 Free body diagram
for spring–mass–damper
system (d'Alembert's inertial
force, $f = m\ddot{x}$, is included)

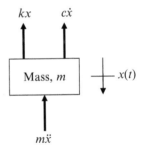

To obtain the differential equation of motion for the spring–mass–damper
system, let's draw the free body diagram. Figure 2.3 shows three forces acting on
the single degree of freedom mass:

- the spring force, kx
- the damping force, $c\dot{x}$
- the "inertial force," $m\ddot{x}$.

This transformation of the accelerating body into an equivalent static system by
the addition of the inertial force is referred to as *d'Alembert's principle*. It enables
us to simply sum the forces and set the result equal to zero (as in the static case).
For Fig. 2.3, this "static" force sum is $\sum f_x = -m\ddot{x} - c\dot{x} - kx = 0$, where a positive
force is assumed in the $+x$ direction (down in Fig. 2.3). Rewriting gives the familiar
form for the equation of motion.

$$m\ddot{x} + c\dot{x} + kx = 0 \qquad (2.3)$$

Note that this gives the same result as if we had used $\sum f_x = m\ddot{x}$ and not
considered the inertial force. In this case, we would have obtained $\sum f_x = m\ddot{x} =$
$-c\dot{x} - kx$ (where the positive force direction is again down in Fig. 2.3), or
$\sum f_x = m\ddot{x} + c\dot{x} + kx = 0$.

To begin our analysis of this equation of motion, let's neglect damping for now
so that the new equation is:

$$m\ddot{x} + kx = 0. \qquad (2.4)$$

IN A NUTSHELL   We will see that the inclusion of damping
typically makes the results surprisingly more complex. For that
reason, it is usually easiest to start with the simpler undamped case.

This equation is similar to the "differential equation of harmonic motion"
we saw in Eq. 1.10. We have already seen in Sect. 1.5 that the exponential function

can be used to describe harmonic motion through Euler's formula. Therefore, let's assume a (harmonic) solution to Eq. 2.4 of the form:

$$x(t) = Xe^{st}, \tag{2.5}$$

where $s = i\omega$ is the Laplace variable. We can then take the first and second time derivatives of Eq. 2.5 to determine the velocity and acceleration, respectively.

$$\dot{x}(t) = sXe^{st} \tag{2.6}$$

$$\ddot{x}(t) = s^2Xe^{st} \tag{2.7}$$

 IN A NUTSHELL   The solution to a differential equation is a function which makes it true. The solution to the differential equation described by Eq. 2.4 is a function of time, $x(t)$, such that the sum of its second derivative multiplied by m and the function itself multiplied by k is zero. One solution procedure for ordinary linear differential equations with constant coefficients (like Eq. 2.4) begins with making an assumption of the solution form. Equation 2.5 shows a typical assumption. One instructor somewhere in our past said we should assume $x(t) = Be^{st}$ because that form is the "best" assumption. This may sound corny, but it provides a nice mnemonic. In fact, it turns out that this assumption provides all of the solutions that exist for this differential equation.

We can now substitute for x and $\ddot{x}$ in Eq. 2.4 using Eqs. 2.5 and 2.7.

$$m(s^2Xe^{st}) + k(Xe^{st}) = 0 \tag{2.8}$$

Grouping terms gives $Xe^{st}(ms^2 + k) = 0$. There are two possibilities for this equation to be true. Since it is a product of two terms, at least one of the two terms must be equal to zero. If $Xe^{st} = 0$, this means that no motion has occurred. Our assumed form leads to all possible solutions and this is a valid solution. However, it is not a very interesting outcome, so it is referred to as the *trivial solution*. We are interested in the alternative:

$$ms^2 + k = 0. \tag{2.9}$$

This is used to obtain our vibration solution and is referred to as the *character-istic equation* for the system. From this equation, we can identify the *natural frequency*, or the frequency at which the system will vibrate if disturbed from equilibrium and released. Because we are not considering damping in this discussion, it is referred to as the undamped natural frequency, ω_n.

Let's solve for s from Eq. 2.9. We see that $s^2 = -\frac{k}{m}$. Taking the square root of both sides gives $s = \pm\sqrt{-\frac{k}{m}} = \pm i\sqrt{\frac{k}{m}}$. We have already stated that $s = i\omega$. Comparing these two equations for s gives the natural frequency of vibration for our undamped single degree of freedom system.

$$\omega_n = \sqrt{\frac{k}{m}} \tag{2.10}$$

Here, we have selected the positive root to give a positive natural frequency. If the stiffness units are N/m and the mass units are kg, then the units of natural frequency are:

$$\sqrt{\frac{\frac{N}{m}}{kg}} = \sqrt{\frac{\frac{kg\frac{m}{s^2}}{m}}{kg}} = \sqrt{\frac{1}{s^2}} = \frac{rad}{s}.$$

Note that the unit radian does not strictly require a unit symbol and is typically omitted in mathematics literature. It has been added here for clarity and to emphasize the description of the rotating vector in the Argand diagram. If we wish to express the natural frequency in units of Hz (or cycles/s), we simply apply Eq. 1.4. In this case, we use the variable f_n to indicate the alternate units of the natural frequency.

$$f_n = \frac{\omega_n}{2\pi} \tag{2.11}$$

 IN A NUTSHELL   Equation 2.10 makes intuitive sense. We would expect systems that have a high natural frequency to be stiff and have a low mass. The highest pitch guitar strings are thin and tight. We would also expect systems that have a low natural frequency to be more massive and flexible. The lowest pitch guitar strings are the loosest and often have a second string wound around them to increase their mass.

The total solution to the equation of motion is the sum of the two solutions determined from the roots of the characteristic equation: $s_1 = +i\omega_n$ and $s_2 = -i\omega_n$. Using Eq. 2.5, the two solutions are summed to obtain:

$$x(t) = X_1 e^{s_1 t} + X_2 e^{s_2 t} = X_1 e^{i\omega_n t} + X_2 e^{-i\omega_n t}. \tag{2.12}$$

In the right-hand side of this equation, the first term, $X_1 e^{i\omega_n t}$, is a counter-clockwise rotating vector in the complex plane and the second term, $X_2 e^{-i\omega_n t}$, is a clockwise rotating vector. To determine the coefficients X_1 and X_2, we use the *initial conditions*. These are typically applied at time $t = 0$ to the system that was originally at equilibrium in order to disturb it from this state. We can describe these initial conditions (or disturbances) as: (1) $x(0) = x_0$, the initial displacement;

Fig. 2.4 Complex plane
representation of $X_1 = Ae^{i\beta}$

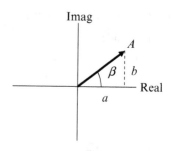

and (2) $\dot{x}(0) = \dot{x}_0$, the initial velocity. If we let $t = 0$ in Eq. 2.12, we obtain $x(0) = x_0 = X_1 e^0 + X_2 e^0$. Because $e^0 = 1$, we can write this as:

$$x_0 = X_1 + X_2. \tag{2.13}$$

Next, we take the time derivative of Eq. 2.12 to obtain $\frac{dx}{dt} = \dot{x}(t) = i\omega_n X_1 e^{i\omega_n t} - i\omega_n X_2 e^{-i\omega_n t}$. Setting $t = 0$ and substituting the initial velocity gives $\dot{x}(0) = \dot{x}_0 = i\omega_n X_1 e^0 - i\omega_n X_2 e^0$. This result can be rewritten as:

$$\dot{x}_0 = i\omega_n X_1 - i\omega_n X_2. \tag{2.14}$$

We can solve for X_1 by multiplying Eq. 2.13 by $i\omega_n$ and summing this result with Eq. 2.14. The X_2 terms cancel and we are left with $X_1 = \frac{i\omega_n x_0 + \dot{x}_0}{2i\omega_n}$. Let's rationalize the right-hand side of this equation by multiplying both the numerator and denominator by i.

$$X_1 = \frac{i^2 \omega_n x_0 + i\dot{x}_0}{2i^2 \omega_n} = \frac{-\omega_n x_0 + i\dot{x}_0}{-2\omega_n} = \frac{\omega_n x_0 - i\dot{x}_0}{2\omega_n} \tag{2.15}$$

We can now solve for X_2 using Eq. 2.13: $X_2 = x_0 - X_1 = \frac{2\omega_n x_0}{2\omega_n} - X_1$. See Eq. 2.16.

$$X_2 = \frac{2\omega_n x_0 - \omega_n x_0 + i\dot{x}_0}{2\omega_n} = \frac{\omega_n x_0 + i\dot{x}_0}{2\omega_n} \tag{2.16}$$

Because X_1 and X_2 are identical except for the sign of their imaginary parts, they are referred to as *complex conjugates*. We can also write them in exponential form as described in Sect. 1.5. If the real and imaginary parts of X_1 are a and b, then we have that $X_1 = a + ib$, where $a = \frac{\omega_n x_0}{2\omega_n} = \frac{x_0}{2}$ and $b = \frac{-\dot{x}_0}{2\omega_n}$. In exponential notation, the equivalent expression is $X_1 = Ae^{i\beta}$, where $A = \sqrt{a^2 + b^2}$ and $\beta = \tan^{-1}\left(\frac{b}{a}\right)$. These relationships are depicted in Fig. 2.4.

Again using a and b, we can express the second complex coefficient as $X_2 = a - ib = Ae^{-i\beta}$. The corresponding graphical relationships between a, b, A, and β are shown in Fig. 2.5.

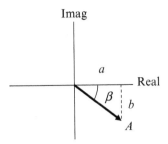

Fig. 2.5 Complex plane representation of $X_2 = Ae^{-i\beta}$

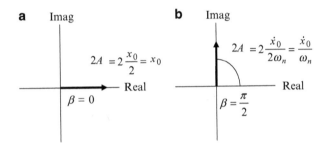

Fig. 2.6 Vector representation of $x(t) = 2A\cos(\omega_n t + \beta)$ at $t = 0$ for: (**a**) $\dot{x}_0 = 0$ and $x_0 \neq 0$; and (**b**) $x_0 = 0$ and $\dot{x}_0 \neq 0$

Substituting for X_1 and X_2 in Eq. 2.12 gives:

$$x(t) = X_1 e^{i\omega_n t} + X_2 e^{-i\omega_n t} = Ae^{i\beta}e^{i\omega_n t} + Ae^{-i\beta}e^{-i\omega_n t}$$
$$x(t) = A\left(e^{i(\omega_n t + \beta)} + e^{-i(\omega_n t + \beta)}\right) \tag{2.17}$$

Rewriting Eq. 1.12, we have that $2\cos(\theta) = e^{i\theta} + e^{-i\theta}$, where θ is an arbitrary argument. Using this relationship, we can rewrite Eq. 2.17 as $x(t) = A(2\cos(\omega_n t + \beta)) = 2A\cos(\omega_n t + \beta)$. As we have seen, A and β depend on ω_n (i.e., the model parameters k and m) and the initial conditions. Specifically,

$$A = \sqrt{\left(\frac{x_0}{2}\right)^2 + \left(\frac{-\dot{x}_0}{2\omega_n}\right)^2} \text{ and } \beta = \tan^{-1}\left(\frac{\frac{-\dot{x}_0}{2\omega_n}}{\frac{x_0}{2}}\right) = \tan^{-1}\left(\frac{-\dot{x}_0}{\omega_n x_0}\right).$$

We can plot $x(t) = 2A\cos(\omega_n t + \beta)$ for two distinct cases: (1) $\dot{x}_0 = 0$ and $x_0 \neq 0$; and (2) $x_0 = 0$ and $\dot{x}_0 \neq 0$. For the first case, $\beta = 0$ and the representation of the counterclockwise rotating vector x in the complex plane at $t = 0$ is shown in Fig. 2.6a. For the second case, $\beta = \frac{\pi}{2}$ and x is shown at $t = 0$ in Fig. 2.6b. The time-domain representations of these two cases (i.e., the projection of the vectors on the real axis) are included in Fig. 2.7a (first case) and 2.7b (second case).

We can express undamped free vibration (i.e., the solution to Eq. 2.4) in several forms. Using somewhat careless notation (the A in the following list is not the same A we just discussed), these forms can be generically written as:

- $x(t) = A\cos(\omega_n t + B)$
- $x(t) = C\sin(\omega_n t + D)$

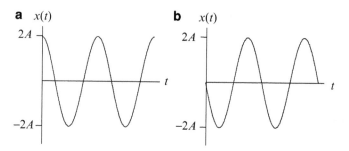

Fig. 2.7 Time-domain representation of $x(t) = 2A\cos(\omega_n t + \beta)$ for: **(a)** $\dot{x}_0 = 0$ and $x_0 \neq 0$; and **(b)** $x_0 = 0$ and $\dot{x}_0 \neq 0$

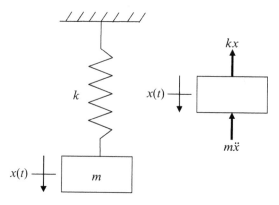

Fig. 2.8 *By the Numbers 2.1* – the example spring–mass system and corresponding free body diagram are shown

- $x(t) = E\sin(\omega_n t) + F\cos(\omega_n t)$
- $x(t) = Ge^{i\omega_n t} + He^{-i\omega_n t}$,

where A through F are real-valued (no imaginary part) and G and H are complex.

IN A NUTSHELL   It might seem like a long (mathematical) way to go, but the end result is that free vibration of a single degree of freedom system with no damping is sinusoidal and occurs at the natural frequency of the system. The initial conditions might make the vibration larger or smaller or make it look more like a sine or a cosine. However, the initial conditions do not change the frequency of the motion. The natural frequency is a fundamental property of a single degree of freedom system.

By the Numbers 2.1

Consider the spring–mass system shown in Fig. 2.8, where $k = 5 \times 10^7$ N/m and $m = 0.1$ kg. The free body diagram is also shown. Based on this diagram, the force balance is $\sum f_x = 0 = -m\ddot{x} - kx$. The equation of motion is therefore $m\ddot{x} + kx = 0$, as we have already seen.

Let's consider the spring stiffness and try to make some sense of this value. We can rewrite $k = 5 \times 10^7$ N/m as $k = \frac{5,000}{100} \frac{N}{\mu m}$. In other words, it requires 5,000 N to deflect the spring by 100 μm. To put 5,000 N in more physical terms, let's convert it to pounds-force by dividing by 4.448. The result is $\frac{5,000}{4.448} = 1,124$ lb$_f$. This is approximately 2.55 Kawasaki Ninja 500R motorcycles (curb weight 440 lb$_f$). If we placed these 2.55 motorcycles on top of the spring, it would deflect by 100 μm, which is approximately the diameter of a human hair. This seems like a very stiff spring, but this is actually a typical k value for many mechanical systems, such as the bending stiffness for an endmill clamped in a spindle on a milling machine.

Let's next discuss the mass of 0.1 kg. We can convert this to newtons by multiplying Earth's gravitational acceleration value of 9.81 m/s^2. The result is $f = 9.81(0.1) = 0.98$ N. Dividing this result by 4.448 to convert it to pounds-force, we obtain $f = \frac{0.98}{4.448} = 0.22$ lb$_f$. This is about the same as the precooked weight of the beef patty on a McDonald's Quarter Pounder. Relative to the motorcycles from the force description, this is not much weight.

Given the stiff spring and low mass, what does your intuition tell you about the corresponding natural frequency? You would probably expect that this combination would yield a high natural frequency, i.e., if we did deflect the mass (supported by the spring) from its equilibrium position and then let it vibrate, a high oscillating frequency would not be a surprise. Using Eq. 2.10, we find that $\omega_n = \sqrt{\frac{5 \times 10^7}{0.1}} = 22,361$ rad/s. We can express this result in Hz by applying Eq. 2.11: $f_n = \frac{\omega_n}{2\pi} = \frac{22,361}{2\pi} = 3,559$ Hz. Therefore, if we were to disturb this system from its equilibrium position, it would complete 3,559 cycles of oscillation per second. For comparison purposes, the highest note on a standard piano has a frequency of 4,186 Hz, where the sound we hear is due to the free vibration of a taut, thin wire which is excited by striking it with a "hammer" (the padded hammer motion is initiated by the pressing of the piano key).

To solve the equation of motion for this harmonic vibration, let's select initial conditions of $x_0 = 0.002$ m and $\dot{x}_0 = 0$. This means that we pulled the mass down by 0.002 m and then released it. Although many forms are available to us, let's use $x(t) = E \sin(\omega_n t) + F \cos(\omega_n t)$ for the vibration description. To apply both initial conditions and solve for E and F, we will need the time derivative of the position:

$$\frac{dx}{dt} = \dot{x}(t) = \omega_n E \cos(\omega_n t) - \omega_n F \sin(\omega_n t). \tag{2.18}$$

We can now apply the two initial conditions by setting $t = 0$.

$$x(0) = x_0 = E \sin(0) + F \cos(0) = F \tag{2.19}$$

$$\dot{x}(0) = \dot{x}_0 = \omega_n E \cos(0) - \omega_n F \sin(0) = \omega_n E \tag{2.20}$$

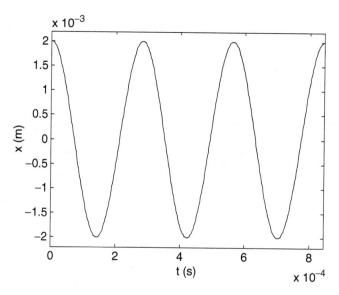

Fig. 2.9 *By the Numbers 2.1* – the free vibration response to an initial displacement of 0.002 m is shown

From Eqs. 2.19 and 2.20, we see that $F = x_0$ and $E = \frac{\dot{x}_0}{\omega_n}$ so that we can generically write:

$$x(t) = \frac{\dot{x}_0}{\omega_n} \sin(\omega_n t) + x_0 \cos(\omega_n t). \tag{2.21}$$

IN A NUTSHELL   We can always use the solution form provided by Eq. 2.21 for single degree of freedom free vibration problems when the initial conditions are known.

Substituting our initial conditions and natural frequency, Eq. 2.21 simplifies to $x(t) = 0.002 \cos(22,361t)$ m. The resulting vibration is shown in Fig. 2.9. Note that the period of vibration for this example is $\tau = \frac{1}{f_n} = \frac{2\pi}{\omega_n} = \frac{2\pi}{22,361} = 2.81 \times 10^{-4}$, s $= 0.281$ ms.

In practice, we would expect that the free vibration magnitude would decay over time and eventually stop, not persist with a constant magnitude as we see in Fig. 2.9. This is because all physical systems exhibit some level of damping. However, let's continue this undamped example by considering a second form for the harmonic vibration solution $x(t) = A \cos(\omega_n t + B)$. The derivative is:

$$\dot{x}(t) = -\omega_n A \sin(\omega_n t + B). \tag{2.22}$$

We apply the initial conditions to obtain:

$$x(0) = x_0 = A \cos(0 + B) = A \cos(B) \tag{2.23}$$

and

$$\dot{x}(0) = \dot{x}_0 = -\omega_n A \sin(0 + B) = -\omega_n A \sin(B). \tag{2.24}$$

We can divide Eq. 2.24 by Eq. 2.23 to obtain:

$$\frac{-\omega_n A \sin(B)}{A \cos(B)} = \frac{\dot{x}_0}{x_0}, \tag{2.25}$$

which can be rewritten as $-\omega_n \tan(B) = \frac{\dot{x}_0}{x_0}$. Solving for the phase angle, B, gives:

$$B = \tan^{-1}\left(-\frac{\dot{x}_0}{\omega_n x_0}\right). \tag{2.26}$$

Given B, we can solve for A using Eq. 2.23:

$$A = \frac{x_0}{\cos(B)}. \tag{2.27}$$

For this example, we find that $B = \tan^{-1}\left(-\frac{0}{22,361(0.002)}\right) = 0$ and $A = \frac{0.002}{\cos(0)} = 0.002$ m. Substituting into $x(t) = A \cos(\omega_n t + B)$ naturally gives the same result we found previously: $x(t) = 0.002 \cos(22,361t)$ m. We can follow the identical approach for any of the harmonic vibration forms provided.

2.2 Energy-Based Approach

As an alternative to drawing a free body diagram to identify the equation of motion for a vibratory system, we can use an energy-based approach. In this method, we recognize that oscillating systems constantly switch between kinetic and potential energy. For our spring–mass system, we can express the *kinetic energy* of the moving mass as $K = \frac{1}{2}mv^2$ and the *potential energy* of the spring as $P = \frac{1}{2}kx^2$. As a check on these equations, let's take a look at the units. For K, we have $kg\left(\frac{m}{s}\right)^2 = \frac{kg \cdot m}{s^2}m = N \cdot m$. Similarly, P gives $\frac{N}{m}(m)^2 = N \cdot m$. As expected, these units describe energy/work.

The sum of the kinetic and potential energies is a constant value in time: $K + P = $ constant. Substituting for K and P considering our spring–mass system and taking the time derivative of this sum gives:

$$\frac{d}{dt}\left(\frac{1}{2}mv^2 + \frac{1}{2}kx^2\right) = 0. \tag{2.28}$$

Fig. 2.10 A cylinder rolls
without slipping on a concave
cylindrical surface

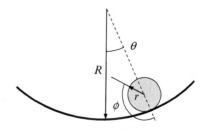

We now need to calculate the time derivatives of the two terms in Eq. 2.28. Rewriting the velocity, v, as $\dot{x}$, its derivative is:

$$\frac{d}{dt}\left(\frac{1}{2}m\dot{x}^2\right) = \frac{1}{2}m(2\dot{x})\ddot{x} = m\ddot{x}(\dot{x}). \tag{2.29}$$

Similarly, the potential energy derivative is:

$$\frac{d}{dt}\left(\frac{1}{2}kx^2\right) = \frac{1}{2}k(2x)\dot{x} = kx(\dot{x}). \tag{2.30}$$

Substituting Eqs. 2.29 and 2.30 into Eq. 2.28 gives:

$$\dot{x}(m\ddot{x} + kx) = 0, \tag{2.31}$$

which is the same result obtained from the free body diagram approach for this spring–mass system: $m\ddot{x} + kx = 0$. The energy method is particularly useful when the "mass" and "spring" elements are difficult to identify. Let's consider an example (Thomson and Dahley 1998).

A cylinder with mass, m, rolls without slipping on a concave cylindrical surface as shown in Fig. 2.10. Let's determine the cylinder's differential equation of motion for small oscillations about the lowest point for the concave surface (this is the equilibrium position). For no slipping, the two arc lengths, $R\theta$ and $r\phi$, are equal: $R\theta = r\phi$. Their time derivatives are also equal:

$$R\dot{\theta} = r\dot{\phi}. \tag{2.32}$$

To apply the energy method, we need to write equations for both K and P. For the kinetic energy, there is both translation and rotation of the rolling cylinder. The translational velocity, v, is equal to the product of the radius from the concave surface center to the cylinder center, $R - r$, and the angular velocity about the surface center, $\dot{\theta}$:

$$v = (R - r)\dot{\theta} \tag{2.33}$$

Fig. 2.11 Translational velocity, v, for the rolling cylinder

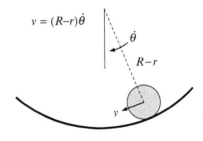

Fig. 2.12 Rotational velocity, $\dot{\Phi}$, for the rolling cylinder

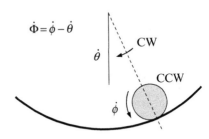

as shown in Fig. 2.11. The rotational velocity for the cylinder itself is $\dot{\Phi} = \dot{\phi} - \dot{\theta}$. We require this difference because $\dot{\phi}$ is counterclockwise (CCW), while $\dot{\theta}$ is clockwise (CW) as the cylinder rolls from right to left as shown in Fig. 2.12. We can express the cylinder's rotational velocity in terms of $\dot{\theta}$ only, however, by applying the no-slipping condition represented by Eq. 2.32. Solving this equation for $\dot{\phi}$ gives $\dot{\phi} = \frac{R}{r}\dot{\theta}$. Substitution in $\dot{\phi} - \dot{\theta}$ yields:

$$\dot{\Phi} = \frac{R}{r}\dot{\theta} - \dot{\theta} = \left(\frac{R-r}{r}\right)\dot{\theta}. \tag{2.34}$$

Given these velocity expressions, we can now write the kinetic energy equation. We again have the $\frac{1}{2}mv^2$ term for the translational kinetic energy, but we also have a $\frac{1}{2}J\dot{\Phi}^2$ term for the rotational kinetic energy, where J is the mass moment of inertia. The kinetic energy equation is:

$$K = \frac{1}{2}m\left((R-r)\dot{\theta}\right)^2 + \frac{1}{2}J\left(\left(\frac{R-r}{r}\right)\dot{\theta}\right)^2. \tag{2.35}$$

Simplifying Eq. 2.35 gives:

$$K = \frac{1}{2}m\left((R-r)^2\dot{\theta}^2 + \frac{r^2}{2}\left(\frac{R-r}{r}\right)^2\dot{\theta}^2\right)$$

$$K = \frac{1}{2}m(R-r)^2\left(\dot{\theta}^2 + \frac{\dot{\theta}^2}{2}\right) = \frac{3}{4}m(R-r)^2\dot{\theta}^2, \tag{2.36}$$

where we substituted for the uniform cylinder's mass moment of inertia: $J = \frac{mr^2}{2}$.

Fig. 2.13 Cylinder height as
a function of the angle θ

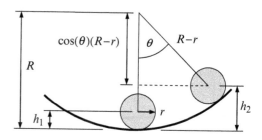

Because there are no springs in this system, the potential energy is based only
on the height of the cylinder's center above the lowest position. We can express this
in terms of the angle θ as shown in Fig. 2.13, where h_1 is the lowest position height
and h_2 is the height that varies with θ. These two heights are:

$$h_1 = r \tag{2.37}$$

and

$$h_2 = R - \cos(\theta)(R - r). \tag{2.38}$$

The gravitational potential energy is then:

$$P = mg(h_2 - h_1) = mg(R - \cos(\theta)(R - r) - r)$$
$$P = mg(R - r)(1 - \cos(\theta)). \tag{2.39}$$

Finally, we sum the kinetic and potential energy and calculate the time derivative of this sum. Alternatively, we can first calculate the derivatives of each, sum
them, and then set this result equal to zero. Note that this sum is set equal to zero
because the derivative of a constant is zero, i.e., $\frac{d}{dt}(K + P = \text{constant})$ gives
$\frac{dK}{dt} + \frac{dP}{dt} = 0$. The kinetic and potential energy derivatives are:

$$\frac{d}{dt}(K) = \frac{d}{dt}\left(\frac{3}{4}m(R - r)^2\dot{\theta}^2\right) = \frac{3}{4}m(R - r)^2\left(2\dot{\theta}\right)\ddot{\theta} \tag{2.40}$$

and

$$\frac{d}{dt}(P) = \frac{d}{dt}(mg(R - r)(1 - \cos(\theta))) = mg(R - r)(\sin(\theta))\dot{\theta}. \tag{2.41}$$

Their sum is $m\dot{\theta}\left(\frac{3}{2}(R - r)^2\ddot{\theta} + g(R - r)\sin(\theta)\right) = 0$. For small angles, we can
approximate $\sin(\theta)$ as θ and rewrite this equation as:

$$\ddot{\theta} + \frac{2}{3}\frac{g}{(R - r)}\theta = 0. \tag{2.42}$$

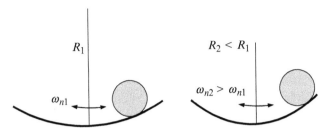

Fig. 2.14 Natural frequency dependence on R for the rolling cylinder example

This is the system equation of motion written in the same form as Eq. 1.10, which is the *standard form* for the *differential equation of harmonic motion*. Using this form, $\frac{d^2x}{dt^2} + \omega^2 x = 0$, we can immediately identify the natural frequency for the system:

$$\omega_n = \sqrt{\frac{2}{3}\frac{g}{(R-r)}}. \tag{2.43}$$

This equation tells us that a larger radius, R, of the concave surface should give a lower natural frequency. Using Fig. 2.14, we see that this makes intuitive sense.

Let's return to the spring–mass system and show how to quickly identify the natural frequency using the energy expressions. First, we can recognize that the kinetic energy is maximum when the potential energy is zero since their sum is a constant. In equation form, we have $K_{max} + 0 =$ constant. For the undamped, single degree of freedom, lumped parameter model, this maximum kinetic energy is identified when the oscillating mass is passing through its $x = 0$ position (where the velocity is maximum, $\dot{x}_{max}$). Naturally, the potential energy is zero at $x = 0$ because the spring extension is zero. Second, we can also see that the potential energy is maximum when the kinetic energy is zero: $0 + P_{max} =$ constant. The maximum potential energy occurs at the maximum spring extension, x_{max}, where the mass velocity is zero. Because K_{max} and P_{max} are both equal to the same constant, we can equate them.

$$\frac{1}{2}m\dot{x}_{max}^2 = \frac{1}{2}kx_{max}^2 \tag{2.44}$$

To determine the $\dot{x}_{max}$ and x_{max} values, we need to select a form for the harmonic solution of the differential equation of motion from our previous list. Let's use $x(t) = A\cos(\omega_n t + B)$. The maximum value is $x_{max} = A$. The velocity is $\dot{x}(t) = -\omega_n A \sin(\omega_n t + B)$ and its maximum value is $\dot{x}_{max} = \omega_n A$. Substituting in Eq. 2.44 gives:

$$\frac{1}{2}m(\omega_n^2 A^2) = \frac{1}{2}k(A^2). \tag{2.45}$$

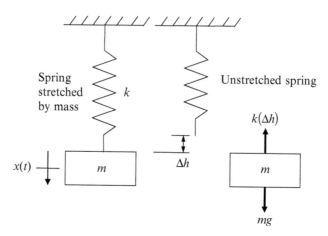

Fig. 2.15 Static free body diagram for spring–mass system

Rewriting Eq. 2.45 gives $\omega_n^2 = \frac{k}{m}$. This result is the same as Eq. 2.10, which is based on the free body diagram analysis. Solving for the natural frequency yields $\omega_n = \sqrt{\frac{k}{m}}$.

Given the rolling cylinder and spring–mass examples we have considered in this section, you may be asking the question: "Why didn't we include the gravitational potential energy for the spring–mass example?" This is a great question and its answer resides in how we are defining the $x = 0$ location. In all spring–mass($-$damper) examples in this text, we will set x to be equal to zero at the static equilibrium position of the mass. This means that the spring is actually deflected from its free length by the gravitational force (mass) already. The force balance from the static free body diagram in Fig. 2.15 is $k(\Delta h) = mg$, where Δh is the spring extension due to the weight of the lumped parameter mass, m.

Now let's consider some other position while the mass is vibrating. In this case, the spring force is $k(x + \Delta h)$ and the force balance from the free body diagram in Fig. 2.16 is:

$$\sum f_x = mg - k(x + \Delta h) - m\ddot{x} = 0. \tag{2.46}$$

Expanding this equation and substituting $k(\Delta h)$ for mg gives $k(\Delta h) - kx - k(\Delta h) - m\ddot{x} = 0$. Canceling and rewriting gives the expected equation of motion $m\ddot{x} + kx = 0$. Our conclusion is that the gravity potential is canceled by the deflected spring force, so we do not need to consider either.[1]

---

[1] This also describes why there is no gravitational force, mg, in Fig. 2.3.

Fig. 2.16 Free body diagram
for vibrating spring–mass
system including the
gravitational force and static
spring deflection

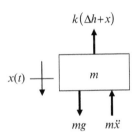

IN A NUTSHELL   The gravity term does not appear in the
spring–mass system because the equations are written about
the equilibrium position. This kind of system vibrates with the same
frequency and in the same way whether it is aligned with gravity,
perpendicular to gravity, or even in the absence of gravity. In the
case of the rolling cylinder, it is gravity that supplies the spring component.
The cylinder example does not vibrate in the absence of gravity.

2.3   Additional Information

Before continuing with the spring–mass–damper (i.e., the damped harmonic
oscillator) analysis in the next section, there are three topics that we should explore.

2.3.1   Equivalent Springs

Figure 2.17 shows two possibilities for combining two springs, k_1 and k_2.
In Fig. 2.17a, the springs are arranged in *parallel*. What we'd like to do is replace
these two parallel springs with a single, *equivalent spring*, k_{eq}. We can find this
equivalent spring constant using Fig. 2.18. The total force, f, required to deflect the
two springs balances the sum of the two spring forces, f_1 and f_2, in the static case.
Because the deflection is the same for both springs, we can write:

$$f = f_1 + f_2 = k_1 x + k_2 x = (k_1 + k_2)x. \tag{2.47}$$

By inspection, we see that $k_{eq} = k_1 + k_2$. For n springs in parallel with constants
$k_1, k_2, k_3, \ldots, k_n$, the equivalent spring constant is:

$$k_{eq} = \sum_{j=1}^{n} k_j. \tag{2.48}$$

Fig. 2.17 Two springs arranged in (a) parallel and (b) series

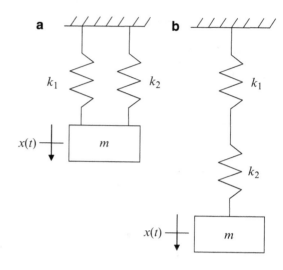

Fig. 2.18 Static free body diagram for two springs in parallel

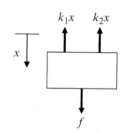

For the springs in *series* shown in Fig. 2.17b, we recognize that a force applied to the end of the two springs will cause a deflection, x, and that the sum of the deflections for the individual springs, x_1 and x_2, must equal the total deflection. Additionally, we know that the individual deflections depend on the force, f, applied at the free end; see Fig. 2.19. These deflections are $x_1 = \frac{f}{k_1}$ for the top spring and $x_2 = \frac{f}{k_2}$ for the bottom spring. The total deflection is therefore $x = x_1 + x_2 = \frac{f}{k_1} + \frac{f}{k_2}$. We find the equivalent spring constant using Eq. 2.49.

$$k_{eq} = \frac{f}{x} = \frac{f}{\frac{f}{k_1} + \frac{f}{k_2}} = \frac{1}{\frac{1}{k_1} + \frac{1}{k_2}} \tag{2.49}$$

In general, for n springs in series with constants $k_1, k_2, k_3, \ldots, k_n$, the equivalent spring constant is calculated using:

$$\frac{1}{k_{eq}} = \sum_{j=1}^{n} \frac{1}{k_j}. \tag{2.50}$$

Fig. 2.19 Deflections for two springs in series

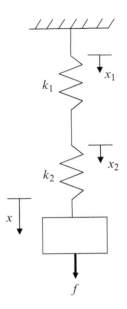

IN A NUTSHELL   The process of determining an equivalent spring means that we apply a "test force" and measure the resulting displacement. This is generally true both in our idealized models and in practice. It is useful because springs very rarely look like our idealized textbook picture. This concept gives us an idea for a physical test we can use to construct an idealized model of a physical system.

2.3.2   Torsional Systems

For single degree of freedom *torsional systems*, the independent coordinate is the rotational angle, θ, rather than position. We will again assume a lumped parameter system as shown in Fig. 2.20. A massless rod, which serves as a torsional spring, K, is attached to a disk with a mass moment of inertia, J. The spring constant for the rod can be expressed as:

$$K = \frac{GI}{l}, \qquad (2.51)$$

where G is the shear modulus, I is the polar moment of inertia, and l is the rod's length. The shear modulus can be written as a function of the elastic modulus, E, and Poisson's ratio, v: $G = \frac{E}{2(1+v)}$. The polar moment of inertia for the rod with diameter, d, is $I = \frac{\pi d^4}{32}$.

Fig. 2.20 Lumped parameter model for single degree of freedom torsional system

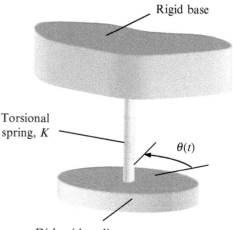

Rigid base

Torsional spring, K

$\theta(t)$

Disk with radius, r, mass, m, and mass moment of inertia, J

Fig. 2.21 Free body diagram for torsional system (top view through the rigid base in Fig. 2.20). The d'Alembert inertial torque is included

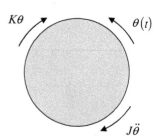

$K\theta$

$\theta(t)$

$J\ddot{\theta}$

The free body diagram, which includes the inertial torque, is shown in Fig. 2.21. Summing the torques (counterclockwise is assumed positive) gives $\sum T = -J\ddot{\theta} - K\theta = 0$. We can rewrite this to obtain the standard form for the torsional differential equation of harmonic motion:

$$\ddot{\theta} + \frac{K}{J}\theta = 0. \tag{2.52}$$

We can now directly identify the natural frequency: $\omega_n = \sqrt{\frac{K}{J}}$.

Let's take a look at the SI units for this system. For the spring constant, G has units of Pa or N/m^2, I has units of m^4, and l has units of m. Combining these, we obtain:

$$\frac{\frac{N}{m^2} m^4}{m} = N \cdot m. \tag{2.53}$$

However, we can see that the stiffness is multiplied by the rotation angle to give the final torque. Therefore, we can include units of radians in the stiffness denominator: $\frac{\text{N·m}}{\text{rad}}$. Remember that in mathematical "writing," we can include or exclude radians as desired.

We also need units for the disk's mass moment of inertia. As we saw with the rolling cylinder problem, $J = \frac{mr^2}{2} = \frac{md^2}{8}$, where r indicates radius, d is diameter, and m is the disk mass. The units are kg · m², but we will again include radians in the denominator for compatibility with the equation of motion: $\frac{\text{kg·m}^2}{\text{rad}}$. Let's now verify the natural frequency units.

$$\omega_n = \sqrt{\frac{\frac{\text{N·m}}{\text{rad}}}{\frac{\text{kg·m}^2}{\text{rad}}}} = \sqrt{\frac{\text{kg}\frac{\text{m}}{\text{s}^2}\text{m}}{\text{kg}\cdot\text{m}^2}} = \sqrt{\frac{1}{\text{s}^2}} = \frac{\text{rad}}{\text{s}}. \tag{2.54}$$

2.3.3   Nonlinear Springs

In this text, we are only considering vibration of linear systems. However, there are instances where a nonlinear model better describes the system behavior. One way to incorporate nonlinear behavior is through the use of nonlinear springs, or springs where the force is not linear with displacement as we showed in Fig. 2.2. A common nonlinear spring model, referred to as a *Duffing spring*, includes a cubic nonlinearity:

$$f = k_0 x + k_1 x^3. \tag{2.55}$$

For the Duffing spring, k_0 is the linear spring constant ($k_0 > 0$) and k_1 is the nonlinear spring constant. If k_1 is greater than zero, the spring is a hardening spring. If k_1 is less than zero, it is a softening spring. The (undamped) Duffing differential equation of motion is:

$$m\ddot{x} + k_0 x + k_1 x^3 = 0, \tag{2.56}$$

where the linear spring in Fig. 2.8 has simply been replaced by the Duffing spring. Analytical solutions for such nonlinear differential equations are difficult to obtain, but we do have some tools to help us understand their behavior. Let us explore the Duffing spring in MATLAB® MOJO 2.1. In this example, we will compare the displacement–force relationships for three cases: (1) $k_1 = 0$ (linear); (2) $k_1 = 2.5 \times 10^4$ N/m³ (hardening); and (3) $k_1 = -2.5 \times 10^4$ N/m³ (softening). In all instances, $k_0 = 1 \times 10^3$ N/m. The results are provided in Fig. 2.22. Note that these values were selected for plotting convenience and do not represent a particular physical system.

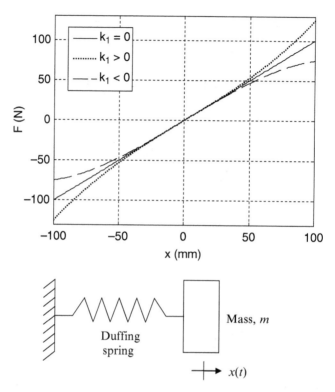

Fig. 2.22 Duffing spring responses for linear, hardening, and softening cases

MATLAB® MOJO 2.1

```
% matlab_mojo_2_1.m

clc
clear all
close all

% define variables
x = -0.1:0.001:0.1;          % m
k0 = 1000;                    % N/m

% define force
F = k0*x;
k1 = 25000;                  % N/m^3

Fstiff = F + k1*x.^3;        % stiffening spring
Fsoft = F - k1*x.^3;         % softening spring

figure(1)
plot(x*1e3, F, 'k-', x*1e3, Fstiff, 'k:', x*1e3, Fsoft, 'k--')
set(gca,'FontSize', 14)
xlabel('x (mm)')
ylabel('F (N)')
legend('k_1 = 0', 'k_1 > 0', 'k_1 < 0')
grid
axis([-100 100 -130 130])
```

2.4   Damped Harmonic Oscillator

Let's now include damping in the single degree of freedom spring–mass lumped parameter model. As we mentioned previously, all systems exhibit damping or "losses." There are three main types of damping used to model physical systems.

2.4.1   Viscous Damping

As we discussed previously, the viscous damping model relates force to velocity as shown in Eq. 2.2. In this equation, the proportionality constant c is the viscous damping coefficient with SI units of N-s/m. Physically, this model adequately describes the retarding force on a body that is moving at a moderate speed through a fluid. Because it is convenient to implement mathematically, it is the most common selection for modeling vibratory systems.

2.4.2   Coulomb Damping

This type of damping represents the energy dissipation due to dry sliding between two surfaces (or friction). In this case, the force, which always opposes the direction of motion, is described by Eq. 2.57. In this equation, μ is the friction coefficient and N is the normal force between the two bodies, which is perpendicular (normal) to the contacting surfaces.

$$f = \mu N \tag{2.57}$$

2.4.3   Solid Damping

Solid, or structural, damping occurs due to internal energy dissipation within the material of the vibrating body. Consider a steel beam floating in space. If an astronaut were to tap this beam at its end, it would begin to rotate about its center of mass – this is referred to as *rigid body motion* – and it would also begin vibrating. Since nothing is touching the beam and there is no obvious resistance to motion, why does the vibrating motion eventually stop? This is solid damping and we will describe it using a complex elastic modulus for the beam material, E_s in Sect. 8.3. See Eq. 2.58, where η is the material-dependent solid damping factor.

$$E_s = E(1 + i\eta) \tag{2.58}$$

2.4.4 Damped System Behavior

Since viscous damping is the most common modeling choice, we will consider it now to represent our single degree of freedom spring–mass–damper system. As shown in Figs. 2.1 and 2.3 from Sect. 2.1, the equation of motion is $m\ddot{x} + c\dot{x} + kx = 0$ (see Eq. 2.3). Because the vibratory response when the system is disturbed from its equilibrium position is again harmonic, we can select a solution of the form $x(t) = Xe^{st}$ with $\dot{x}(t) = sXe^{st}$ and $\ddot{x}(t) = s^2Xe^{st}$. Substituting in the equation of motion and grouping terms gives:

$$\left(ms^2 + cs + ks\right)Xe^{st} = 0. \tag{2.59}$$

As with the undamped model, there are two options for Eq. 2.59. If $Xe^{st} = 0$, no motion has occurred and this is referred to as the trivial solution. The characteristic equation is therefore:

$$ms^2 + cs + ks = 0. \tag{2.60}$$

The characteristic equation is quadratic in s and has two roots. Dividing by m gives $s^2 + \frac{c}{m}s + \frac{k}{m} = 0$ and the two roots can be determined using the quadratic equation.

$$s_{1,2} = \frac{-\frac{c}{m} \pm \sqrt{\left(\frac{c}{m}\right)^2 - 4(1)\frac{k}{m}}}{2(1)} = -\frac{c}{2m} \pm \sqrt{\left(\frac{c}{2m}\right)^2 - \frac{k}{m}} \tag{2.61}$$

The total solution for the system vibration is the sum of the two harmonic responses defined by the two roots: $x(t) = X_1 e^{s_1 t} + X_2 e^{s_2 t} = X_1 e^{\left(-\frac{c}{2m} + \sqrt{\left(\frac{c}{2m}\right)^2 - \frac{k}{m}}\right)t} + X_2 e^{\left(-\frac{c}{2m} - \sqrt{\left(\frac{c}{2m}\right)^2 - \frac{k}{m}}\right)t}$. This equation can be rewritten as shown in Eq. 2.62, where the first term in the product (i.e., the exponential term) describes the damping envelope that bounds the decaying oscillation and the second term (in parentheses) defines the oscillatory part.

$$x(t) = e^{\left(-\frac{c}{2m}\right)t}\left(X_1 e^{\left(\sqrt{\left(\frac{c}{2m}\right)^2 - \frac{k}{m}}\right)t} + X_2 e^{\left(-\sqrt{\left(\frac{c}{2m}\right)^2 - \frac{k}{m}}\right)t}\right) \tag{2.62}$$

The system behavior depends on the value of the radical expression, $\sqrt{\left(\frac{c}{2m}\right)^2 - \frac{k}{m}}$, and there are three possibilities.

1. If $\left(\frac{c}{2m}\right)^2 - \frac{k}{m} < 0$, the characteristic equation will have complex roots (i.e., s_1 and s_2 will have both real and imaginary parts). In this case, the response is vibratory (see Eq. 2.62).

2. If $\left(\frac{c}{2m}\right)^2 - \frac{k}{m} > 0$, the characteristic equation will have real-valued roots and there is no oscillation. The viscous damping is large enough to prevent vibration.

3. If $\left(\frac{c}{2m}\right)^2 - \frac{k}{m} = 0$, the two roots are real and equal, $s_{1,2} = -\frac{c}{2m}$, and there is just no oscillation. This case is referred to as *critical damping*. If a system with critical damping is displaced, it will return to its equilibrium position as quickly as possible without ever passing through the equilibrium position.

We can determine the damping coefficient required to achieve critical damping, c_c, by rearranging the radical equation for case 3: $\left(\frac{c_c}{2m}\right)^2 - \frac{k}{m} = 0$. We have that $\left(\frac{c_c}{2m}\right)^2 = \frac{k}{m}$ or $c_c = 2m\sqrt{\frac{k}{m}}$. Finally, by simplifying we obtain:

$$c_c = 2\sqrt{km}. \tag{2.63}$$

We can now define a dimensionless parameter, referred to as the *damping ratio*, using the critical damping coefficient. The damping ratio, ζ, is:

$$\zeta = \frac{c}{c_c} = \frac{c}{2\sqrt{km}}. \tag{2.64}$$

IN A NUTSHELL   It is often convenient to describe the damping of a system in comparison to the damping that would just prevent vibration. This is the damping ratio. A damping ratio of, say, 0.3 means that the system has 30% of the damping required to prevent vibration. Unless designers make special efforts, most mechanical structures have very low damping ratios – almost always less than 10% and more typically on the order of 1–5%.

Let's now see how we can rewrite the equation of motion to use the damping ratio. We have already seen the form $s^2 + \frac{c}{m}s + \frac{k}{m} = 0$. We can write the final term, $\frac{k}{m}$, as ω_n^2, but what about the $\frac{c}{m}$ coefficient on s? This can be expressed as $\frac{c}{m} = 2\zeta\omega_n$. Solving for ζ gives $\zeta = \frac{c}{2m\omega_n}$. This can be rewritten as $\zeta = \frac{c}{2\sqrt{m^2}\sqrt{\frac{k}{m}}} = \frac{c}{2\sqrt{\frac{km^2}{m}}} = \frac{c}{2\sqrt{km}}$, which validates the $\frac{c}{m} = 2\zeta\omega_n$ equation. We can therefore rewrite the equation of motion in the form shown in Eq. 2.65. We will use this form in many instances as we move forward.

$$s^2 + 2\zeta\omega_n s + \omega_n^2 = 0 \tag{2.65}$$

In the same way as we discussed for the radical expression value, there are three possibilities for the damping ratio.

1. If $\zeta < 1$ $(c < c_c)$, the response is vibratory with a magnitude that decays exponentially over time. The system that exhibits this behavior is referred to as *underdamped*. This is the situation for most mechanical systems.
2. If $\zeta > 1$ $(c > c_c)$, there is no vibration. This is the *overdamped* case.
3. If $\zeta = 1$ $(c = c_c)$, the system is *critically damped*.

2.4.5 Underdamped System

Let's explore the underdamped case in a little more detail. As stated earlier, the two roots of the characteristic equation will be complex-valued. These roots are calculated using Eq. 2.61, where the term under the radial is negative. Let's verify that the radial term is indeed negative for an underdamped system. We need to show that $\left(\frac{c}{2m}\right)^2 - \frac{k}{m} < 0$. This equation can be rewritten as $\frac{c}{2m} < \sqrt{\frac{k}{m}}$ (or $c < 2m\omega_n$) and simplified to obtain $c < 2\sqrt{m^2}\sqrt{\frac{k}{m}}$, or $c < 2\sqrt{km}$. Because this is equivalent to $c < c_c$, which indicates underdamped behavior, we have validated our assertion.

Let's rewrite the radical term as $-\left(\frac{k}{m} - \left(\frac{c}{2m}\right)^2\right)$. Using the undamped natural frequency and damping ratio, we can then redefine this expression as $-\left(\omega_n^2 - (\zeta\omega_n)^2\right) = -\left(\omega_n^2(1 - \zeta^2)\right)$. We can substitute this result in Eq. 2.61 to obtain a new form for the roots equation as shown in Eq. 2.66.

$$s_{1,2} = -\zeta\omega_n \pm \sqrt{-\left(\omega_n^2(1 - \zeta^2)\right)}$$
$$s_{1,2} = -\zeta\omega_n \pm i\omega_n\sqrt{1 - \zeta^2} \tag{2.66}$$

We can now introduce the *damped natural frequency*, ω_d, which describes the frequency of free vibration when damping is present:

$$\omega_d = \omega_n\sqrt{1 - \zeta^2}. \tag{2.67}$$

Because the damping ratio for mechanical systems is typically small ($\zeta < 0.1$), the value of the damped natural frequency is only slightly less than the undamped natural frequency. For example, if $\zeta = 0.05$ (5% viscous damping), then $\omega_d = \omega_n\sqrt{1 - 0.05^2} = 0.999\omega_n$.

The roots in Eq. 2.66 can now be written in more compact notation: $s_{1,2} = -\zeta\omega_n \pm i\omega_d$. Substitution in the assumed harmonic form enables us to rewrite Eq. 2.62 as shown in Eq. 2.68, where the first term in the product again describes the damping envelope that bounds the oscillation (at the damped natural frequency) defined by the second term in the product. See Fig. 2.23.

$$x(t) = e^{-\zeta\omega_n t}\left(X_1 e^{i\omega_d t} + X_2 e^{-i\omega_d t}\right) \tag{2.68}$$

For the exponential term in Eq. 2.68, a higher damping ratio gives a more rapid decay rate. For the oscillatory part, if the period of vibration is τ (expressed in seconds), the damped natural frequency (in Hz) is:

$$f_d = \frac{1}{\tau}. \tag{2.69}$$

Also, the relationship between the damped natural frequency in rad/s, ω_d, and Hz is $\omega_d = 2\pi f_d$, as we saw for the undamped case.

Fig. 2.23 Damped free
vibration response

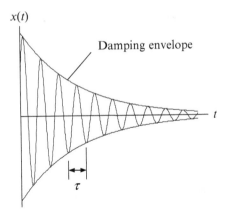

By the Numbers 2.2

Let's calculate the free vibration response for a spring–mass–damper system with
the following parameters:

- $m = 1$ kg
- $c = 20$ N-s/m
- $k = 1 \times 10^4$ N/m
- the initial displacement is $x_0 = 25$ mm
- the initial velocity is $\dot{x}_0 = 1{,}000$ mm/s.

The undamped natural frequency is $\omega_n = \sqrt{\frac{k}{m}} = \sqrt{\frac{1 \times 10^4}{1}} = 100$ rad/s.
Converting to Hz, we obtain $f_n = \frac{\omega_n}{2\pi} = 15.9$ Hz. The damping ratio is $\zeta = \frac{c}{2\sqrt{km}} = $
$\frac{20}{2\sqrt{1 \times 10^4(1)}} = 0.1$ (or 10% damping). The damped natural frequency is therefore
$\omega_d = \omega_n \sqrt{1 - \zeta^2} = 100\sqrt{1 - 0.1^2} = 99.5$ rad/s.

For this underdamped case, we can use the form for the solution provided in
Eq. 2.68. In order to apply the initial conditions, we also need the velocity, which
we obtain by calculating the time derivative of Eq. 2.68. See Eq. 2.70.

$$\dot{x}(t) = e^{-\zeta\omega_n t}(i\omega_d)\left(X_1 e^{i\omega_d t} - X_2 e^{-i\omega_d t}\right) - \zeta\omega_n e^{-\zeta\omega_n t}\left(X_1 e^{i\omega_d t} + X_2 e^{-i\omega_d t}\right) \quad (2.70)$$

Substituting $t = 0$ into Eqs. 2.68 and 2.70 gives:

$$x(0) = x_0 = X_1 + X_2 \quad (2.71)$$

and

$$\dot{x}(0) = \dot{x}_0 = i\omega_d(X_1 - X_2) - \zeta\omega_n(X_1 + X_2). \quad (2.72)$$

Fig. 2.24 X_1 and X_2 complex conjugates

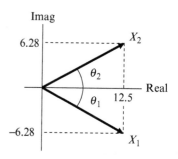

Using Eq. 2.71, we can substitute for $(X_1 + X_2)$ in Eq. 2.72. This gives $\dot{x}_0 = i\omega_d(X_1 - X_2) - \zeta\omega_n x_0$. Solving this equation for $(X_1 - X_2)$ yields $X_1 - X_2 = \frac{\dot{x}_0 + \zeta\omega_n x_0}{i\omega_d}$. We can rationalize this equation by multiplying both the numerator and denominator by the imaginary variable i:

$$X_1 - X_2 = \frac{\dot{x}_0 + \zeta\omega_n x_0}{i\omega_d}\frac{i}{i} = \frac{-i(\dot{x}_0 + \zeta\omega_n x_0)}{\omega_d}. \tag{2.73}$$

We can now add Eqs. 2.71 and 2.73 to eliminate X_2. The result is:

$$X_1 = \frac{x_0}{2} - i\frac{(\dot{x}_0 + \zeta\omega_n x_0)}{2\omega_d}. \tag{2.74}$$

Using Eq. 2.71, we can then determine X_2.

$$X_2 = \frac{x_0}{2} + i\frac{(\dot{x}_0 + \zeta\omega_n x_0)}{2\omega_d} \tag{2.75}$$

The coefficients X_1 and X_2 are complex conjugates. This is always the case for this form of the underdamped harmonic motion solution. Substituting the values for this example into Eqs. 2.74 and 2.75 gives:

$$X_1 = \frac{25}{2} - i\frac{(100 + 0.1(100)25)}{2(99.5)} = 12.5 - i6.28$$

and

$$X_2 = 12.5 + i6.28.$$

Let's rewrite these expressions in vector notation before substituting in the response equation (Eq. 2.68). The vector representations of these complex conjugates are plotted in Fig. 2.24. Using this figure, we can determine the magnitude and phase for each vector. The magnitude, which is the square root of

Fig. 2.25 Phase calculation
for X_2

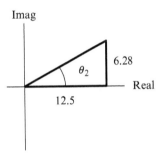

the sum of the squares of the real (Re) and imaginary (Im) parts, is the same for
both vectors and is given by:

$$|X_{1,2}| = \sqrt{Re^2 + Im^2} = \sqrt{12.5^2 + 6.28^2} = 13.99.$$

The phase is the inverse tangent of the ratio of the imaginary part to the real
part; this is evident from Fig. 2.25 which highlights the "triangle" formed by
the real and imaginary parts of the X_2 vector. The corresponding phase is
$\theta_2 = \tan^{-1}\left(\frac{Im}{Re}\right) = \tan^{-1}\left(\frac{6.28}{12.5}\right) = 0.466$, rad $= 26.67°$. For the X_1 vector, the phase
is $\theta_1 = \tan^{-1}\left(\frac{Im}{Re}\right) = \tan^{-1}\left(\frac{-6.28}{12.5}\right) = -0.466$, rad $= -26.67°$.
The vector notations for X_1 and X_2 are then:

$$X_1 = 13.99e^{-i0.466}$$
$$X_2 = 13.99e^{i0.466}.$$

We can now substitute in Eq. 2.68, where $\zeta\omega_n = 0.1(100) = 10$ rad/s and
$\omega_d = 99.5$ rad/s.

$$x(t) = e^{-\zeta\omega_n t}\left(X_1 e^{i\omega_d t} + X_2 e^{-i\omega_d t}\right) = e^{-10t}\left(13.99e^{-i0.466}e^{i99.5t} + 13.99e^{i0.466}e^{-i99.5t}\right)$$
$$x(t) = 13.99e^{-10t}\left(e^{i(99.5t-0.466)} + e^{-i(99.5t-0.466)}\right)$$

Using Eq. 1.12 (derived from Euler's formula), we can rewrite this equation as:

$$x(t) = 2(13.99)e^{-10t}\cos(99.5t - 0.466) = 27.98e^{-10t}\cos(99.5t - 0.466) \text{ mm},$$

where the magnitude is 27.98 mm, the phase is -0.466 rad, and the decay rate is
described by the exponential term e^{-10t}. This damped free vibration response is shown
in Fig. 2.26, where the initial value $x(0) = 27.98e^{-10(0)}\cos(99.5(0) - 0.466) =$
$27.98\cos(-0.466) = 25$ mm matches the initial displacement, $x_0 = 25$ mm, and
the *period of vibration* is:

$$\tau = \frac{1}{f_d} = \frac{2\pi}{\omega_d} = \frac{2\pi}{\omega_n\sqrt{1 - \zeta^2}} = \frac{2\pi}{100\sqrt{1 - 0.1^2}} = 0.063 \text{ s}.$$

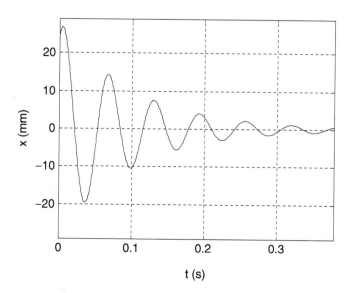

Fig. 2.26 *By the Numbers 2.2* – free vibration response for damped system

Figure 2.26 was produced using the code provided in MATLAB® MOJO 2.2.

MATLAB® MOJO 2.2

```
% matlab_mojo_2_2.m

clc
clear all
close all

% define variables
A = 27.98;                      % mm
omegan = 100;                   % rad/s
zeta = 0.1;
omegad = omegan*sqrt(1-zeta^2);
fd = omegad/2/pi;               % Hz
tau = 1/fd;                     % s
phi = 0.466;                    % rad
t = 0:tau/100:6*tau;            % s

% define displacement
x = A*exp(-zeta*omegan*t).*cos(omegad*t - phi);        % mm

figure(1)
plot(t, x, 'k-')
set(gca,'FontSize', 14)
xlabel('t (s)')
ylabel('x (mm)')
grid
axis([0 max(t) -29 29])
```

By the Numbers 2.3

Let's now consider the free vibration response for a spring–mass–damper system with an initial displacement $x(0) = x_0$ and zero initial velocity for three cases: (1) $\zeta = 2$ (overdamped); (2) $\zeta = 0.2$ (underdamped); and (3) $\zeta = 1$ (critically damped).

1. Overdamped ($\zeta = 2$)

   Because there are two real-valued roots for the characteristic equation, we can write the response in the following form:

$$x(t) = X_1 e^{\left(-\zeta+\sqrt{\zeta^2-1}\right)\omega_n t} + X_2 e^{\left(-\zeta-\sqrt{\zeta^2-1}\right)\omega_n t}. \tag{2.76}$$

   The corresponding velocity equation is:

$$\dot{x}(t) = \left(-\zeta+\sqrt{\zeta^2-1}\right)\omega_n X_1 e^{\left(-\zeta+\sqrt{\zeta^2-1}\right)\omega_n t} + \left(-\zeta-\sqrt{\zeta^2-1}\right)\omega_n X_2 e^{\left(-\zeta-\sqrt{\zeta^2-1}\right)\omega_n t}. \tag{2.77}$$

   For $\zeta = 2$, the exponent term $-\zeta \pm \sqrt{\zeta^2-1}$ is $-2 \pm \sqrt{2^2-1} = -2 \pm \sqrt{3}$. Substituting gives:

$$x(t) = X_1 e^{\left(-2+\sqrt{3}\right)\omega_n t} + X_2 e^{\left(-2-\sqrt{3}\right)\omega_n t}$$

   for position and:

$$\dot{x}(t) = \left(-2+\sqrt{3}\right)\omega_n X_1 e^{\left(-2+\sqrt{3}\right)\omega_n t} + \left(-2-\sqrt{3}\right)\omega_n X_2 e^{\left(-2-\sqrt{3}\right)\omega_n t}.$$

   for velocity.

   We can now apply the initial conditions $x(0) = x_0$ and $\dot{x}(0) = 0$. Substituting gives: $x(0) = \dot{x}_0 = X_1 + X_2$ and $\dot{x}(0) = 0 = \left(-2+\sqrt{3}\right)\omega_n X_1 + \left(-2-\sqrt{3}\right)\omega_n X_2$. The velocity equation can be simplified to be $\dot{x}(0) = 0 = \left(-2+\sqrt{3}\right)X_1 + \left(-2-\sqrt{3}\right)X_2$ by dividing both sides by ω_n. We now have a system of two linear equations with two unknowns. We could solve for X_1 and X_2 by a variety of methods, but let's use *matrix inversion* here since we will begin writing our equations of motion in matrix form for the two degree of freedom systems in Sect. 4.1. Our two equations in matrix form are:

$$\begin{bmatrix} 1 & 1 \\ -2+\sqrt{3} & -2-\sqrt{3} \end{bmatrix} \begin{bmatrix} X_1 \\ X_2 \end{bmatrix} = \begin{bmatrix} x_0 \\ 0 \end{bmatrix},$$

where the position equation is the top row of the 2×2 (rows $\times$ columns) matrix multiplied by the 2×1 column vector $\begin{bmatrix} X_1 \\ X_2 \end{bmatrix}$ and set equal to the top, or $(1,1)$, entry, x_0, in the 2×1 column vector on the right-hand side of the equal sign $\begin{bmatrix} x_0 \\ 0 \end{bmatrix}$.

We perform this term-by-term multiplication to obtain $(1)X_1 + (1)X_2 = x_0$. Similarly, the second row gives the velocity equation. We multiply the $(2,1)$ entry, $-2+\sqrt{3}$, in the 2×2 matrix by X_1 and the $(2,2)$ entry, $-2-\sqrt{3}$, by X_2 and set the result equal to 0, the $(2,1)$ entry in the column vector on the right-hand side of the equation. If we rewrite this matrix equation as $A\vec{X} = \vec{B}$, then algebraically we know that we can find $\vec{X}$ by moving A to the right-hand side of the equation. For scalar values, we can simply divide both sides by A or, equivalently, multiply both sides by the inverse of A. We can write this as $\vec{X} = A^{-1}\vec{B}$. In our matrix problem, we can perform the same operation, but determining the inverse of the 2×2 A matrix requires a little work.[2] Let's first write the $\vec{X} = A^{-1}\vec{B}$ equation explicitly.

$$\begin{bmatrix} X_1 \\ X_2 \end{bmatrix} = \begin{bmatrix} 1 & 1 \\ -2+\sqrt{3} & -2-\sqrt{3} \end{bmatrix}^{-1} \begin{bmatrix} x_0 \\ 0 \end{bmatrix}$$

Inverting the 2×2 A matrix to determine A^{-1} requires three steps.

1. Switch the on-diagonal terms. This means that we replace the $(1,1)$ entry, 1, by the $(2,2)$ entry, $-2-\sqrt{3}$, and the $(2,2)$ entry by the $(1,1)$ entry.
2. Change the signs of the off-diagonal terms. The $(1,2)$ entry becomes -1 and the $(2,1)$ entry becomes $-(-2+\sqrt{3}) = 2 - \sqrt{3}$.
3. Divide each entry in the matrix by the *determinant*. For a 2×2 matrix, the determinant is the difference between the product of the on-diagonal terms and the product of the off-diagonal terms, i.e., $(1,1)(2,2) - (1,2)(2,1)$. For our matrix, this is:

$$(1)\left(-2-\sqrt{3}\right) - (1)\left(-2+\sqrt{3}\right) = -2 - \sqrt{3} + 2 - \sqrt{3} = -2\sqrt{3}.$$

The result for the matrix inversion is:

$$\begin{bmatrix} 1 & 1 \\ -2+\sqrt{3} & -2-\sqrt{3} \end{bmatrix}^{-1} = \frac{1}{-2\sqrt{3}} \begin{bmatrix} -2-\sqrt{3} & -1 \\ 2-\sqrt{3} & 1 \end{bmatrix}$$

$$= \frac{1}{2\sqrt{3}} \begin{bmatrix} 2+\sqrt{3} & 1 \\ -2+\sqrt{3} & -1 \end{bmatrix}.$$

---

[2] These matrix manipulations are a subset of the topics covered in a *linear algebra* course.

Matrix inversion can also be completed in MATLAB® using the inv command. From the command prompt (≫), we first define the 2×2 A matrix. The semicolon indicates the end of the first row in the matrix description.

```
>> A = [1 1;-2+sqrt(3) -2-sqrt(3)]

A =

    1.0000    1.0000
   -0.2679   -3.7321
```

We next determine the matrix inverse. The answer (ans) shown below is identical to the result we obtained using the three-step inversion procedure.

```
>> inv(A)

ans =

    1.0774    0.2887
   -0.0774   -0.2887
```

The complete matrix equation is now:

$$\begin{bmatrix} X_1 \\ X_2 \end{bmatrix} = \frac{1}{2\sqrt{3}} \begin{bmatrix} 2+\sqrt{3} & 1 \\ -2+\sqrt{3} & -1 \end{bmatrix} \begin{bmatrix} x_0 \\ 0 \end{bmatrix}.$$

We write the X_1 equation using the top row:

$$X_1 = \frac{1}{2\sqrt{3}}\left((2+\sqrt{3})x_0 + (1)0\right) = \frac{2+\sqrt{3}}{2\sqrt{3}}x_0 \approx 1.08x_0.$$

We determine X_2 from the bottom row:

$$X_2 = \frac{1}{2\sqrt{3}}\left((-2+\sqrt{3})x_0 + (-1)0\right) = \frac{-2+\sqrt{3}}{2\sqrt{3}}x_0 \approx -0.08x_0.$$

Substituting in Eq. 2.76 gives:

$$x(t) = 1.08x_0 e^{(-2+\sqrt{3})\omega_n t} - 0.08x_0 e^{(-2-\sqrt{3})\omega_n t} \text{ or}$$

$$\frac{x(t)}{x_0} = 1.08e^{(-0.27)\omega_n t} - 0.08e^{(-3.73)\omega_n t}.$$

The ratio is plotted as a function of $\omega_n t$ in Fig. 2.27.

2. Underdamped ($\zeta = 0.2$)

We can write the response for the underdamped system with two complex conjugate roots from the characteristic equation as shown in Eq. 2.68:

$$x(t) = e^{-\zeta\omega_n t}\left(X_1 e^{i\sqrt{1-\zeta^2}\omega_n t} + X_2 e^{-i\sqrt{1-\zeta^2}\omega_n t}\right)$$

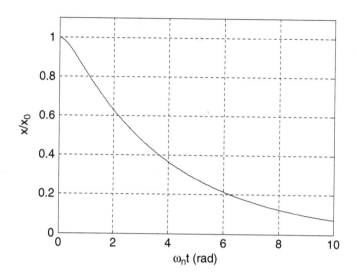

Fig. 2.27 *By the Numbers 2.3* – response for overdamped system with initial displacement only

or, equivalently, as:

$$x(t) = Xe^{-\zeta\omega_n t} \sin\left(\sqrt{1 - \zeta^2}\omega_n t + \phi\right). \tag{2.78}$$

The velocity equation is:

$$\dot{x}(t) = -\zeta\omega_n Xe^{-\zeta\omega_n t} \sin\left(\sqrt{1 - \zeta^2}\omega_n t + \phi\right)$$
$$+ \sqrt{1 - \zeta^2}\omega_n Xe^{-\zeta\omega_n t} \cos\left(\sqrt{1 - \zeta^2}\omega_n t + \phi\right). \tag{2.79}$$

In this example, $x(0) = x_0$, $\dot{x}(0) = 0$, and $\sqrt{1 - \zeta^2} = \sqrt{1 - 0.2^2} = \sqrt{0.96}$. Substituting gives:

$$x(0) = x_0 = X\sin(\phi) \text{ and}$$

$$\dot{x}(0) = 0 = -0.2\omega_n X\sin(\phi) + \sqrt{0.96}\omega_n X\cos(\phi).$$

We can use the velocity equation to determine ϕ. Rewriting gives:

$$0.2\omega_n X\sin(\phi) = \sqrt{0.96}\omega_n X\cos(\phi) \text{ or } \frac{X\sin(\phi)}{X\cos(\phi)} = \tan(\phi) = \frac{\sqrt{0.96}\omega_n}{0.2\omega_n}.$$

This yields $\phi = 1.37$ rad $= 78.5°$. Using the position equation, we solve for X:

$$X = \frac{x_0}{\sin(\phi)} = \frac{x_0}{\sqrt{0.96}}.$$

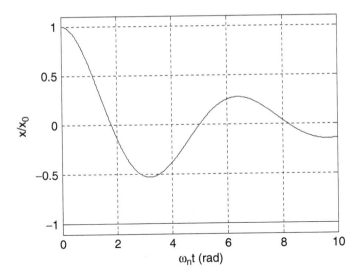

Fig. 2.28 *By the Numbers 2.3* – response for underdamped system with initial displacement only

Substituting in Eq. 2.78 gives:

$$x(t) = \frac{x_0}{\sqrt{0.96}} e^{-0.2\omega_n t} \sin\left(\sqrt{0.96}\omega_n t + 1.37\right) \text{ or}$$

$$\frac{x(t)}{x_0} = \frac{1}{\sqrt{0.96}} e^{-0.2\omega_n t} \sin\left(\sqrt{0.96}\omega_n t + 1.37\right).$$

This ratio is plotted in Fig. 2.28. Note that the response is now oscillatory for the underdamped system.

3. Critically damped ($\zeta = 1$)

The response for this case is written as:

$$x(t) = (X_1 + X_2 t)e^{-\omega_n t} \tag{2.80}$$

because there are two repeated, real-valued roots. The corresponding velocity equation is:

$$\dot{x}(t) = -\omega_n(X_1 + X_2 t)e^{-\omega_n t} + X_2 e^{-\omega_n t}. \tag{2.81}$$

Substituting the initial conditions, $x(0) = x_0$ and $\dot{x}(0) = 0$, into Eqs. 2.80 and 2.81 gives $x(0) = x_0 = X_1$ and $\dot{x}(0) = 0 = -\omega_n(X_1) + X_2$. From the velocity equation, we obtain $X_2 = \omega_n X_1 = \omega_n x_0$. We can now replace X_1 and X_2 in Eq. 2.80 to find the position as a function of ω_n, x_0, and t. The ratio $\frac{x(t)}{x_0}$ is plotted versus $\omega_n t$ in Fig. 2.29. As a check on this figure, we can see that the value is 1 at $\omega_n t = 0$ as expected, $\frac{x}{x_0} = (1+0)e^0 = 1$.

$$x(t) = (x_0 + \omega_n x_0 t)e^{-\omega_n t} \text{ or } \frac{x(t)}{x_0} = (1 + \omega_n t)e^{-\omega_n t}$$

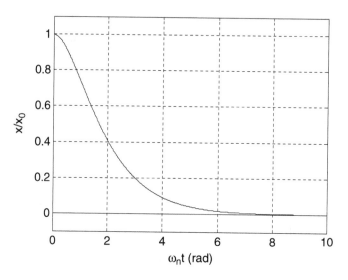

Fig. 2.29 *By the Numbers 2.3* – response for critically damped system with initial displacement only

2.4.6 *Damping Estimate from Free Vibration Response*

We have said that all physical systems include damping and we have seen that the amount of damping influences the resulting free vibration response. We can now "reverse engineer" this analysis. We will use the behavior during free oscillation to determine the amount of damping in a system. Quantifying damping is important because it is difficult to predict using models. Designers can use the material properties and dimensions of the components in a structure to predict the natural frequencies using finite element analysis, for example. However, the magnitude of the vibrations that occur when the structure is excited is more challenging to predict based only on first principles. Therefore, damping is typically identified experimentally.

We have already seen in Eq. 2.78 that the free vibration response of an underdamped single degree of freedom system can be written as $x(t) = Xe^{-\zeta\omega_n t} \sin\left(\sqrt{1 - \zeta^2}\omega_n t + \phi\right)$. The general response is shown in Fig. 2.30, where the response value is identified at two time instants, t_1 and t_2, separated by the *damped period of vibration*:

$$\tau_d = \frac{1}{f_d} = \frac{2\pi}{\omega_d} = \frac{2\pi}{\omega_n\sqrt{1 - \zeta^2}}. \tag{2.82}$$

Fig. 2.30 General response
for the free vibration of an
underdamped single degree of
freedom system

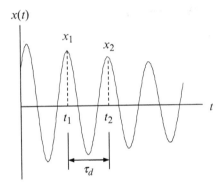

We see that the response decreases from x_1 to x_2 over the period, τ_d. For the
viscously damped system, we also know that the decay is exponential and depends
on the damping ratio, ζ. We can therefore define the *logarithmic decrement*, δ,
to describe this amplitude reduction:

$$\delta = \ln\left(\frac{x_1}{x_2}\right). \tag{2.83}$$

We can now substitute for x_1 and x_2 in Eq. 2.83 using the expression $x(t) =
Xe^{-\zeta\omega_n t}\sin\left(\sqrt{1-\zeta^2}\omega_n t + \phi\right)$ evaluated at the times, t_1 and t_2:

$$\delta = \ln\left(\frac{Xe^{-\zeta\omega_n t_1}\sin\left(\sqrt{1-\zeta^2}\omega_n t_1 + \phi\right)}{Xe^{-\zeta\omega_n t_2}\sin\left(\sqrt{1-\zeta^2}\omega_n t_2 + \phi\right)}\right). \tag{2.84}$$

We can simplify Eq. 2.84 by recognizing that $t_2 = t_1 + \tau_d$.

$$\delta = \ln\left(\frac{Xe^{-\zeta\omega_n t_1}\sin\left(\sqrt{1-\zeta^2}\omega_n t_1 + \phi\right)}{Xe^{-\zeta\omega_n (t_1+\tau_d)}\sin\left(\sqrt{1-\zeta^2}\omega_n (t_1 + \tau_d) + \phi\right)}\right) \tag{2.85}$$

Because the sine function is periodic, we know that $\sin(t_1) = \sin(t_1 + \tau_d)$. This
enables us to rewrite Eq. 2.85 because the sine terms in the numerator and
denominator cancel.

$$\delta = \ln\left(\frac{e^{-\zeta\omega_n t_1}}{e^{-\zeta\omega_n(t_1+\tau_d)}}\right) = \ln\left(\frac{e^{-\zeta\omega_n t_1}}{e^{-\zeta\omega_n t_1}e^{-\zeta\omega_n\tau_d}}\right) = \ln\left(e^{\zeta\omega_n\tau_d}\right) = \zeta\omega_n\tau_d \tag{2.86}$$

We can now experimentally determine the damping ratio by following three steps.

1. Substitute for τ_d in Eq. 2.86 using Eq. 2.82.

$$\delta = \zeta\omega_n\tau_d = \zeta\omega_n \frac{2\pi}{\omega_n\sqrt{1-\zeta^2}} = \frac{2\pi\zeta}{\sqrt{1-\zeta^2}} \qquad (2.87)$$

2. Measure x_1 and x_2 during free vibration and calculate δ using Eq. 2.83.
3. Solve for ζ from Eq. 2.87. Rewriting gives $\delta\sqrt{1-\zeta^2} = 2\pi\zeta$. By squaring both sides, we obtain $\delta^2(1-\zeta^2) = 4\pi^2\zeta^2$. Combining terms yields $\zeta^2 = \frac{\delta^2}{4\pi^2+\delta^2}$, which we can write as:

$$\zeta = \sqrt{\frac{\delta^2}{4\pi^2 + \delta^2}}. \qquad (2.88)$$

Alternately, we can recognize that $\sqrt{1-\zeta^2} \approx 1$ for small ζ so that we can rewrite Eq. 2.87 as $\delta \simeq 2\pi\zeta$. Solving for ζ gives:

$$\zeta \simeq \frac{\delta}{2\pi}. \qquad (2.89)$$

IN A NUTSHELL   Equation 2.89 provides a very useful result. In a physical single degree of freedom system, we know how to measure the mass, m, using a balance, for example. We can apply a static force and measure the resulting deflection to identify the static stiffness, k. Damping is more difficult to quantify. We often are not even sure about the source of the damping. Is it viscous? Did it come from friction? An easy way to determine the equivalent viscous damping is to initiate vibration and then measure the height of two successive vibration peaks. The natural logarithm of the ratio of the peak heights divided by 2π closely approximates the damping ratio. Experimentally, when the damping ratio is low, the difference in two successive peak heights may be small and, therefore, difficult to measure accurately. However, because this ratio of peak heights holds for any two successive peaks, the accuracy of the estimate may be improved by considering the heights of peaks separated by several cycles, N. That is, $\frac{x_0}{x_N} = \left(\frac{x_0}{x_1}\right)\left(\frac{x_1}{x_2}\right)\left(\frac{x_2}{x_3}\right)...\left(\frac{x_{N-1}}{x_N}\right) = \left(\frac{x_0}{x_1}\right)^N$, so $\ln\left(\frac{x_0}{x_N}\right) = N\ln\left(\frac{x_0}{x_1}\right) = N\delta$. This means that in order to calculate the damping ratio using the peak height change over several cycles, we compute the natural logarithm of the ratio, divide by 2π, and divide this result by the number of cycles. Additionally, this approach can be used to determine how many cycles of the motion would be required for the vibration to fall below a predetermined amplitude. If the damping ratio is known, then we can solve for N.

2.4.7   Damping Estimate Uncertainty

For any measurement, there is always *uncertainty* associated with the result. For the logarithmic decrement, the combined standard uncertainty in the damping ratio, $u_c(\zeta)$, depends on the standard uncertainties in the measurements of x_1 and x_2, $u(x_1)$ and $u(x_2)$ (Taylor and Kuyatt 1994). To determine how these uncertainties are related, we can perform a first-order Taylor series expansion of Eq. 2.89 after substituting for δ: $\zeta \simeq \frac{1}{2\pi} \ln\left(\frac{x_1}{x_2}\right) = \frac{1}{2\pi}(\ln(x_1) - \ln(x_2))$. If we neglect any potential relationship (or correlation) between the x_1 and x_2 values, represented by the *covariance*, we can find $u_c(\zeta)$ using:

$$u_c^2(\zeta) = \left(\frac{\partial \zeta}{\partial x_1}\right)^2 u^2(x_1) + \left(\frac{\partial \zeta}{\partial x_2}\right)^2 u^2(x_2), \tag{2.90}$$

where the partial derivatives are $\frac{\partial \zeta}{\partial x_1} = \frac{1}{2\pi}\frac{1}{x_1}$ and $\frac{\partial \zeta}{\partial x_2} = \frac{1}{2\pi}\frac{1}{x_2}$. Substituting gives:

$$u_c(\zeta) = \frac{1}{2\pi} \sqrt{\frac{u^2(x_1)}{x_1^2} + \frac{u^2(x_2)}{x_2^2}}, \tag{2.91}$$

where the average (or mean) values of x_1 and x_2 are used to evaluate the combined standard uncertainty.

By the Numbers 2.4

The tip displacement for a freely vibrating cantilever beam was measured as shown in Fig. 2.31. The displacement values, x_1 and x_2, were recorded at two times, separated by the damped vibration period of 0.01 s; these values were 0.93 mm and 0.82 mm. The manufacturer-specified measurement uncertainty for the *linear variable differential transformer*[3] (LVDT) used to perform the displacement measurement was 0.01 mm. Let's determine the damped natural frequency, mean value of the damping ratio, and the associated uncertainty in the damping ratio for this measurement activity.

First, we use Eq. 2.82 to find the damped natural frequency, $f_d = \frac{1}{\tau_d} = \frac{1}{0.01} = 100$ Hz. Second, for the mean damping ratio, we use Eq. 2.83 to calculate the logarithmic decrement, $\delta = \ln\left(\frac{x_1}{x_2}\right) = \ln\left(\frac{0.93}{0.82}\right) = 0.13$. The damping ratio is then

---

[3] An LVDT is a transformer with three coils placed next to one another around a tube. A ferromagnetic core slides in and out of the tube; this cylindrical core usually serves as the moving probe. An alternating current is passed through the center coil and, as the core moves, voltages are induced in the outer coils. These voltages are used to determine the core displacement (http://en.wikipedia.org/wiki/Linear_variable_differential_transformer).

Fig. 2.31 *By the Numbers*
2.4 – measurement of free
oscillation for a cantilever
beam using an LVDT

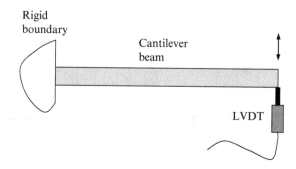

$\zeta \simeq \frac{0.13}{2\pi} = 0.02$, or 2%. Third, the uncertainty in this value is determined using
Eq. 2.71.

$$u_c(\zeta) = \frac{1}{2\pi}\sqrt{\frac{0.01^2}{0.93^2} + \frac{0.01^2}{0.82^2}} = 0.002$$

This combined standard uncertainty represents one standard deviation of the
mean value of ζ. We can interpret it this way. For a normal (or *Gaussian*) distribution of ζ values, we would expect subsequent measurements to appear within the
interval of ± 0.002 about the mean value 68.2% of the time. For (approximately) a
95% *level of confidence*, we would expand this interval to $\pm 2u_c(\zeta) = 0.004$.

Before moving to Sect. 2.5 and unstable behavior, let's return to the Argand
diagram and compare the undamped and damped cases. The undamped free vibration response described by $x(t) = X_u \cos(\omega_n t + \phi_u)$ is shown in Fig. 2.32, where
$\phi_u = \tan^{-1}\left(\frac{-\dot{x}_0}{\omega_n x_0}\right)$ and $X_u = \frac{x_0}{\cos(\phi_u)}$ (the u subscript represents "undamped"). With
the addition of damping, the response is $x(t) = X_d e^{-\zeta \omega_n t} \cos(\omega_d t + \phi_d)$, where
$\phi_d = \tan^{-1}\left(\frac{-\zeta \omega_n x_0 - \dot{x}_0}{\omega_d x_0}\right)$ and $X_d = \frac{x_0}{\cos(\phi_d)}$ (the d subscript represents "damped").
The damped case is pictured in Fig. 2.33 for comparison purposes. Because the
response decays over time, the vector length decreases and the tip traces a "spiral"
pattern.

2.5   Unstable Behavior

For the single degree of freedom damped oscillator, we detailed the solution to
the equation of motion $m\ddot{x} + c\dot{x} + kx = 0$ with the initial conditions $x(0)$ and $\dot{x}(0)$.
We discussed four possible scenarios based on the damping ratio: $\zeta < 1$
(underdamped); $\zeta = 0$ (no damping); $\zeta = 1$ (critically damped); and $\zeta > 1$
(overdamped). For $\zeta < 1$, we saw responses of the type shown in Fig. 2.34. These
decaying responses are called *asymptotically stable* because they exponentially

Fig. 2.32 Argand diagram
for undamped free
vibration response
$x(t) = X_u \cos(\omega_n t + \phi_u)$

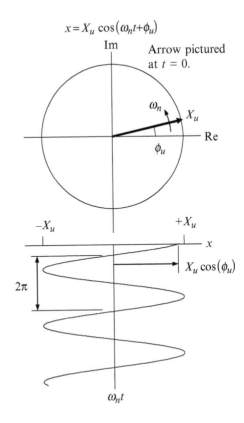

approach zero (the equilibrium position) as time progresses. For the $\zeta = 0$ response
in Fig. 2.35, the behavior is *marginally stable*; the response neither grows nor
decays. The $\zeta = 1$ and $\zeta > 1$ cases are also stable. Like the $\zeta < 1$ case, these
responses approach the equilibrium position as time increases.

What if the response does not approach the equilibrium position? This is called
unstable behavior. We will discuss two types: flutter instability (or self-excited
vibration) and divergent instability.

2.5.1   Flutter Instability

Let's again consider the equation of motion $m\ddot{x} + c\dot{x} + kx = 0$. For positive mass
values ($m > 0$), unstable behavior is obtained if c or k are less than zero. This is
referred to as negative damping or stiffness, respectively. This unstable behavior
leads to oscillatory motion that grows, rather than decays, over time. It is referred to
as *flutter instability* or *self-excited vibration*.

$$x = X_d e^{-\zeta \omega_n t} \cos(\omega_d t + \phi_d)$$

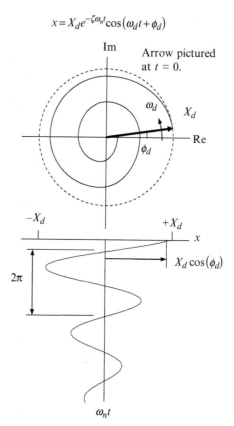

Fig. 2.33 Argand diagram for damped free vibration response $x(t) = X_d e^{-\zeta \omega_n t} \cos(\omega_d t + \phi_d)$

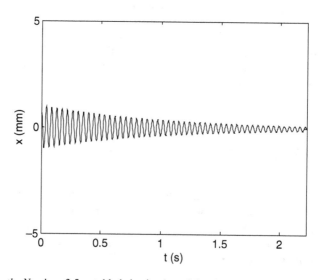

Fig. 2.34 *By the Numbers 2.5 – stable behavior* $(c = 1.1 c_{\mathrm{lim}})$

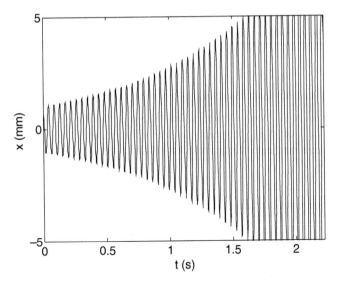

Fig. 2.35 *By the Numbers 2.5* – flutter instability ($c = 0.9c_{\text{lim}}$)

Consider a structure moving through a fluid, such as an aircraft wing moving through the air, that can be modeled as a single degree of freedom system with a velocity-dependent aerodynamic force, $f = \gamma\dot{x}$, acting on it (Inman 2001). The equation of motion is:

$$m\ddot{x} + c\dot{x} + kx = \gamma\dot{x}, \tag{2.92}$$

where we will specify that m, c, k, and γ are all positive. Rewriting this equation yields $m\ddot{x} + (c - \gamma)\dot{x} + kx = 0$. Now we have two possibilities for the effective damping coefficient, $c - \gamma$. If $c - \gamma > 0$, then $\zeta = \frac{c-\gamma}{2\sqrt{km}} > 0$ and we obtain stable behavior. However, if $c - \gamma < 0$, then $\zeta < 0$ and the response grows exponentially over time (flutter). If $c - \gamma = 0$, the system is marginally stable.

While we have only considered analytical solutions to the differential equations of motion so far, let's now solve Eq. 2.92 numerically. We will implement *Euler integration* in a time-domain simulation to determine the system behavior for various $c - \gamma$ values.

The simulation is carried out in small *time steps*, *dt*. Because we are numerically integrating the system equation of motion to determine the resulting vibration, care must be exercised in selecting *dt*. If the value is too large, inaccurate results are obtained. As a rule of thumb, it is generally acceptable to set *dt* to be at least ten times smaller than the period corresponding to the highest natural frequency in the system's dynamic model.

Given the equation of motion $m\ddot{x} + (c - \gamma)\dot{x} + kx = 0$, the acceleration in the current simulation time step is determined using:

$$\ddot{x} = \frac{-(c - \gamma)\dot{x} - kx}{m},$$

where the velocity, $\dot{x}$, and position, x, from the previous time step are used (for the first time step, they are set equal to the initial conditions). The velocity for the current time step is then determined by Euler integration:

$$\dot{x} = \dot{x} + \ddot{x} \cdot dt,$$

where the velocity on the right-hand side of the equation is retained from the previous time step and used to update the current value (on the left-hand side of the equation). The current velocity is then applied to determine the current displacement according to:

$$x = x + \dot{x} \cdot dt.$$

Again, the displacement on the right-hand side of the equation is retained from the previous time step. Finally, the time-dependent displacement can be written to a vector, y, as:

$$y_n = x,$$

where the n subscript on y indicates the time step. This is used to simplify the "book keeping" in the MATLAB® program; see By the Numbers 2.5 and MATLAB® MOJO 2.3. The corresponding time is $t_n = n \cdot dt$.

By the Numbers 2.5

Let's choose the following parameters for Eq. 2.92: $m = 500$ kg, $k = 1 \times 10^7$ N/m, and $\gamma = 1 \times 10^4$ N-s/m with initial conditions of $x(0) = 1 \times 10^{-3}$ m and $\dot{x}(0) = 0$. At the limit of stability, $c - \gamma = 0$. Therefore, for a given γ value, the associated limiting c value is $c_{lim} = \gamma$. In this case, $c_{lim} = 1 \times 10^4$ N-s/m. For $c > c_{lim}$, the system is stable. If $c < c_{lim}$, flutter occurs. Using the code provided in MATLAB® MOJO 2.3, the results shown in Figs. 2.34 and 2.35 were obtained. In Fig. 2.34, the system is stable with $c = 1.1c_{lim}$. Unstable results are seen in Fig. 2.35, where $c = 0.9c_{lim}$. Exponentially increasing oscillatory behavior is observed. For these figures, the response is plotted over 50 periods of vibration, $\tau = \frac{1}{f_n} = 2\pi\sqrt{\frac{m}{k}} = 0.044$ s, using a time step of $dt = \frac{\tau}{20} = 0.0022$ s.

MATLAB® MOJO 2.3
```
% matlab_mojo_2_3.m

clc
close all
clear all

% Define parameters
m = 500;                          % kg
k = 1e7;                          % N/m
gamma = 1e4;                      % N-s/m
clim = gamma;                     % N-s/m
fn = 1/(2*pi)*sqrt(k/m);          % Hz
tau = 1/fn;                       % s
c = 0.9*clim;

% Define simulation variables
dt = tau/20;                      % sec/step
steps = round(50*tau/dt);

% Euler integration initial conditions
x = 1e-3;                         % m
dx = 0;                           % m/s

% Initialize final position and time vectors
y = zeros(1, steps);
time = zeros(1, steps);

for cnt = 1:steps
    ddx = (-(c - gamma)*dx - k*x)/m;   % m/s^2
    dx = dx + ddx*dt;                  % m/s
    x = x + dx*dt;                     % m

    % Write results to vectors
    y(cnt) = x;                        % m
    time(cnt) = cnt*dt;                % s
end

figure(1)
plot(time, y*1e3, 'k')
xlim([0 max(time)])
ylim([-5 5])
set(gca,'FontSize', 14)
xlabel('t (s)')
ylabel('x (mm)')
```

2.5.2   Divergent Instability

In order to model and simulate divergent instability, we will use the *inverted pendulum*, composed of a mass, m, supported by a massless rod of length, l, that rotates about the frictionless pivot, O. The rod is held in its vertical equilibrium position by springs and dampers as shown in Fig. 2.36. The free body diagram for

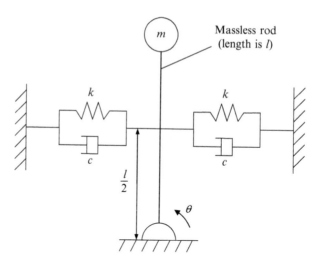

Fig. 2.36 Inverted pendulum

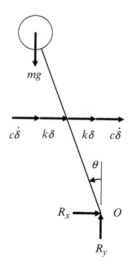

Fig. 2.37 Free body diagram for inverted pendulum

a rotation, θ, of the mass/rod counterclockwise from the equilibrium position is provided in Fig. 2.37. The forces that need to be considered include:

- the gravity force, mg
- the two spring forces, $k\delta$, where $\delta = \frac{l}{2} \sin(\theta)$ is the horizontal deflection due to rotation of the rod as shown in Fig. 2.38
- the two viscous damping forces, $c\dot{\delta}$, where $\dot{\delta} = \frac{d\delta}{dt} = \frac{l}{2} \cos(\theta) \cdot \dot{\theta}$ is the horizontal velocity
- the reaction forces at the pivot in the horizontal, R_x, and vertical, R_y, directions.

In order to sum moments, M, about O, the forces can be redrawn to give the components perpendicular and parallel to the massless rod. Only the perpendicular

Fig. 2.38 Relationship
between δ and θ

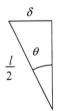

Fig. 2.39 Free body diagram
with force components
perpendicular and parallel to
the massless rod. The
d'Alembert inertial moment
is also included

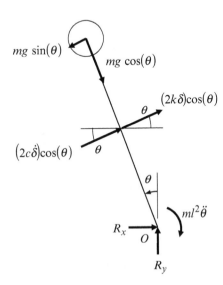

components need to be considered in the moment sum and, because we are
summing about O, the reaction components may be neglected. The free body
diagram is shown in Fig. 2.39, where d'Alembert's inertial moment, $J\ddot{\theta} = ml^2\ddot{\theta}$,
with mass moment of inertia, J, is also included so that $\sum M_O = 0$.

Using Fig. 2.39, the moment sum is:

$$ml^2\ddot{\theta} + \left(2c\dot{\delta}\right)\cos(\theta)\frac{l}{2} + (2k\delta)\cos(\theta)\frac{l}{2} - mg\sin(\theta)l = 0.$$

Substituting for δ and $\dot{\delta}$ gives:

$$ml^2\ddot{\theta} + \left(2c\frac{l}{2}\cos(\theta)\cdot\dot{\theta}\right)\cos(\theta)\frac{l}{2} + \left(2k\frac{l}{2}\sin(\theta)\right)\cos(\theta)\frac{l}{2} - mg\sin(\theta)l = 0.$$

For small rotations, $\sin(\theta) \approx \theta$ and $\cos(\theta) \approx 1$. Substituting these approximations and combining terms yields:

$$ml^2\ddot{\theta} + c\frac{l^2}{2}\dot{\theta} + \left(k\frac{l^2}{2} - mgl\right)\theta = 0.$$

If $k\frac{l^2}{2} - mgl < 0$, the effective stiffness is negative and divergent unstable behavior
is obtained. In this situation, the motion grows without bound and no oscillation
occurs when the pendulum is disturbed from its equilibrium position. Physically, this

means that the spring force is insufficient to counteract gravity and the pendulum simply falls over for nonzero initial conditions. However, if $k\frac{l}{2} - mgl > 0$, the pendulum oscillates around its equilibrium position when disturbed from equilibrium and the response eventually decays to zero due to the viscous dampers. The limiting spring stiffness is found using $k_{lim}\frac{l}{2} - mgl = 0$. Rewriting gives $k_{lim} = \frac{2mg}{l}$. If $k < k_{lim}$, divergent instability occurs.

By the Numbers 2.6

Consider the inverted pendulum in Fig. 2.36 with $m = 0.5$ kg, $c = 1$ N-s/m, and $l = 0.3$ m. The limiting spring stiffness is $k_{lim} = \frac{2(0.5)9.81}{0.3} = 32.7$ N/m. For initial conditions of $\theta(0) = 5°$ and $\dot{\theta}(0) = 0$, let's determine the response $\theta(t)$ using Euler integration. As with the aircraft wing example, the first step is to solve the equation of motion for the acceleration term.

$$\ddot{\theta} = \frac{c\frac{l}{2}\dot{\theta} + \left(k\frac{l}{2} - mgl\right)\theta}{ml^2}$$

The angular velocity of the pendulum for the current time step is then determined by:

$$\dot{\theta} = \dot{\theta} + \ddot{\theta} \cdot dt,$$

where the angular velocity on the right-hand side of the equation is the value from previous time step. This new velocity is then used to calculate the current angle.

$$\theta = \theta + \dot{\theta} \cdot dt$$

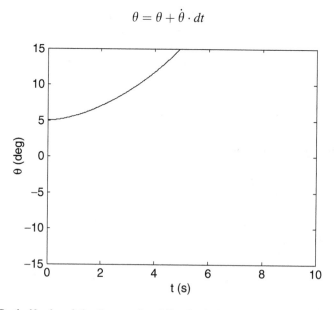

Fig. 2.40 *By the Numbers 2.6* – divergent instability for the inverted pendulum with $k = 0.99k_{lim}$

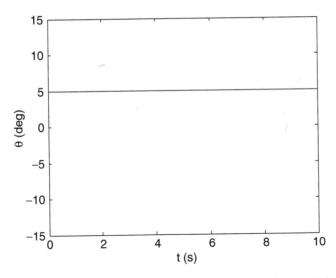

Fig. 2.41 *By the Numbers 2.6* – marginal stability for the inverted pendulum with $k = k_{\lim}$

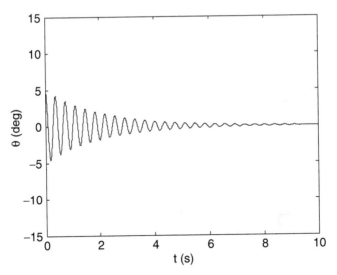

Fig. 2.42 *By the Numbers 2.6* – asymptotic stability for the inverted pendulum with $k = 10k_{\lim}$

Using the code provided in MATLAB® MOJO 2.4, the results displayed in Figs. 2.40–2.42 were obtained for three k values:

- $k = 0.99k_{\lim}$
- $k = k_{\lim}$
- $k = 10k_{\lim}$.

In Fig. 2.40, the system exhibits divergent instability with $k = 0.99k_{lim}$. The pendulum mass simply falls in the direction of the initial angular offset. Marginal stable results are seen in Fig. 2.41, where $k = k_{lim}$. There is no mass rotation because the spring restoring force exactly offsets the gravity force. Exponentially decaying oscillatory behavior is observed in Fig. 2.42, where $k = 10k_{lim}$. In this asymptotically stabile case, the pendulum's angular position oscillates about the equilibrium position (zero rotation). For these figures, the response is plotted over 2,000 steps using a time increment of $dt = 0.005$ s.

MATLAB® MOJO 2.4

```
% matlab_mojo_2_4.m

clc
close all
clear all

% Define parameters
m = 0.5;                          % kg
c = 1;                            % N-s/m
g = 9.81;                         % m/s^2
l = 0.3;                          % m
klim = 2*m*g/l;                   % N/m
k = 0.99*klim;

% Define simulation variables
dt = 0.005;        % sec/step
steps = 2000;

% Euler integration initial conditions
theta = 5*pi/180;                 % rad
dtheta = 0;                       % rad/s

% Initialize final theta vector and time vector
th = zeros(1, steps);
time = zeros(1, steps);

for cnt = 1:steps
    ddtheta = (-(c*l^2/2)*dtheta-(k*l^2/2-m*g*l)*theta)/(m*l^2);   % rad/s^
    dtheta = dtheta + ddtheta*dt;                 % rad/s
    theta = theta + dtheta*dt;                    % rad

    % Write results to vectors
    th(cnt) = theta;                              % rad
    time(cnt) = cnt*dt;                           % s
end

figure(1)
plot(time, th*180/pi, 'k')
xlim([0 max(time)])
ylim([-15 15])
set(gca,'FontSize', 14)
xlabel('t (s)')
ylabel('\theta (deg)')
```

2.6 Free Vibration Measurement

To conclude this chapter, let's introduce the *beam experimental platform (BEP)* that we will use to demonstrate various concepts throughout this text. The design dimensions and materials are provided in Appendix A. A photograph of the measurements setup is shown in Fig. 2.43. The 12.7-mm diameter steel rod is clamped in the base with an overhang length of 125 mm. An *accelerometer* is attached to the cantilever beam's free end. This is a piezoelectric measurement transducer that gives a voltage which is proportional to acceleration. We will discuss it in more detail in Sect. 7.3.3. The rod was disturbed from equilibrium by a light tap from a small hammer and the resulting vibration was recorded. A plot of the acceleration, a, versus time, t, is provided in Fig. 2.44, where the hammer impact was applied at 0.005 s. We see that the response resembles the underdamped free vibration results we have already studied; it decays exponentially over time. However, there are also some differences. It does not uniformly decay to zero; in fact, it seems to grow and decrease periodically within the overall damping envelope. This is because there are actually multiple natural frequencies of the continuous beam excited simultaneously by the hammer tap. To model this behavior, we need to consider multiple degrees of freedom. As we move forward, we will discuss both multiple degree of freedom systems and modeling techniques for continuous beams.

Fig. 2.43 Photograph of free vibration measurement using the BEP

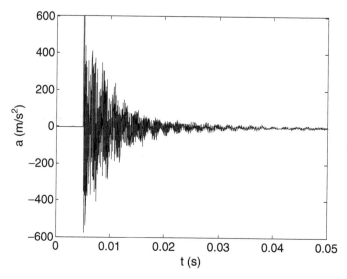

Fig. 2.44 Tip acceleration for the BEP due to a hammer tap disturbance at $t = 0.005$ s

Chapter Summary

- The lumped parameter model, composed of a mass (located at each coordinate) that is supported by massless springs and/or dampers, can be used to represent physical systems.
- D'Alembert's principle can be used to represent a dynamic model as a static system by including the inertial force(s) or moment(s) in the free body diagram.
- A system's characteristic equation is used to determine its natural frequency(s).
- Initial conditions are used to solve the differential equation of motion and determine the time-domain response for free vibration.
- When applying an energy-based approach, the expressions for kinetic and potential energy are used to identify the equation of motion for a system.
- An equivalent spring can be developed for springs in series and springs in parallel.
- The Duffing spring, which includes a cubic nonlinearity, can be used to describe the behavior of nonlinear systems.
- Three primary damping models are viscous, Coulomb, and solid damping.
- The dimensionless damping ratio, ζ, is used to describe the behavior of damped systems.
- For underdamped systems, where $\zeta < 1$, the response is vibratory with a magnitude that decays exponentially over time.
- For overdamped systems, where $\zeta > 1$, there is no vibration.
- Matrix inversion can be used to solve linear systems of equations.
- The logarithmic decrement, which is determined from measurements of free vibration, can be used to estimate the damping ratio.

- All measurement results include uncertainty. Given an equation for the value in question (expressed as a function of other measured inputs), the uncertainty of that value can be determined by a Taylor series expansion of the equation.
- Two types of instability are: (1) flutter or self-excited vibration; and (2) divergent instability.
- The equation of motion can be solved numerically using Euler integration.
- An accelerometer can be used to measure the vibration of structures.

Exercises

1. For a single degree of freedom spring–mass system with $m = 1$ kg and $k = 4 \times 10^4$ N/m, complete the following for the case of free vibration.

   (a) Determine the natural frequency in Hz and the corresponding period of vibration.
   (b) Given an initial displacement of 5 mm and zero initial velocity, write an expression for the time response of free vibration using the following form:

   $$x(t) = X_1 e^{i\omega_n t} + X_2 e^{-i\omega_n t} \text{ mm}$$

   (c) Plot the first ten cycles of motion for the result from part (b).

2. For a single degree of freedom spring–mass system, complete the following.

   (a) If the free vibration is described as $x(t) = A\cos(\omega_n t + \Phi_c)$, determine expressions for A and Φ_c if the initial displacement is x_0 and the initial velocity is $\dot{x}_0$.
   (b) If the free vibration is described as $x(t) = A\cos(\omega_n t) + B\sin(\omega_n t)$, determine expressions for A and B if the initial displacement is x_0 and the initial velocity is $\dot{x}_0$.

3. The differential equation of motion for a cylinder rolling on a concave cylindrical surface is $\ddot{\theta} + \frac{2}{3}\frac{g}{R-r}\theta = 0$, where g is the gravitational constant.

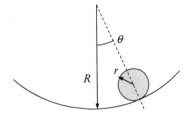

Fig. P2.3 Cylinder rolling on a concave cylindrical surface

(a) Write the expression for the natural frequency, ω_n (rad/s).
(b) If the free vibration is described as $\theta(t) = A\sin(\omega_n t + \Phi_s)$, determine expressions for A and Φ_s if the initial angle is θ_0 and the initial angular velocity is $\dot{\theta}_0$.
(c) If $R = 200$ mm, $r = 10$ mm, $\theta_0 = 5° = 0.087$ rad, and $\dot{\theta}_0 = 0$, plot $\theta(t)$ (deg) using the function from part b for the time interval from $t = 0$ to 5 s in steps of 0.005 s.

4. For a single degree of freedom spring–mass–damper system with $m = 1$ kg, $k = 4\times10^4$ N/m, and $c = 10$ N-s/m, complete the following for the case of free vibration.

(a) Calculate the natural frequency (in rad/s), damping ratio, and damped natural frequency (in rad/s).
(b) Given an initial displacement of 5 mm and zero initial velocity, write the expression for the underdamped, free vibration in the form $x(t) = e^{-\zeta\omega_n t}(A\cos(\omega_d t) + B\sin(\omega_d t))$ mm.
(c) Plot the first ten cycles of motion.
(d) Calculate the viscous damping value, c (in N-s/m), to give the critically damped case for this system.

5. For a single degree of freedom spring–mass–damper system with $m = 0.2$ lb$_m$, $k = 2.5\times10^3$ lb$_f$/in., $c = 10.92$ lb$_f$-s/ft, $x_0 = 0.1$ in., and $\dot{x}_0 = 0$, complete the following for the case of free vibration.

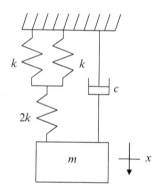

Fig. P2.5 Single degree of freedom spring–mass–damper system under free vibration

(a) Determine the equivalent spring constant (in lb$_f$/in) for the spring configuration shown in the figure.
(b) Determine the force (in lb$_f$) required to cause the initial displacement of 0.1 in. (assume the system was at static equilibrium prior to introducing the initial displacement).
(c) Calculate the damping ratio. You will need the units correction factor: (32.2 ft-lb$_m$)/(lb$_f$-s^2). Is this system underdamped or overdamped?
(d) Calculate the damped natural frequency (in Hz).

6. For a single degree of freedom spring–mass–damper system under free vibration, determine the values for the mass, m (kg), viscous damping coefficient, c (N-s/m), and spring constant, k (N/m), given the following information:

   • the damping ratio is 0.1
   • the undamped natural frequency is 100 Hz
   • the initial displacement is 1 mm
   • the initial velocity is 5 mm/s
   • if the system was critically damped, the value of the damping coefficient would be 586.1 N-s/m.

7. For a single degree of freedom spring–mass–damper system under free vibration, the following information is known: $m = 2$ kg, $k = 1 \times 10^6$ N/m, $c = 500$ N-s/m, $x_0 = 4$ mm, and $\dot{x}_0 = 0$ mm/s.

   Determine the corresponding expression for velocity (mm/s) if position is given in the form $x(t) = Ae^{-\zeta \omega_n t} \cos(\omega_d t + \varphi_c)$. Numerically evaluate all coefficients and constant terms in your final expression.

8. The requirement for small features on small parts has led to increased demands on measuring systems. One approach for determining the size of features (such as a hole's diameter) is to use a probe to touch the surface at several locations (e.g., points on the hole wall) and then use these coordinates to calculate the required dimension. To probe small features, small probes are required. However, at small size and force scales, intermolecular forces can dominate.

   An example is the interaction between very small, flexible probes and surfaces. As shown in Fig. P2.8, a 72-μm diameter probe tip comes into contact

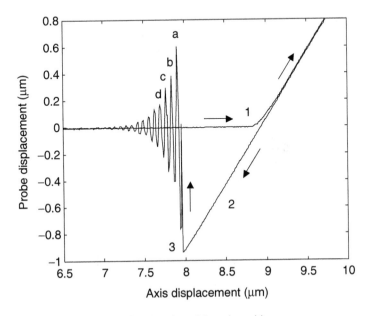

Fig. P2.8 Probe displacement as a function of the axis position

with a measurement surface (1) and the probe begins to deflect. The attractive (van der Waals) force between the probe and surface causes the tip to "stick" to the surface as it is retracted after contact (2). The motion is provided by an axis which moves the probe relative to the surface. Once the retraction force on the probe overcomes the attractive force, the probe is released from the surface (3) and it oscillates under free vibration conditions. Given the probe's free vibration response, determine the damping ratio using the logarithmic decrement approach. The peaks from the free vibration response are provided in Table P2.8. (The data in Fig. P2.8 is courtesy of IBS Precision Engineering, Eindhoven, The Netherlands.)

Table P2.8 Peak values for probe-free vibration

Peak label	Peak value (μm)
a	0.60
b	0.38
c	0.29
d	0.16

9. If the free vibration of a single degree of freedom spring–mass system is described as $x(t) = A \sin(\omega_n t + \Phi_s)$, determine expressions for A and Φ_s if the initial displacement is x_0 and the initial velocity is $\dot{x}_0$.

10. For a single degree of freedom spring–mass–damper system, the free vibration response shown in the Fig. P2.10a was obtained due to an initial displacement with no initial velocity.

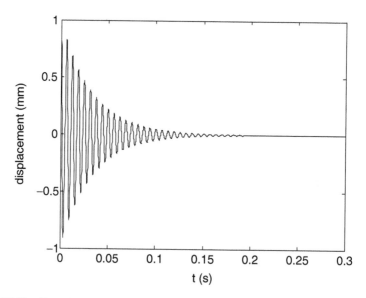

Fig. P2.10a Free vibration response for a single degree of freedom spring–mass–damper system

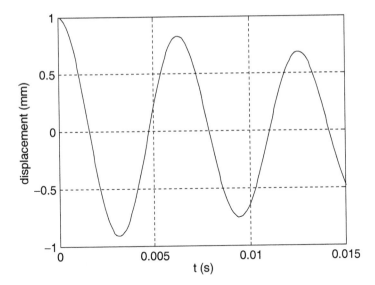

Fig. P2.10b First few cycles of free vibration response for a single degree of freedom spring–mass–damper system

(a) Determine the damping ratio using the logarithmic decrement. Figure P2.10b, which shows just the first few cycles of oscillation, is provided to aid in this calculation.
(b) What was the initial displacement for this system?
(c) Determine the period of oscillation and corresponding damped natural frequency (in Hz).
(d) If the system mass is 1 kg, determine the spring constant (in N/m).

References

http://en.wikipedia.org/wiki/Linear_variable_differential_transformer
Inman D (2001) Engineering vibration, 2nd edn. Prentice Hall, Upper Saddle River
Taylor B, Kuyatt C (1994) NIST technical note 1297, 1994 edn. Guidelines for evaluating and expressing the uncertainty of NIST measurement results, National Institute of Standards and Technology, Gaithersburg
Thomson W, Dahley M (1998) Theory of vibrations with applications, 5th edn. Prentice Hall, Upper Saddle River

Chapter 3
Single Degree of Freedom Forced Vibration

Imagination decides everything.

– Blaise Pascal

3.1 Equation of Motion

Let's continue our study of the lumped parameter spring–mass–damper model, but now consider *forced vibration*. While the oscillation decays over time for a damped system under free vibration, the vibratory motion is maintained at a constant magnitude and frequency when an external energy source (i.e., a forcing function) is present. In Fig. 3.1, a harmonic input force has been added to the model, $f(t) = Fe^{i\omega t}$, where ω is the forcing frequency.

 IN A NUTSHELL   We have already seen that the complex exponential notation can be used to represent sine and cosine functions. The addition of the input force in Fig. 3.1 simply means that the system is being excited by a sinusoidal force. Additionally, because any periodic signal can be expressed by a sum of sine and cosine functions with different frequencies and amplitudes and because the system is linear and superposition can be applied, the discussion that follows applies to systems excited by any periodic force.

The free-body diagram (including the inertial force) is also provided in Fig. 3.1. By summing the forces in the x direction, $\sum f_x = 0$, the equation of motion is determined:

$$m\ddot{x} + c\dot{x} + kx = Fe^{i\omega t}, \tag{3.1}$$

where viscous damping is included in the model.

T.L. Schmitz and K.S. Smith, *Mechanical Vibrations: Modeling and Measurement*,
DOI 10.1007/978-1-4614-0460-6_3, © Springer Science+Business Media, LLC 2012

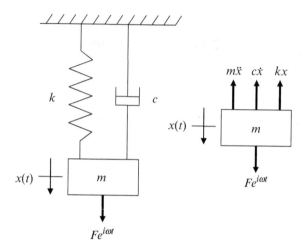

Fig. 3.1 Spring–mass–damper model with a harmonic input force. The free-body diagram is included

3.2   Frequency Response Function

The total solution to the forced vibration equation of motion (Eq. 3.1) has two parts: (1) the homogeneous, or *transient*, solution; and (2) the particular, or *steady-state*, solution. The transient portion is the free vibration response, $m\ddot{x} + c\dot{x} + kx = 0$, and, as we saw in Chap. 2, it rapidly decays for damped systems. The steady-state portion remains after the transient has attenuated and it persists as long as the force is acting on the system. The particular solution takes the same form as the forcing function after the transients are damped out. The resulting vibration has the same frequency as the harmonic force. Specifically, given the force $f(t) = Fe^{i\omega t}$, the corresponding steady-state response can be written as $x(t) = Xe^{i\omega t}$. Given this form for the position, the velocity is $\dot{x}(t) = i\omega Xe^{i\omega t}$ and the acceleration is $\ddot{x}(t) = (i\omega)^2 Xe^{i\omega t} = -\omega^2 Xe^{i\omega t}$. Substituting these expressions in Eq. 3.1, we obtain:

$$-m\omega^2 Xe^{i\omega t} + i\omega c Xe^{i\omega t} + kXe^{i\omega t} = Fe^{i\omega t}. \tag{3.2}$$

Grouping terms gives:

$$\left(-m\omega^2 + i\omega c + k\right)Xe^{i\omega t} = Fe^{i\omega t}. \tag{3.3}$$

This relates the force to the resulting vibration as a function of the forcing frequency, ω. Both sides of Eq. 3.3 include $e^{i\omega t}$, so we can eliminate it. We are now considering the response in the *frequency domain* rather than the time domain since t no longer appears in the equation. Let's rewrite Eq. 3.3 so that we have the ratio of the output (the complex-valued vibration, X) to the input (the real-valued force, F). This is referred to as the *frequency response function* (FRF) for physical systems.

$$\frac{X}{F}(\omega) = G(\omega) = \frac{1}{-m\omega^2 + i\omega c + k} \tag{3.4}$$

IN A NUTSHELL   We see in Eq. 3.4 that the relationship between the exciting force and the resulting vibration depends not only on the system parameters (m, k, and c) and amplitude of the exciting force (F), but also on the frequency of the excitation (ω).

In the *Laplace domain* ($s = \sigma + i\omega$), frequencies from negative infinity to positive infinity ($-\infty \le \omega \le +\infty$) are considered. In this case, the ratio $\frac{X}{F}(s)$ is referred to as the *transfer function* for the system. We can obtain the transfer function from the forced vibration equation of motion, $m\ddot{x} + c\dot{x} + kx = f(t)$, using the Laplace transforms of $x(t)$ and $f(t)$. These are $\mathcal{L}x(t) = X(s) = \int_0^\infty e^{-st}x(t)dt$ and $\mathcal{L}f(t) = F(s) = \int_0^\infty e^{-st}f(t)dt$. We also need the velocity:

$$\mathcal{L}\dot{x}(t) = sX(s) - x(0)$$

and acceleration:

$$\mathcal{L}\ddot{x}(t) = s^2X(s) - sx(0) - \dot{x}(0).$$

Substituting into the equation of motion gives:

$$m\big(s^2X(s) - sx(0) - \dot{x}(0)\big) + c\big(sX(s) - x(0)\big) + kX(s) = F(s).$$

For the transfer function, we are considering the steady-state response so we can neglect the transients and let $x(0) = \dot{x}(0) = 0$. This yields:

$$\big(ms^2 + cs + k\big)X(s) = F(s),$$

which can be rewritten as the transfer function:

$$\frac{X}{F}(s) = \frac{1}{ms^2 + cs + k}.$$

For our purposes, however, we are interested in the measurement and subsequent modeling of physical systems. Therefore, we will limit our discussions to the FRF, which considers only positive frequencies and the system-specific damping.

IN A NUTSHELL   For those with a background in controls, the FRF is a special case of the transfer function. The transfer function is a surface (like a tent) above the s plane (i.e., the plane with a σ axis and an $i\omega$ axis). "Poles" (like tent poles) are where that surface rises very high (to ∞) and "zeroes" are where the tent touches the ground (s plane). The FRF is a slice through the tent fabric along the axis where $\sigma = 0$ (the frequency axis).

Let's check the zero frequency (static or DC) case for Eq. 3.4. Substituting $\omega = 0$, we obtain:

$$\frac{X}{F}(0) = G(0) = \frac{1}{-m(0)^2 + i(0)c + k} = \frac{1}{k}.$$

This is simply Hooke's law that we discussed in Sect. 2.1. For the static case, we see that $F = kX$. In this instance, the response X is not complex; it has no imaginary component. Let's now rewrite the FRF to express it as a function of the natural frequency, ω_n, and damping ratio, ζ, rather than m, c, and k.

$$G(\omega) = \frac{1}{k - m\omega^2 + i\omega c} = \frac{1}{m}\left(\frac{1}{\left(\frac{k}{m} - \omega^2\right) + i\frac{c}{m}\omega}\right)$$

From Chap. 2, we know that $\frac{k}{m} = \omega_n^2$ and $\frac{c}{m} = 2\zeta\omega_n$. Substituting gives:

$$G(\omega) = \frac{1}{m}\left(\frac{1}{(\omega_n^2 - \omega^2) + i2\zeta\omega_n\omega}\right).$$

Multiply this result by $\frac{k}{k}$ to obtain:

$$G(\omega) = \frac{1}{k}\frac{k}{m}\left(\frac{1}{(\omega_n^2 - \omega^2) + i2\zeta\omega_n\omega}\right) = \frac{1}{k}\left(\frac{\omega_n^2}{(\omega_n^2 - \omega^2) + i2\zeta\omega_n\omega}\right).$$

It is common to rewrite this equation using the frequency ratio, $r = \frac{\omega}{\omega_n}$, which expresses how close the excitation frequency is to the natural frequency of the system. Dividing the numerator and denominator of the term in parentheses by ω_n^2, the new FRF form is:

$$G(\omega) = \frac{1}{k}\left(\frac{1}{\left(1 - \left(\frac{\omega}{\omega_n}\right)^2\right) + i2\zeta\left(\frac{\omega}{\omega_n}\right)}\right). \tag{3.5}$$

Substituting the frequency ratio r for $\frac{\omega}{\omega_n}$ yields the more compact FRF equation:

$$G(r) = \frac{1}{k}\left(\frac{1}{(1 - r^2) + i2\zeta r}\right). \tag{3.6}$$

In Eq. 3.6, $r = 1$ represents a special case. In this instance:

$$G(1) = \frac{1}{k}\left(\frac{1}{\left(1 - (1)^2\right) + i2\zeta(1)}\right) = \frac{1}{k}\left(\frac{1}{i2\zeta}\right).$$

This gives the largest value of $G(r)$ for a fixed value of ζ. It is called *resonance*. Physically, this means that when a system is forced at its natural frequency ($\omega = \omega_n$), the steady-state response is largest and its amplitude depends on the system stiffness and damping ratio. Intuitively, a larger stiffness or damping ratio gives a smaller response. Note that the resonant response is purely imaginary.

IN A NUTSHELL   The resonant response, $G(r = 1)$, is not imaginary in the sense that it does not exist. "Purely imaginary" means that if the excitation is represented by a sine function, then the resulting vibration at resonance is represented by a cosine function. The force and vibration reach their maximum values at different times, and, at resonance, the force is maximum when the displacement is zero.

Rather than leaving the FRF in the form shown in Eq. 3.6, let's rationalize by multiplying the numerator and denominator by the complex conjugate of the denominator.

$$G(r) = \frac{1}{k}\left(\frac{1}{(1 - r^2) + i2\zeta r} \cdot \frac{(1 - r^2) - i2\zeta r}{(1 - r^2) - i2\zeta r}\right) = \frac{1}{k}\left(\frac{(1 - r^2) - i2\zeta r}{(1 - r^2)^2 + (2\zeta r)^2}\right) \quad (3.7)$$

This function has both real and imaginary parts. The *real part* is:

$$\text{Re}(G(r)) = \frac{1}{k}\left(\frac{(1 - r^2)}{(1 - r^2)^2 + (2\zeta r)^2}\right) \quad (3.8)$$

and the *imaginary part* is:

$$\text{Im}(G(r)) = \frac{1}{k}\left(\frac{-2\zeta r}{(1 - r^2)^2 + (2\zeta r)^2}\right). \quad (3.9)$$

We can also express Eq. 3.7 in terms of *magnitude* and *phase*. This relates to the vector description we have discussed previously (see Figs. 2.4 and 2.5, for example). Figure 3.2 demonstrates the relationships between the real and imaginary parts and the magnitude and phase. The magnitude is calculated according to:

$$|G(r)| = \sqrt{(\text{Re}(G(r)))^2 + (\text{Im}(G(r)))^2} = \frac{1}{k}\sqrt{\frac{1}{(1 - r^2)^2 + (2\zeta r)^2}}, \quad (3.10)$$

Fig. 3.2 Vector description of the FRF's magnitude and phase

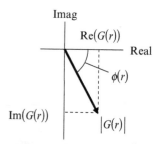

and the phase[1] is determined using:

$$\phi(r) = \tan^{-1}\left(\frac{\text{Im}(G(r))}{\text{Re}(G(r))}\right) = \tan^{-1}\left(\frac{-2\zeta r}{1 - r^2}\right). \tag{3.11}$$

Physically, the magnitude gives the size of the vibration response and the phase describes how much the vibration lags the harmonic, oscillating force. To better understand the phase lag, consider the *Slinky*® – a helical spring toy invented by Richard James in the 1940s (http://en.wikipedia.org/wiki/Slinky). By holding the Slinky® vertically and attaching several coils at the bottom together (using a rubber band, for example), you can approximate a spring–mass–damper system (although the damping is quite low). Now begin moving your hand up and down very slowly. You will observe that the "mass" at the bottom basically follows your hand's motion. As you increase the frequency of your hand's oscillation, however, you will see that the mass exhibits a different behavior; now, its motion directly opposes your hand's motion. For the low forcing frequency, the phase lag is near zero. For the higher frequency, the phase lag is close to 180° (i.e., $\phi \approx -180°$ and the mass's motion is out of phase with your hand's motion). The sharp transition between "in phase" and "out of phase" vibration is observed due to the low damping in the system.

As an alternative to Eqs. 3.10 and 3.11, the magnitude and phase can be expressed in terms of the model parameters m, k, and c. See Eqs. 3.12 and 3.13.

$$|G(r)| = \sqrt{\frac{1}{(k - m\omega^2)^2 + (c\omega)^2}} \tag{3.12}$$

$$\phi = \tan^{-1}\left(\frac{-c\omega}{k - m\omega^2}\right) \tag{3.13}$$

---

[1] Note that the tangent function exhibits quadrant dependence in the complex plane. In MATLAB® the atan2 function can be used to respect this quadrant-dependent behavior.

3.3 Evaluating the Frequency Response Function

Let's now plot the FRF as a function of the frequency ratio, r (or, equivalently, the forcing frequency, ω). Figure 3.3 shows the FRF magnitude from Eq. 3.10. Results are provided for ζ values of 0.01, 0.05, and 0.1 with $k = 1 \times 10^6$ N/m. We see that the peak height is reduced with increased damping. This "sharpness" of the magnitude peak is sometimes described as the system Q (or *quality factor*). A tall, sharp peak (low damping) represents a system with high Q. The Q can be related to ζ as shown in Eq. 3.14.

$$Q = \frac{1}{2\zeta} \tag{3.14}$$

In Fig. 3.3, the $r = 0$ magnitude is $|G(r = 0)| = \left|\frac{X}{F}\right| = \frac{1}{k} = 1 \times 10^{-6}$ m/N. This DC result is independent of the damping ratio. For the resonant case where $r = 1$:

$$|G(r = 1)| = \frac{1}{2k\zeta} = \frac{1}{2(1 \times 10^6)\zeta} = \frac{5 \times 10^{-7}}{\zeta}.$$

The three peak values are $\{5 \times 10^{-5}, 1 \times 10^{-5}, \text{and } 5 \times 10^{-6}\}$ m/N.

The phase plot is typically provided in conjunction with the magnitude plot to fully describe the FRF. The phase for the same system used in Fig. 3.3 is shown in Fig. 3.4; the code used to produce Figs. 3.3 and 3.4 is included in MATLAB® MOJO 3.1. In Fig. 3.4, the phase at $r = 0$ is $\phi = 0$. This is the static result where there is no phase lag between the displacement and force. At $r = 1$ (resonance), the phase is

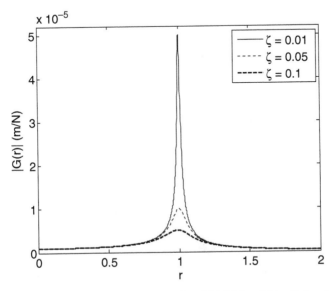

Fig. 3.3 Magnitude plot for underdamped systems ($\zeta = 0.01, 0.05,$ and 0.1 with $k = 1 \times 10^6$ N/m)

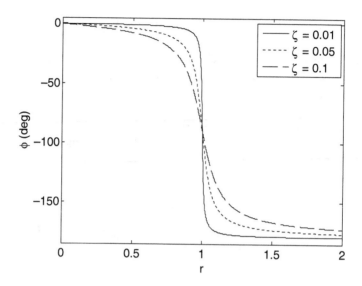

Fig. 3.4 Phase versus r plot for underdamped systems ($\zeta = 0.01$, 0.05, and 0.1 with $k = 1 \times 10^6$ N/m)

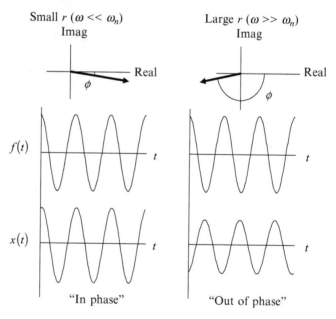

Fig. 3.5 Time-domain representations of phase between force and displacement for: (a) small r and (b) large r ($r \gg 1$)

$\phi = \tan^{-1}\left(\frac{2\zeta}{0}\right) = -\frac{\pi}{2}$ rad $= -90°$. For $r \gg 1$, the phase approaches $-\pi$ rad ($-180°$). This is the "out of phase" condition where the force reaches its minimum value while the displacement reaches its maximum value. See the time-domain representations of the force and displacement in Fig. 3.5.

MATLAB® MOJO 3.1
```
% matlab_mojo_3_1.m

clc
clear all
close all

% define variables
r = 0:0.001:2;
k = 1e6;                    % N/m

% define function
zeta1 = 0.01
mag1 = 1/k*(1./((1-r.^2).^2 + (2*zeta1*r).^2)).^0.5;
phase1 = atan2(-2*zeta1*r, (1-r.^2));

zeta2 = 0.05
mag2 = 1/k*(1./((1-r.^2).^2 + (2*zeta2*r).^2)).^0.5;
phase2 = atan2(-2*zeta2*r, (1-r.^2));

zeta3 = 0.1
mag3 = 1/k*(1./((1-r.^2).^2 + (2*zeta3*r).^2)).^0.5;
phase3 = atan2(-2*zeta3*r, (1-r.^2));

figure(1)
plot(r, mag1, 'k-', r, mag2, 'k:', r, mag3, 'k--')
set(gca,'FontSize', 14)
xlabel('r')
ylabel('|G(r)| (m/N)')
axis([0 2 0 5.2e-5])
legend('\zeta = 0.01', '\zeta = 0.05', '\zeta = 0.1')

figure(2)
plot(r, phase1*180/pi, 'k-', r, phase2*180/pi, 'k:', r, phase3*180/pi,
  'k--')
  set(gca,'FontSize', 14)
  xlabel('r')
  ylabel('\phi (deg)')
  axis([0 2 -185 5])
  legend('\zeta = 0.01', '\zeta = 0.05', '\zeta = 0.1')
```

IN A NUTSHELL   The magnitude and phase representations show how the forced vibration is related to the exciting force over a range of frequencies. Let's look at the vibration that results for a given force amplitude as the frequency of the force changes. In all cases, the frequency of the vibration is the same as the frequency of the exciting force. When the excitation frequency is low, the force and displacement go together and reach their maximum values at almost the same time. They are "in phase." As the frequency of the excitation frequency increases, the amplitude of the displacement increases and the displacement begins to fall behind the force (it reaches its peak value a little later than the force reaches its peak value). At resonance (where the excitation frequency equals the natural frequency), the amplitude of the displacement is at its largest value and the displacement reaches its maximum when the force is zero. The force and displacement are phase

shifted by 90°. As the excitation frequency continues to increase, the amplitude of
the displacement begins to decrease and the displacement lags even farther behind the
force. When the excitation frequency is very high, the amplitude of the displacement
becomes much smaller and the displacement reaches a positive maximum when the
force reaches a negative maximum. They are phase shifted by 180° ("out of phase").

Next, let's plot the real and imaginary parts of the FRF again as a function of
the frequency ratio. Using Eq. 3.8, we obtain Fig. 3.6 for $\zeta = 0.01, 0.05$, and 0.1
with $k = 1 \times 10^6$ N/m. We see that, similar to the magnitude plot in Fig. 3.3, the
peak-to-peak height decreases with increasing damping. For the imaginary part
(Eq. 3.9), a similar trend is observed in Fig. 3.7. The code used to produce Figs. 3.3
and 3.4 is provided in MATLAB® MOJO 3.2.

MATLAB® MOJO 3.2

```
% matlab_mojo_3_2.m

clc
clear all
close all

% define variables
r = 0:0.001:2;
k = 1e6;                    % N/m

% define function
zeta1 = 0.01
real1 = 1/k*(1-r.^2)./((1-r.^2).^2 + (2*zeta1*r).^2);
imag1 = 1/k*(-2*zeta1*r)./((1-r.^2).^2 + (2*zeta1*r).^2);

zeta2 = 0.05
real2 = 1/k*(1-r.^2)./((1-r.^2).^2 + (2*zeta2*r).^2);
imag2 = 1/k*(-2*zeta2*r)./((1-r.^2).^2 + (2*zeta2*r).^2);

zeta3 = 0.1
real3 = 1/k*(1-r.^2)./((1-r.^2).^2 + (2*zeta3*r).^2);
imag3 = 1/k*(-2*zeta3*r)./((1-r.^2).^2 + (2*zeta3*r).^2);

figure(1)
plot(r, real1, 'k-', r, real2, 'k:', r, real3, 'k--')
set(gca,'FontSize', 14)
xlabel('r')
ylabel('Re(G(r))  (m/N)')
axis([0 2 -2.7e-5 2.7e-5])
legend('\zeta = 0.01', '\zeta = 0.05', '\zeta = 0.1')

figure(2)
plot(r, imag1, 'k-', r, imag2, 'k:', r, imag3, 'k--')
set(gca,'FontSize', 14)
xlabel('r')
ylabel('Im(G(r))  (m/N)')
axis([0 2 -5.5e-5 5e-6])
legend('\zeta = 0.01', '\zeta = 0.05', '\zeta = 0.1')
```

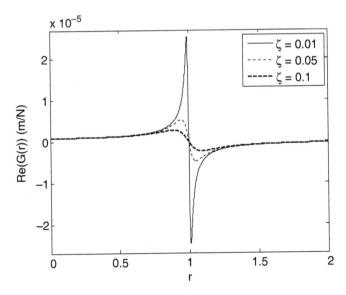

Fig. 3.6 FRF real part versus r plot for underdamped systems ($\zeta = 0.01$, 0.05, and 0.1 with $k = 1 \times 10^6$ N/m)

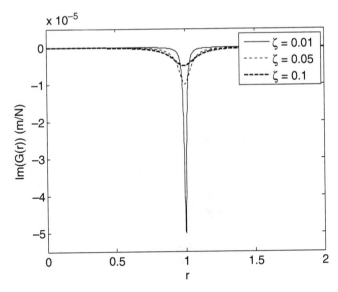

Fig. 3.7 FRF imaginary part versus r plot for underdamped systems ($\zeta = 0.01, 0.05$, and 0.1 with $k = 1 \times 10^6$ N/m)

For the single degree of freedom FRF that we are considering now, several important points can be identified directly from the real and imaginary plots. As we have already discussed, the zero frequency ($r = 0$) response gives a value of $\frac{1}{k}$ for the real part and zero for the imaginary part. At resonance ($r = 1$), the real part is

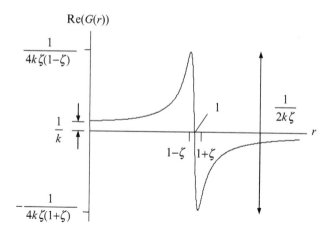

Fig. 3.8 Summary of important points on FRF real part

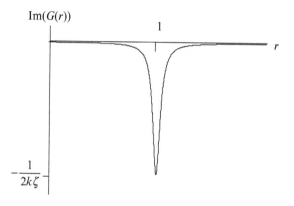

Fig. 3.9 Summary of important points on FRF imaginary part

zero and the imaginary part reaches its minimum value of $-\frac{1}{2k\zeta}$. The maximum real part value of approximately $\frac{1}{4k\zeta(1-\zeta)}$ occurs at a frequency of $r = 1 - \zeta$ ($\omega = \omega_n(1-\zeta)$). At a frequency of $r = 1 + \zeta$ ($\omega = \omega_n(1+\zeta)$), the minimum real part is observed with a value of approximately $-\frac{1}{4k\zeta(1+\zeta)}$. We can also note that the peak-to-peak value of the real part is the same as for the imaginary part: $\frac{1}{2k\zeta}$. Summing up the absolute values of the real part maximum and minimum peaks, we obtain:

$$\frac{1}{4k\zeta(1-\zeta)} + \frac{1}{4k\zeta(1+\zeta)} = \frac{(1+\zeta)+(1-\zeta)}{4k\zeta(1-\zeta)(1+\zeta)} = \frac{2}{4k\zeta(1-\zeta^2)} \approx \frac{1}{2k\zeta}.$$

This approximation is valid for small ζ values, which is typical for mechanical structures. For example, even with 10% damping ($\zeta = 0.1$), ζ^2 is only 0.01. These frequencies and peaks are identified in Figs. 3.8 and 3.9.

Fig. 3.10 *By the Numbers*
3.1 – the example spring-
mass-damper system and
single degree of freedom
model parameters are shown

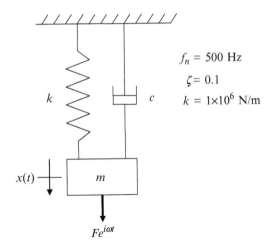

$f_n = 500$ Hz

$\zeta = 0.1$

$k = 1 \times 10^6$ N/m

By the Numbers 3.1

Consider the lumped parameter spring–mass–damper system displayed in Fig. 3.10.
The harmonic force, $f(t) = Fe^{i\omega t}$, is acting on the system to produce forced
vibration. The system natural frequency is 500 Hz, the damping ratio is 0.1, and
the stiffness is 1×10^6 N/m.

First, given the system parameters f_n, ζ, and k, we can determine m and c.
We obtain the mass using:

$$m = \frac{k}{(2\pi f_n)^2} = \frac{1 \times 10^6}{(2\pi \cdot 500)^2} = 0.1 \text{ kg.}$$

The viscous damping ratio is $c = 2\zeta m\omega_n = 2(0.1)0.1(2\pi \cdot 500) = 63$ N-s/m.
Let's now sketch the magnitude and phase for this system. For the magnitude
plot, we can directly identify the magnitude at forcing frequency values of $\omega = 0$
($r = 0$) and $\omega = \omega_n$ ($r = 1$). At DC, the magnitude is $\frac{1}{k} = \frac{1}{1\times10^6} = 1 \times 10^{-6}$m/N
100 μm/100 N. This means that we'd get a deflection of approximately the diame-
ter of a human hair if we apply a constant force of 100 N (or 22.5 lb$_f$). At resonance,
where the forcing frequency is equal to the natural frequency, the magnitude is
$\frac{1}{2k\zeta} = \frac{1}{2(1\times10^6)0.1} = 5 \times 10^{-6}$N/m. Note that the *dynamic flexibility*[2] is five times
higher than the *static flexibility* for this system. At a higher frequency, say $r = 5$
where the forcing frequency is $f = 5f_n = 5 \cdot 500 = 2{,}500$ Hz, the magnitude is:

$$|G(5)| = \frac{1}{1 \times 10^6} \sqrt{\frac{1}{(1 - 5^2)^2 + (2(0.1)5)^2}} = 4.2 \times 10^{-8} \text{ N/m} = 4.2 \mu\text{m}/100 \text{ N.}$$

---

[2] Flexibility, or compliance, is the inverse of stiffness.

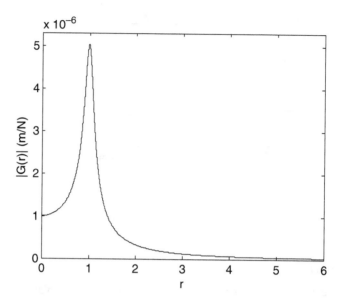

Fig. 3.11 *By the Numbers 3.1* – magnitude plot

Now the dynamic flexibility is ~24 times smaller than the static flexibility and ~120 times smaller than the resonant case. This emphasizes the strong dependence of the steady-state response for dynamic systems on the (harmonic) forcing frequency. The magnitude plot is provided in Fig. 3.11.

For the phase plot (Fig. 3.12), we know that the phase is $\phi = 0$ at $\omega = 0$ ($r = 0$), it is $-90°$ at $\omega = \omega_n$ ($r = 1$), and it approaches $-180°$ for large forcing frequencies (r values $\gg 1$). Specifically, we calculate the following.

- For $r = 0$, $\phi(r = 0) = \tan^{-1}\left(\frac{-2\zeta r}{1-r^2}\right) = \tan^{-1}\left(\frac{0}{1}\right) = 0$.
- For $r = 1$, $\phi(r = 1) = \tan^{-1}\left(\frac{-2\zeta}{0}\right) = -90°$. Recognizing that the phase is the inverse tangent of the ratio of the imaginary part to the real part, Fig. 3.13 demonstrates this result. Because the real part is zero and the imaginary part is negative, the displacement lags the (real-valued) force by 90°.
- For $r = 5$, $\phi(r = 5) = \tan^{-1}\left(\frac{-2(0.1)5}{1-5^2}\right)$. Here we have to exercise caution in calculating the phase. If we simply input this ratio into a calculator and use the $\tan^{-1}$ key/function, we would find that $\phi = 2.4°$ (or 0.04 rad). A positive phase indicates that the displacement has somehow anticipated[3] the force and is leading it in time. This is not possible for our system, so this must not be the correct phase. The error is due to the quadrant dependence of the inverse tangent. Figure 3.14 shows that the correct phase is $\phi = -(180 - 2.4) = -177.6° = -3.1$ rad. In MATLAB®, the correct result of -3.1 rad is obtained using $\texttt{atan2(-2*0.1*5, 1-5^2)}$.

---

[3] Such anticipatory behavior would be exhibited by a noncausal system (Kamen 1990).

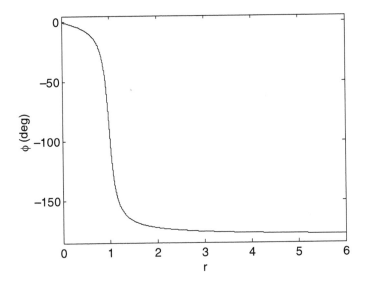

Fig. 3.12 *By the Numbers 3.1 – phase plot*

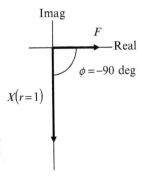

Fig. 3.13 *By the Numbers 3.1 – the displacement lags the force by 90° for $r = 1$*

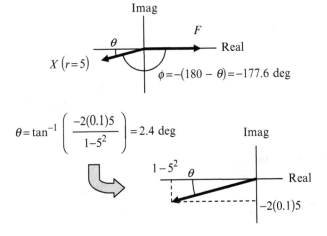

$$\theta = \tan^{-1}\left(\frac{-2(0.1)5}{1-5^2}\right) = 2.4 \text{ deg}$$

Fig. 3.14 *By the Numbers 3.1 – the displacement lags the force by 177.6° for $r = 5$*

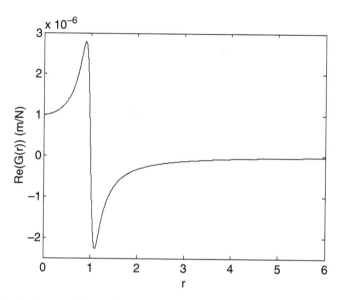

Fig. 3.15 *By the Numbers 3.1* – real part of the FRF

Next, let's look at the real and imaginary parts of the FRF. The real plot is provided in Fig. 3.15. At $\omega = 0$ ($r = 0$), the value is:

$$\mathrm{Re}(G(0)) = \frac{1}{1 \times 10^6} \left(\frac{(1 - 0^2)}{(1 - 0^2)^2 + (2(0.1)0)^2} \right) = 1 \times 10^{-6}\,\mathrm{m/N}.$$

At $\omega = \omega_n$ ($r = 1$), the value of the real part is:

$$\mathrm{Re}(G(1)) = \frac{1}{1 \times 10^6} \left(\frac{(1 - 1^2)}{(1 - 1^2)^2 + (2(0.1)1)^2} \right) = 0.$$

At $\omega = 5\omega_n$ ($r = 5$), the real part is:

$$\mathrm{Re}(G(5)) = \frac{1}{1 \times 10^6} \left(\frac{(1 - 5^2)}{(1 - 5^2)^2 + (2(0.1)5)^2} \right) = -4.2 \times 10^{-8}\,\mathrm{m/N}.$$

Let's also determine the real part value at its peaks. For the maximum peak, which occurs approximately at $r = 1 - \zeta = 1 - 0.1 = 0.9$, the value is:

$$\mathrm{Re}(G(0.9)) = \frac{1}{1 \times 10^6} \left(\frac{(1 - 0.9^2)}{(1 - 0.9^2)^2 + (2(0.1)0.9)^2} \right) = 2.8 \times 10^{-6}\,\mathrm{m/N}.$$

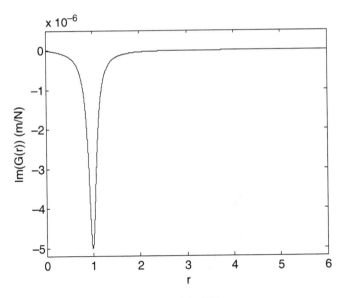

Fig. 3.16 *By the Numbers 3.1* – imaginary part of the FRF

Note that this result can be approximated using $\frac{1}{4k\zeta(1-\zeta)}$. For the minimum (negative) peak, which occurs at a frequency ratio of approximately $1+\zeta = 1+0.1 = 1.1$, the value is:

$$\text{Re}(G(1.1)) = \frac{1}{1 \times 10^6}\left(\frac{(1-1.1^2)}{(1-1.1^2)^2 + (2(0.1)1.1)^2}\right) = -2.3 \times 10^{-6}\text{m/N}.$$

Similarly, this result can be approximated using $-\frac{1}{4k\zeta(1+\zeta)}$.

To complete this example, let's plot the Argand diagram or the real part of the FRF versus the imaginary part of the FRF. The result is provided in Fig. 3.17, where the points for $r=0$, $r=1-\zeta=1-0.1=0.9$, $r=1$, and $r=1+\zeta=1+0.1=1.1$ are identified by the open circles. For these points, the values of the real and imaginary parts are given in Table 3.1.

We can see in Fig. 3.17 that the points for $r=0.9$ and 1.1 do not appear at the Argand "circle" quadrants as we might have expected. They are, after all, supposed to identify the maximum and minimum real part values. The reason for this discrepancy is that using the frequency ratios $r=1-\zeta$ and $r=1+\zeta$, respectively, to identify the maximum and minimum real part values is an approximation. We can determine the actual r values by differentiating Eq. 3.8 with respect to r and setting this result equal to zero.

$$\frac{d}{dr}\left(\frac{1}{k}\left(\frac{(1-r^2)}{(1-r^2)^2 + (2\zeta r)^2}\right)\right) = 0 \tag{3.15}$$

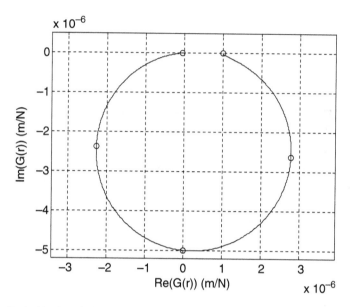

Fig. 3.17 *By the Numbers 3.1* – Argand diagram. The frequency ratios in Table 3.1 are identified by the circles. Clockwise from the top-right circle: $r = 0, 0.9, 1, 1.1$, and 5

Table 3.1 *By the Numbers 3.1* – values of the FRF real and imaginary parts for various frequency ratios (see Fig. 3.17)	r	$\text{Re}(G(r))$ (m/N)	$\text{Im}(G(r))$ (m/N)
	0	1×10^{-6}	0
	0.9	2.8×10^{-6}	-2.8×10^{-6}
	1	0	-5×10^{-6}
	1.1	-2.3×10^{-6}	-2.4×10^{-6}
	5	-4.2×10^{-8}	-1.7×10^{-9}

Computing the derivative in Eq. 3.15 yields:

$$\frac{\left((1-r^2)^2 + (2\zeta r)^2\right)(-2r) - (1-r^2)(2(1-r^2)(-2r) + 2(2\zeta r)(2\zeta))}{\left((1-r^2)^2 + (2\zeta r)^2\right)^2} = 0.$$

(3.16)

Expanding the numerator and canceling terms give an equation that is quadratic in r^2:

$$2r^4 - 4r^2 + (2 - 8\zeta^2) = 0.$$
(3.17)

We can obtain the roots, $r_{1,2}{}^2$, of Eq. 3.17 using the quadratic equation:

$$r_{1,2}{}^2 = \frac{-(-4) \pm \sqrt{(-4)^2 - 4(2)(2 - 8\zeta^2)}}{2(2)} = 1 \pm \frac{\sqrt{64\zeta^2}}{4} = 1 \pm 2\zeta. \quad (3.18)$$

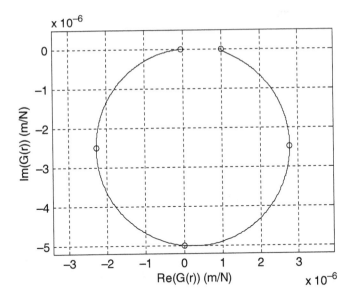

Fig. 3.18 *By the Numbers 3.1* – Argand diagram. The frequency ratios determined using Eq. 3.18 are identified by the open circles. Clockwise from the top-right circle: $r = 0, \sqrt{0.8}, 1, \sqrt{1.2}$, and 5

For our approximation of $r = 1 \pm \zeta$, the square of this expression gives:

$$r_1^2 = 1 + 2\zeta + \zeta^2$$
$$r_2^2 = 1 - 2\zeta + \zeta^2. \tag{3.19}$$

For small ζ, the additional ζ^2 term at the end of the expressions in Eq. 3.19 is negligible. However, as ζ increases, the error in this approximation becomes evident – as seen in Fig. 3.17. Figure 3.18 shows the Argand diagram where the same r values are identified using Eq. 3.18 (rather than the approximation). The code used to generate this figures is provided in MATLAB® MOJO 3.3.

If we look carefully at Fig. 3.18, we see that we are not quite done. In fact, the $r = 1$ point does not exactly identify the minimum value of the FRF imaginary part. This is again an approximation; a good one, but an approximation nonetheless. We can follow the same steps as before, but this time use Eq. 3.9 to determine the proper r value.

$$\frac{d}{dr}\left(\frac{1}{k}\left(\frac{-2\zeta r}{(1 - r^2)^2 + (2\zeta r)^2}\right)\right) = 0 \tag{3.20}$$

Calculating the derivative gives:

$$\frac{\left((1 - r^2)^2 + (2\zeta r)^2\right)(-2\zeta) - (-2\zeta r)(2(1 - r^2)(-2r) + 2(2\zeta r)(2\zeta))}{\left((1 - r^2)^2 + (2\zeta r)^2\right)^2} = 0. \tag{3.21}$$

Expanding the numerator and canceling terms yields:

$$3r^4 + (4\zeta^2 - 2)r^2 - 1 = 0. \tag{3.22}$$

Again using the quadratic equation, we determine the roots to be:

$$r_{1,2}{}^2 = \frac{(2 - 4\zeta^2) \pm 4\sqrt{\zeta^4 - \zeta^2 + 1}}{6}. \tag{3.23}$$

For $\zeta = 0.1$, substitution gives:

$$r_{1,2}{}^2 = \frac{\left(2 - 4(0.1)^2\right) \pm 4\sqrt{(0.1)^4 - (0.1)^2 + 1}}{6} = \frac{1.96}{6} \pm \frac{4\sqrt{0.9901}}{6}.$$

Using $r^2 = \frac{1.96}{6} + \frac{4\sqrt{0.9901}}{6}$, the r value for the minimum imaginary part is 0.995, rather than 1. Note, however, that as ζ approaches zero in Eq. 3.23, r approaches 1 (again using the positive root).

MATLAB® MOJO 3.3

```
% matlab_mojo_3_3.m

clc
clear all
close all

% define variables
r = 0:0.001:6;
k = 1e6;                        % N/m

% define function
zeta1 = 0.1;
real1 = 1/k*(1-r.^2)./((1-r.^2).^2 + (2*zeta1*r).^2);
imag1 = 1/k*(-2*zeta1*r)./((1-r.^2).^2 + (2*zeta1*r).^2);

figure(1)
plot(real1, imag1, 'k-')
set(gca,'FontSize', 14)
xlabel('Re(G(r))  (m/N)')
ylabel('Im(G(r))  (m/N)')
axis([-2.5e-6 3e-6 -5.2e-6 5e-7])
hold on
grid
axis equal

r_points = [0 sqrt(1-2*zeta1) 1 sqrt(1+2*zeta1) 5];

for cnt = 1:length(r_points)
    r1 = r_points(cnt);
    real_points(cnt) = 1/k*(1-r1^2)/((1-r1^2)^2 + (2*zeta1*r1)^2);
    imag_points(cnt) = 1/k*(-2*zeta1*r1)/((1-r1^2)^2 + (2*zeta1*r1)^2);
end

plot(real_points, imag_points, 'ko')
```

Fig. 3.19 *By the Numbers 3.2* – the single degree of freedom spring–mass–damper system and model parameters are shown

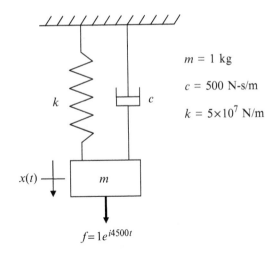

$m = 1$ kg

$c = 500$ N-s/m

$k = 5 \times 10^7$ N/m

$x(t)$ $\quad$ m

$f = 1e^{i4500t}$

By the Numbers 3.2

Let's consider a second example, now with $m = 1$ kg, $c = 500$ N-s/m, $k = 5 \times 10^7$N/m, and $f = Fe^{i\omega t} = 1e^{i4500t}$ N. See Fig. 3.19. The associated natural frequency is $\omega_n = \sqrt{\frac{5 \times 10^7}{1}} = 7,071.1$ rad/s and the damping ratio is $\zeta = \frac{500}{2\sqrt{5 \times 10^7}(1)} = 0.035$ (or 3.5%). The forcing frequency is 4,500 rad/s, so the frequency ratio is $r = \frac{4,500}{7,071.1} = 0.636$.

The complex FRF, $G(r)$, for this damped, single degree of freedom system is:

$$G(0.636) = \frac{1}{5 \times 10^7} \left(\frac{\left(1 - (0.636)^2\right) - i2(0.035)(0.636)}{\left(1 - (0.636)^2\right)^2 + (2(0.035)(0.636))^2} \right).$$

The real part is shown in Fig. 3.20 and the value for $r = 0.636$ is:

$$Re(G(0.636)) = \frac{1}{5 \times 10^7} \left(\frac{\left(1 - (0.636)^2\right)}{\left(1 - (0.636)^2\right)^2 + (2(0.035)(0.636))^2} \right)$$

$$= 3.34 \times 10^{-8} \text{m/N}.$$

The real part is shown in Fig. 3.21 and the value for $r = 0.636$ is:

$$Im(G(0.636)) = \frac{1}{5 \times 10^7} \left(\frac{-2(0.035)(0.636)}{\left(1 - (0.636)^2\right)^2 + (2(0.035)(0.636))^2} \right)$$

$$= -2.5 \times 10^{-9} \text{m/N}.$$

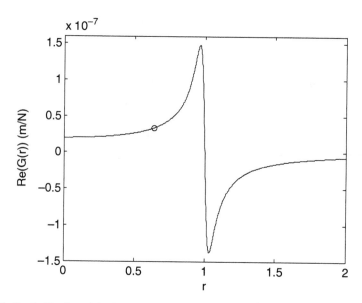

Fig. 3.20 *By the Numbers 3.2* – FRF real part. The $r = 0.636$ point is identified

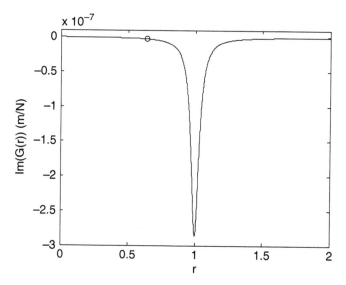

Fig. 3.21 *By the Numbers 3.2* – FRF imaginary part. The $r = 0.636$ point is identified

The real and imaginary values of the frequency-domain displacement, $X(0.636)$, are determined by multiplying the FRF by the force magnitude of $F = 1$ N. The real part is:

$$\mathrm{Re}(X(0.636)) = F \cdot \mathrm{Re}(G(0.636)) = 1 \cdot 3.34 \times 10^{-8} = 3.34 \times 10^{-8} \mathrm{m},$$

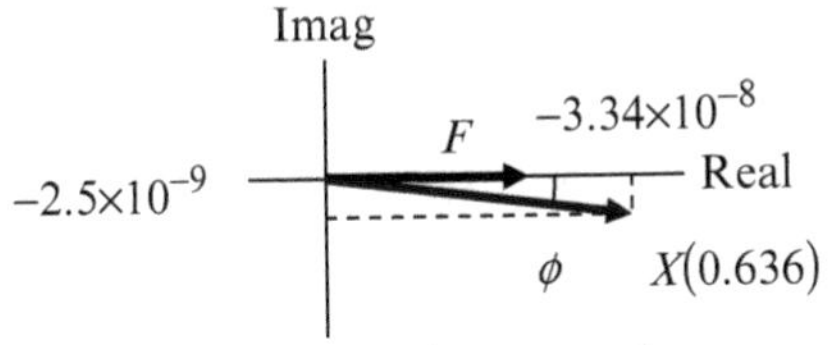

$$\phi(0.636) = \tan^{-1}\left(\frac{-2.5\times10^{-9}}{3.34\times10^{-8}}\right) = -4.28 \text{ deg}$$

$$\left|X(0.636)\right| = \sqrt{\left(3.34\times10^{-8}\right)^2 + \left(-2.5\times10^{-9}\right)^2} = 3.35\times10^{-8} \text{ m}$$

Fig. 3.22 By the Numbers 3.2 – Argand diagram for $r = 0.636$

and the imaginary part is:

$$\text{Im}(X(0.636)) = F \cdot \text{Im}(G(0.636)) = 1 \cdot (-2.5 \times 10^{-9}) = -2.5 \times 10^{-9}\text{m}.$$

We can now plot these components of the response in the complex plane. The force and displacement vectors are represented in Fig. 3.22, where the displacement magnitude is:

$$|X(0.636)| = \sqrt{\text{Re}(G(0.636))^2 + \text{Im}(G(0.636))^2} = 3.35 \times 10^{-8}\text{m},$$

and the phase is:

$$\phi(0.636) = \tan^{-1}\left(\frac{\text{Im}(G(0.636))}{\text{Re}(G(0.636))}\right) = -4.28° = -0.075 \text{ rad}.$$

Of course we would obtain the same results by applying Eqs. 3.10 and 3.11.

3.4   Defining a Model from a Frequency Response Function Measurement

In Figs. 3.8 and 3.9, we saw that key points from the FRF can be quickly identified based on the peaks in the real and imaginary part plots. What if we were able to perform a measurement to determine the FRF for a particular structure? We could then identify a model to represent the measured system using a *peak picking* approach.

While we will not discuss FRF measurement until Chap. 7, we can still describe the steps necessary to use measurement data to define a single degree of freedom

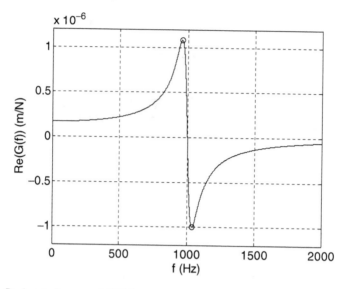

Fig. 3.23 Real part of measured FRF for a test structure

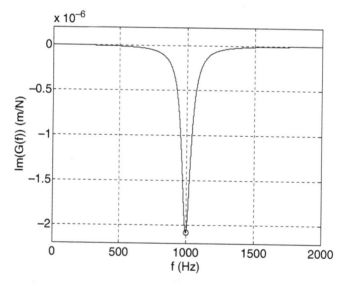

Fig. 3.24 Imaginary part of measured FRF for a test structure

system model. Let's assume that the FRF displayed in Figs. 3.23 (real part) and 3.24 (imaginary part) was obtained from a measurement of a test structure. In the figures, the peaks are identified by circles and the corresponding peak frequencies and values are summarized in Table 3.2.

Table 3.2 Frequencies and values of FRF real and imaginary part peaks	Peak	Frequency (Hz)	Value (m/N)
	Real maximum	959.2	1.085×10^{-6}
	Real minimum	1,039.2	-1.002×10^{-6}
	Imaginary minimum	999.2	-2.084×10^{-6}

Defining a single degree of freedom spring–mass–damper model based on the data in Table 3.2 requires five primary steps.

1. The frequency of the minimum imaginary part peak is taken to be the system natural frequency: $f_n = 999.2$ Hz.
2. The real part maximum and minimum peaks occur at approximately $f = f_n(1 - \zeta)$ and $f = f_n(1 + \zeta)$, respectively. Differencing these two frequencies gives:

$$f_n(1 + \zeta) - f_n(1 - \zeta) = 2\zeta f_n.$$

For the measured FRF, we have $1,039.2 - 959.2 = 80 = 2\zeta f_n$. Since we already know f_n, we can solve for the damping ratio:

$$\zeta = \frac{80}{2(999.2)} = 0.04 \, (4\%).$$

3. We determine the stiffness from the peak value of the minimum imaginary part. Recall from Fig. 3.9 that the minimum value is $-\frac{1}{2k\zeta}$. We found ζ in Step 2, so we can now solve for k.

$$k = \frac{-1}{2(0.04)\left(-2.084 \times 10^{-6}\right)} = 6 \times 10^6 \, \text{N/m}$$

4. Given the natural frequency and stiffness, we can find the model mass.

$$m = \frac{k}{\omega_n^2} = \frac{k}{(2\pi f_n)^2} = \frac{6 \times 10^6}{(2\pi \cdot 999.2)^2} = 0.15 \, \text{kg}$$

5. The viscous damping coefficient is then:

$$c = 2\zeta \sqrt{km} = 2(0.04)\sqrt{(6 \times 10^6)0.15} = 75.9 \, \text{N-s/m}.$$

IN A NUTSHELL   The peak picking strategy is a method of curve fitting. We have a measured FRF and we are trying to choose model parameters for a single degree of freedom system that has an FRF like the one we measured. There are many curve-fitting methods, but the peak picking method uses three easily identifiable points on the real and imaginary parts of the FRF. This is one reason to choose the real/imaginary FRF representation.

Fig. 3.25 Spring–mass–
damper model identified from
a measured FRF (Figs. 3.23
and 3.24) using the peak
picking approach

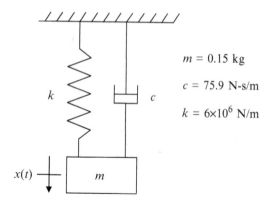

$m = 0.15$ kg

$c = 75.9$ N-s/m

$k = 6{\times}10^6$ N/m

The spring–mass–damper model is shown in Fig. 3.25. Let's now use our model to predict the vibration magnitude if a harmonic force, $f(t) = 1{,}000e^{i\omega_n t}$ N, is applied at the measurement point in the measurement direction. Note that the excitation frequency is equal to the system natural frequency, $\omega_n = 2\pi f_n = 6{,}278$ rad/s. For this resonant condition, the vibration magnitude is:

$$|X| = \frac{F}{2k\zeta} = \frac{1{,}000}{2(6 \times 10^6)0.04} = 2.1 \times 10^{-3}\ \text{m} = 2.1\ \text{mm}$$

If this vibration magnitude is too large, we could attempt to reduce it by:

- Increasing the damping
- Increasing the stiffness – this will serve to not only decrease the magnitude for any forcing function, but will also increase the natural frequency so that the existing force will no longer excite the structure at resonance. If the increase in stiffness is Δk, the new natural frequency will be $\omega_n = \sqrt{\frac{k+\Delta k}{m}}$.

3.5 Rotating Unbalance

A special case of forced vibration is *rotating unbalance*. This occurs when a rotating structure does not possess perfect symmetry in its mass distribution. Common examples include:

- electric motors
- turbines
- automobile wheels
- washing machines.

Fig. 3.26 Single degree of freedom model with rotating unbalance

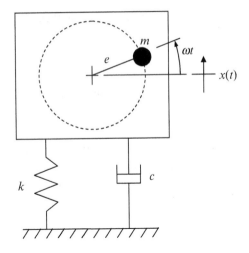

Fig. 3.27 Free-body diagram for the rotating unbalance model in Fig. 3.26

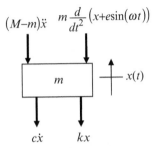

For the final example, you may have witnessed a washing machine "walk across the floor" during the spin cycle due to an uneven distribution of the wet (heavy) clothes in the drum. To describe this behavior in terms of forced vibration, consider the system shown in Fig. 3.26. An unbalanced mass, m, with an eccentricity (the distance of the unbalanced mass from the center of rotation), e, rotates with an angular speed, ω. The vertical displacement of the mass in Fig. 3.26 is:

$$x + e \sin(\omega t),$$

where x is the motion of the support structure with a mass of $M - m$. (The total system mass is the sum of the support structure and unbalanced mass, $M - m + m = M$.) The free-body diagram for the system is shown in Fig. 3.27. Unlike our previous free-body diagrams, this one includes two inertial forces; one for the support structure, $f = (M - m)\ddot{x}$, and one for the rotating unbalanced mass, $f = m \frac{d}{dt^2}(x + e \sin(\omega t))$. Calculating the derivative gives:

$$f = m\left(\ddot{x} - e\omega^2 \sin(\omega t)\right).$$

Summing the forces to zero in the x direction, $\sum f_x = 0$, gives the equation of motion.

$$(M - m)\ddot{x} + m(\ddot{x} - e\omega^2 \sin(\omega t)) + c\dot{x} + kx = 0$$

Simplifying and rewriting this equation yields:

$$M\ddot{x} + c\dot{x} + kx = me\omega^2 \sin(\omega t), \tag{3.24}$$

where the right hand side of this equation is a harmonic forcing function with the frequency-dependent magnitude $me\omega^2$. The force amplitude naturally increases with increasing unbalanced mass and eccentricity, but also grows with the square of the rotating frequency.

The magnitude and phase of the corresponding frequency-domain vibration are:

$$|X| = \frac{me\omega^2}{\sqrt{(k - M\omega^2)^2 + (c\omega)^2}} \tag{3.25}$$

and

$$\phi = \tan^{-1}\left(\frac{c\omega}{k - M\omega^2}\right). \tag{3.26}$$

As before, we can rewrite these equations to be functions of r and ζ. Here, r is the ratio of the unbalanced mass rotating speed to the natural frequency, $\omega_n = \sqrt{\frac{k}{M}}$. Also, the damping ratio is $\zeta = \frac{c}{2\sqrt{kM}}$. See Eqs. 3.27 and 3.28.

$$|X| = \frac{\frac{m}{M}er^2}{\sqrt{(1 - r^2)^2 + (2\zeta r)^2}} \tag{3.27}$$

$$\phi = \tan^{-1}\left(\frac{2\zeta r}{1 - r^2}\right) \tag{3.28}$$

A nondimensionalized magnitude plot is provided in Fig. 3.28. The vertical axis of $\frac{MX}{me}$ was obtained by moving M, m, and e from the numerator of the right-hand side to the left-hand side in Eq. 3.27. Responses are provided for $\zeta = 0.01, 0.05,$ and 0.1.

In some cases, we wish to maintain a vibration magnitude below a certain level to avoid damage to the system, such as bearing damage for a rotating shaft; see Fig. 3.29. In this case, we may need to select a range of acceptable rotating speeds so that this maximum vibration magnitude is not exceeded. In Fig. 3.30, the acceptable rotating speed ranges are $\omega < \omega_1$ and $\omega > \omega_2$. The problem with the second range is that the speed must pass through resonance ($\omega = \omega_n$) in order reach ω_2. This would only be an acceptable alternative if short-term, large vibration magnitudes will not harm the system.

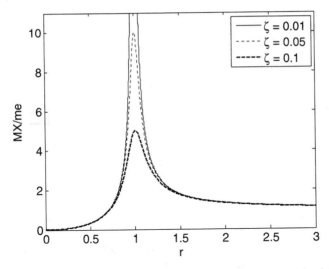

Fig. 3.28 Rotating unbalance frequency-domain vibration response for $\zeta = 0.01$, 0.05, and 0.1

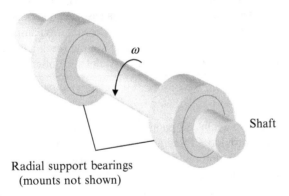

Fig. 3.29 A rotating shaft supported by bearing. Large magnitude rotating unbalance could damage the bearings

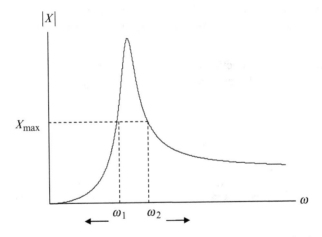

Fig. 3.30 Acceptable speed ranges for a maximum allowable vibration magnitude

Fig. 3.31 *By the Numbers 3.3* – Unbalanced mass used to produce a cell phone's "silent ring"

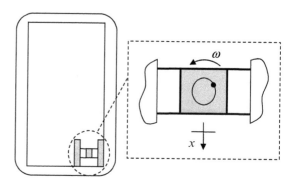

 IN A NUTSHELL   Designers of rotating systems use Eq. 3.27. In some instances (such as a jet engine), the rotational speed often changes and the designers attempt to place the natural frequency of the system far above the rotational frequency of the shaft. In this case, the system never operates at resonance. In other situations (like a turbine in a power plant), the rotational speed is held constant over long periods of time (on the order of months). In this case, designers often place the operating speed far above the resonance so that the vibration amplitude is as small as possible during operation. During start up, the forcing frequency passes through resonance, but this happens quickly and not very often.

By the Numbers 3.3

Let's consider an example. On cell phones, the "silent ring" can be produced using an eccentric mass to provide a force and vibration due to the rotating imbalance. Figure 3.31 shows a small motor with an unbalanced mass supported by four flexible tethers for illustration purposes.[4] The parallel pairs of tethers produce a flexure mechanism that is flexible in the x (vertical) direction but stiff in the orthogonal (horizontal) direction (Smith 2000). For this device, the rotating speed was varied, and the following results were obtained.

1. The maximum magnitude of vibration, X, was 1.0 mm.
2. For high frequencies, the magnitude asymptotically approached 0.02 mm $= 20\,\mu$m (about one fifth of a human hair's diameter).

Using this information, let's estimate the damping ratio, ζ, for the system. First, we can use the magnitude equation, Eq. 3.27, substitute $r = 1$ for resonance and set the right-hand side equal to 1.0 mm.

---

[4] This tether/payload geometry is referred to as a floating element structure in microelectrome-chanical systems (MEMS) design and has been used for shear stress measurement (Xu et al. 2009).

$$|X| = \frac{\frac{m}{M} e(1)^2}{\sqrt{\left(1 - (1)^2\right)^2 + (2\zeta(1))^2}} = \frac{\frac{m}{M} e}{2\zeta} = 1.0 \text{ mm} \qquad (3.29)$$

Second, we can let $r \to \infty$ in Eq. 3.27 and set the right-hand side equal to 0.02 mm.

$$|X| = \frac{\frac{m}{M} e(\infty)^2}{\sqrt{\left(1 - (\infty)^2\right)^2 + (2\zeta(\infty))^2}} = \frac{\frac{m}{M} e(\infty)^2}{(\infty)^2} = \frac{m}{M} e = 0.02 \text{ mm} \qquad (3.30)$$

Substituting $\frac{m}{M} e = 0.02$ mm from Eq. 3.30 into Eq. 3.29 yields:

$$|X| = \frac{0.02}{2\zeta} = 1.0 \text{ mm} \qquad (3.31)$$

Solving for the damping ratio gives $\zeta = 0.01 = 1\%$.

3.6 Base Motion

Base motion is observed when a system (for example, an automobile engine) is excited through elastic supports (such as the engine mounts that connect the engine to the automobile frame). Another example is the motion of an automobile's chassis in response to a wavy road surface.

Let's consider Fig. 3.32, where the motion, $x(t)$, of a single degree of freedom spring–mass–damper system is excited by motion of the structure's base, $y(t)$. The free-body diagram for the mass, which is also included in Fig. 3.32, gives the equation of motion:

$$m\ddot{x} + c(\dot{x} - \dot{y}) + k(x - y) = 0. \qquad (3.32)$$

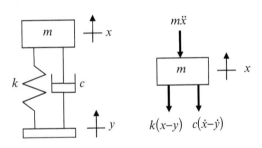

Fig. 3.32 Base motion for a single degree of freedom spring–mass–damper

We can rewrite Eq. 3.32 to isolate the response (x) terms on the left and the input (y) terms on the right.

$$m\ddot{x} + c\dot{x} + kx = c\dot{y} + ky \tag{3.33}$$

For harmonic base motion, $y(t) = Ye^{i\omega t}$, the response is also harmonic, $x(t) = Xe^{i\omega t}$. Calculating the derivatives of these expressions and substituting in Eq. 3.33 gives:

$$\left(-m\omega^2 + ic\omega + k\right)Xe^{i\omega t} = (ic\omega + k)Ye^{i\omega t}. \tag{3.34}$$

We can now calculate the ratio of the response, X, to the base motion input, Y.

$$\frac{X}{Y} = \frac{ic\omega + k}{-m\omega^2 + ic\omega + k} \tag{3.35}$$

Let's rewrite this equation in the r and ζ notation we have used previously.

$$\frac{X}{Y} = \frac{(k) + i(c\omega)}{(k - m\omega^2) + i(c\omega)} = \frac{\left(\frac{k}{m}\right) + i\left(\frac{c}{m}\omega\right)}{\left(\frac{k}{m} - \omega^2\right) + i\left(\frac{c}{m}\omega\right)} = \frac{\left(\omega_n^2\right) + i(2\zeta\omega_n\omega)}{\left(\omega_n^2 - \omega^2\right) + i(2\zeta\omega_n\omega)} \tag{3.36}$$

Dividing the numerator and denominator by ω_n^2, we obtain:

$$\frac{X}{Y} = \frac{(1) + i\left(2\zeta\frac{\omega}{\omega_n}\right)}{\left(1 - \frac{\omega^2}{\omega_n^2}\right) + i\left(2\zeta\frac{\omega}{\omega_n}\right)} = \frac{1 + i(2\zeta r)}{(1 - r^2) + i(2\zeta r)}. \tag{3.37}$$

We can now rationalize Eq. 3.37.

$$\frac{X}{Y} = \frac{1 + i(2\zeta r)}{(1 - r^2) + i(2\zeta r)} \cdot \frac{(1 - r^2) - i(2\zeta r)}{(1 - r^2) - i(2\zeta r)} = \frac{1 + (4\zeta^2 - 1)r^2 - i(2\zeta r^3)}{(1 - r^2)^2 + (2\zeta r)^2} \tag{3.38}$$

The real part of Eq. 3.38 is:

$$\mathrm{Re}\left(\frac{X}{Y}\right) = \frac{1 + (4\zeta^2 - 1)r^2}{(1 - r^2)^2 + (2\zeta r)^2}. \tag{3.39}$$

The imaginary part is:

$$\mathrm{Im}\left(\frac{X}{Y}\right) = \frac{-2\zeta r^3}{(1 - r^2)^2 + (2\zeta r)^2}. \tag{3.40}$$

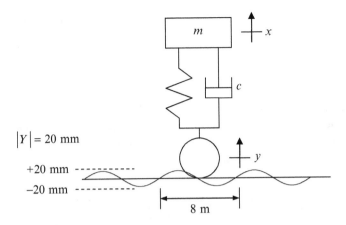

$|Y| = 20$ mm

+20 mm

−20 mm

8 m

Fig. 3.33 *By the Numbers 3.4* – Base motion example for automobile suspension (not to scale)

The magnitude, which is referred to as the *displacement transmissibility*, is the square root of the sum of the squares of the real and imaginary parts. The displacement transmissibility describes the transfer of the base motion to the mass motion. In many instances, we wish this transmissibility to be low so that the base motion is only weakly transmitted to the degree of freedom of interest.

$$\left|\frac{X}{Y}\right| = \left(\frac{\left(1 + (4\zeta^2 - 1)r^2\right)^2 + (2\zeta r^3)^2}{(1 - r^2)^2 + (2\zeta r)^2}\right)^{\frac{1}{2}} \tag{3.41}$$

The corresponding phase lag of the mass motion with respect to the base motion is:

$$\phi = \tan^{-1}\left(\frac{\text{Im}}{\text{Re}}\right) = \tan^{-1}\left(\frac{-2\zeta r^3}{1 + (4\zeta^2 - 1)r^2}\right). \tag{3.42}$$

By the Numbers 3.4

The suspension for an automobile can be modeled as a single degree of freedom spring–mass–damper as shown in Fig. 3.32 with a natural frequency of 3 Hz and a damping ratio of 0.5. The natural frequency is based on the combination of the car's mass and the suspension spring's stiffness. The damping comes primarily from the shock absorber. If $y(t)$ represents the motion of the wheel center as it follows the oscillating road surface depicted in Fig. 3.33, determine the magnitude of the automobile's motion, X, when the car is traveling at 60 mph.

The first step in the solution is to determine the forcing frequency, f (in Hz), due to the automobile's motion across the wavy road. The car's speed, v, is:

$$v = 60\frac{\text{min}}{\text{hr}} \cdot \frac{1}{3,600}\frac{\text{hr}}{\text{s}} \cdot 5,280\frac{\text{ft}}{\text{min}} \cdot 12\frac{\text{in.}}{\text{ft}} \cdot 0.0254\frac{\text{m}}{\text{in.}} = 26.8\frac{\text{m}}{\text{s}}.$$

The forcing frequency is the speed divided by the spatial wavelength of the road's surface:

$$f = \frac{26.8}{8} = 3.35 \, \text{Hz}.$$

The frequency ratio is therefore $r = \frac{f}{f_n} = \frac{3.35}{3} = 1.12$. We determine $|X|$ using Eq. 3.40, where $\zeta = 0.5$. A plot of $|X|$ versus r is provided in Fig. 3.34; the operating point is identified by a circle. The code used to produce Fig. 3.34 is provided in MATLAB® MOJO 3.4.

$$|X| = 20 \left(\frac{\left(1 + \left(4(0.5)^2 - 1 \right) 1.12^2 \right)^2 + \left(2(0.5)1.12^3 \right)^2}{\left(1 - 1.12^2 \right)^2 + \left(2(0.5)1.12 \right)^2} \right)^{\frac{1}{2}} = 26.1 \, \text{mm}$$

We see that the 20-mm road excitation is transmitted as a 26.1-mm magnitude oscillation of the automobile. This amplification occurs because the forcing frequency is near the suspension's natural frequency. At much lower or higher speeds, the vibration level would be less.

MATLAB® MOJO 3.4
```
% matlab_mojo_3_4.m

clc
close all
clear all

% Model dynamics
fn = 3;                    % Hz
zeta = 0.5;
Y_mag = 20;                % mm

r = 0:0.001:2;

Re = (1 + (4*zeta^2 - 1)*r.^2)./((1 - r.^2).^2 + (2*zeta*r).^2);
Im = - (2*zeta*r.^3)./((1 - r.^2).^2 + (2*zeta*r).^2);

X_mag = Y_mag*(Re.^2 + Im.^2).^0.5;

figure(1)
plot(r, X_mag, 'k')
set(gca,'FontSize', 14)
xlabel('r')
ylabel('|X|  (mm)')
hold on

index = find(r == 1.12);

plot(r(index), X_mag(index), 'ko')
```

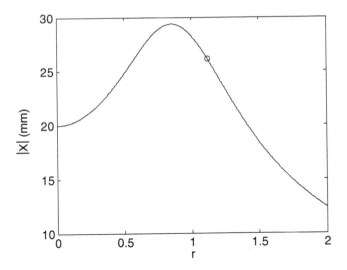

Fig. 3.34 *By the Numbers 3.4* – Automobile response, X, to the wavy road input

---

IN A NUTSHELL   "Rumble strips" on a roadway are designed with a spacing that causes large motion of the car body when the car passes over the bumps. One way to minimize the vibration inside the car would be to drive over the rumble strips at a high rate of speed (moving the excitation frequency far above resonance). We do not recommend experimental verification, however.

3.7   Impulse Response

When describing the equation of motion for a single degree of freedom lumped parameter system under forced vibration in Sect. 3.1, we assumed a harmonic force input of the form $f(t) = Fe^{i\omega t}$, where ω is the forcing frequency. This enabled us to define the corresponding FRF and explore its behavior. However, we recognize that not all forces are best described as a harmonic function at a single frequency. In a given situation, the force may be composed of multiple frequencies (e.g., an earthquake); it may be random (i.e., follows an erratic pattern that must be described statistically), or it may be transient. A common transient example is the impulsive force. In Chap. 7, we will explore the use of an impact hammer to excite a structure in order to measure its FRF. Because this is an important measurement technique, let's determine the response of our spring–mass–damper model from Fig. 3.1 to an impulsive input. This input can be defined as:

$$f(t) = \begin{cases} 0, t \le 0 \\ F, 0 < t < \Delta t \\ 0, t \ge \Delta t \end{cases} \tag{3.43}$$

Fig. 3.35 Graphical impulse
description

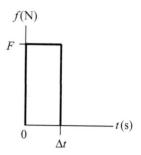

and is depicted in Fig. 3.35. We see that it has a constant value, F, over the
short-time interval from 0 to Δt. The solution to the equation of motion,
$m\ddot{x} + c\dot{x} + kx = f(t)$, is based on Newton's second law, which tells us that the
impulse of the force is equal to the change in momentum of the system that it
excites. We define the impulse of the force as the area under the force curve; this
area is $F\Delta t$ for the function defined in Eq. 3.24, but may be generically described as
the product of the average force value and the time interval over which the force is
applied (provided the time interval is small). If we assume zero initial conditions for
the system in Fig. 3.1, its change in momentum is mv_0, where v_0 is the velocity of
the mass due to the force application and the initial displacement is still zero.
Therefore, $F\Delta t = mv_0$. This is an interesting observation because it enables us to
treat this problem as a case of free vibration with a nonzero initial velocity,
$v_0 = \frac{F\Delta t}{m}$, and a zero initial displacement.

We have already considered the solution of free vibration using initial conditions in Sect. 2.4.5. One form for the general response
is $x(t) = Xe^{-\zeta \omega_n t} \sin\left(\sqrt{1 - \zeta^2}\omega_n t + \phi\right)$, where:

$$X = \frac{\sqrt{(\dot{x}_0 + \zeta\omega_n x_0)^2 + \left(x_0\sqrt{1 - \zeta^2}\omega_n\right)^2}}{\sqrt{1 - \zeta^2}\omega_n} \quad \text{and} \quad \phi = \tan^{-1}\left(\frac{x_0\sqrt{1 - \zeta^2}\omega_n}{\dot{x}_0 + \zeta\omega_n x_0}\right). \tag{3.44}$$

IN A NUTSHELL   Equation 3.44 provides the magnitude and
phase for the free vibration of an underdamped single degree of
freedom system in terms of the initial conditions. If we have a single
degree of freedom system problem and the system is underdamped,
then we can use these values to write the resulting motion as a
function of time if the initial conditions are known.

For the impulse response, $x_0 = 0$ and $\dot{x}_0 = v_0$, so Eq. 3.45 simplifies to:

$$X = \frac{v_0}{\sqrt{1 - \zeta^2}\omega_n} \quad \text{and } \phi = 0. \tag{3.45}$$

The system response to the impulse is therefore:

$$x(t) = \frac{v_0}{\sqrt{1 - \zeta^2}\omega_n} e^{-\zeta\omega_n t} \sin\left(\sqrt{1 - \zeta^2}\omega_n t\right). \tag{3.46}$$

Substituting $v_0 = \frac{F\Delta t}{m}$ gives:

$$x(t) = \frac{F\Delta t}{m\sqrt{1 - \zeta^2}\omega_n} e^{-\zeta\omega_n t} \sin\left(\sqrt{1 - \zeta^2}\omega_n t\right). \tag{3.47}$$

We can rewrite this equation as $x(t) = F\Delta t \cdot h(t)$, where:

$$h(t) = \frac{1}{m\sqrt{1 - \zeta^2}\omega_n} e^{-\zeta\omega_n t} \sin\left(\sqrt{1 - \zeta^2}\omega_n t\right) \tag{3.48}$$

is the *impulse response function* for our underdamped spring–mass–damper system.

By the Numbers 3.5

Let's consider the system shown in Fig. 3.1, but now with the Eq. 3.42 force applied to the mass. Let's select a natural frequency of 500 Hz, a damping ratio of 0.05, and a stiffness of 1×10^6 N/m. The mass is, therefore, $m = \frac{k}{(2\pi f_n)^2} = \frac{1\times10^6}{(2\pi\cdot500)^2} = 0.1$ kg. Also, $F = 100$ N and $\Delta t = 1 \times 10^{-3}$ s. Substituting in Eq. 3.46 yields:

$$x(t) = \frac{100\left(1 \times 10^{-3}\right)}{0.1\sqrt{1 - 0.05^2}500} e^{-0.05(500)t} \sin\left(\sqrt{1 - 0.05^2}500t\right).$$

The vibration response to the impulsive input is displayed in Fig. 3.36. We see that:

- the initial displacement is zero
- the force causes a rapid departure from the equilibrium position to a maximum value of 1.853 mm at $t = 3 \times 10^{-3}$ s
- the oscillating response exponentially decays to zero over time.

We will explore this further in Sect. 7.4.

We can also use the impulse response function to determine the vibration response for our underdamped single degree of freedom system with zero initial conditions due to a general, nonperiodic input force. If we consider the forcing

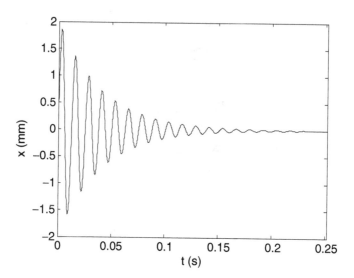

Fig. 3.36 *By the Numbers 3.5* – Impulse response for spring–mass–damper system

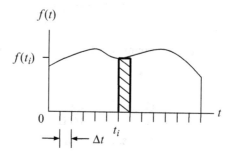

Fig. 3.37 Representation of a general force as a series of impulsive forces

function shown in Fig. 3.37, we see that it is possible to approximate this profile as n separate impulsive forces. In this case, the force is not applied only at $t = 0$. Rather, it also appears at $t = t_i$, where $i = 1...n$ is the impulse index. The impulse response function can now be written as:

$$h(t - t_i) = \frac{1}{m\sqrt{1 - \zeta^2}\omega_n} e^{-\zeta\omega_n(t-t_i)} \sin\left(\sqrt{1 - \zeta^2}\omega_n(t - t_i)\right), \qquad (3.49)$$

and the response to $f(t_i)$ is $x(t_i) = f(t_i)\Delta t \cdot h(t - t_i)$, where Δt is the total time divided by n. The final response for the linear system is simply the sum of all the responses $x(t_i)$. As $\Delta t \to 0$, this sum can be expressed as the *convolution integral*:

$$x(t) = \int_0^t f(\tau)h(t - \tau)d\tau. \qquad (3.50)$$

Substitution of Eq. 3.49 into Eq. 3.50 gives:

$$x(t) = \frac{1}{m\sqrt{1 - \zeta^2}\omega_n} e^{-\zeta\omega_n t} \int_0^t f(\tau) e^{\zeta\omega_n \tau} \sin\left(\sqrt{1 - \zeta^2}\omega_n(t - \tau)\right) d\tau. \quad (3.51)$$

Chapter Summary

- The frequency response function relates the harmonic force applied to a system to the resulting vibration as a function of the forcing frequency, ω.
- The largest value of a system's frequency response function occurs at resonance, when the system is forced at its natural frequency ($\omega = \omega_n$).
- The complex-valued frequency response function can be written as its real and imaginary parts or its magnitude and phase.
- A high Q system has low damping and vice versa.
- The dynamic flexibility, or compliance, of a system depends on the excitation frequency.
- A peak picking approach can be applied to a measured frequency response function in order to identify a model that represents the dynamic behavior of the measured system.
- A special case of forced vibration is rotating unbalance, where the force magnitude depends on the unbalanced mass, its eccentricity, and the rotating frequency.
- Base motion is observed when a system is excited through elastic supports.
- As an alternative to a single-frequency harmonic force model, the force may be described as being composed of multiple frequencies, random, or transient.
- An example of a transient force is the impulsive force.
- The impulse response function can be used to identify the behavior of a system due to an impulsive force input.
- The convolution integral can be used to determine the response of a system with zero initial conditions due to a general, nonperiodic input force.

Exercises

1. An apparatus known as a centrifuge is commonly used to separate solutions of different chemical compositions. It operates by rotating at high speeds to separate substances of different densities.

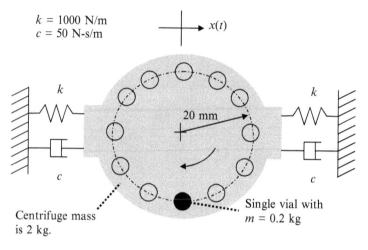

$k = 1000$ N/m
$c = 50$ N-s/m

$x(t)$

k

20 mm

c

k

c

Centrifuge mass
is 2 kg.

Single vial with
$m = 0.2$ kg

Fig. P3.1 Centrifuge model with a single vial

(a) If a single vial with a mass of 0.2 kg is placed in the centrifuge (see
Fig. P3.1) at a distance of 20 mm from the rotating axis, determine the
magnitude of the resulting vibration, X (in mm), of the single degree of
freedom centrifuge structure. The rotating speed is 200 rpm.

(b) Determine the magnitude of the forcing function (in N) due to the single
0.2 kg vial rotating at 200 rpm.

2. For a single degree of freedom spring–mass–damper system with $m = 2.5$ kg,
$k = 6 \times 10^6$ N/m, and $c = 180$ N-s/m, complete the following for the case of
forced harmonic vibration.

(a) Calculate the undamped natural frequency (in rad/s) and damping ratio.

(b) Sketch the imaginary part of the system FRF versus frequency. Identify the
frequency (in Hz) and amplitude (in m/N) of the key features.

(c) Determine the value of the imaginary part of the vibration (in mm) for this
system at a forcing frequency of 1500 rad/s if the harmonic force magnitude is 250 N.

3. A single degree of freedom lumped parameter system has mass, stiffness, and
damping values of 1.2 kg, 1×10^7 N/m, and 364.4 N-s/m, respectively.
Generate the following plots of the frequency response function.

(a) Magnitude (m/N) versus frequency (Hz) and phase (deg) versus frequency
(Hz)

(b) Real part (m/N) versus frequency (Hz) and imaginary part (m/N) versus
frequency (Hz)

(c) Argand diagram, real part (m/N) versus imaginary part (m/N).

4. For the single degree of freedom torsional system under harmonic forced vibration (see Fig. P3.4), complete parts a through c if $J = 40$ kg-m^2/rad, $C = 150$ N-m-s/rad, $K = 5 \times 10^5$ N-m/rad, and $T_0 = 65$ N-m.

Fig. P3.4 Single degree of freedom torsional system under harmonic forced vibration

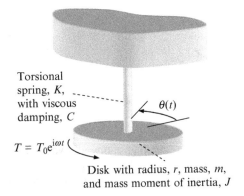

Torsional spring, K, with viscous damping, C

$\theta(t)$

$T = T_0 e^{i\omega t}$

Disk with radius, r, mass, m, and mass moment of inertia, J

(a) Calculate the undamped natural frequency (rad/s) and damping ratio.

(b) Sketch the Argand diagram (complex plane representation) of $\frac{\theta}{T}(\omega)$. Numerically identify key frequencies (rad/s) and amplitudes (rad/N-m).

(c) Given a forcing frequency of 100 rad/s for the harmonic external torque, determine the phase (in rad) between the torque and corresponding steady-state vibration of the system, θ.

5. For a single degree of freedom spring–mass–damper system subject to forced harmonic vibration with $m = 1$ kg, $k = 1 \times 10^6$ N/m, and $c = 120$ N-s/m, complete the following.

(a) Calculate the damping ratio.

(b) Write expressions for the real part, imaginary part, magnitude, and phase of the system frequency response function (FRF). These expressions should be written as a function of the frequency ratio, r, stiffness, k, and damping ratio, ζ.

(c) Plot the real part (in m/N), imaginary part (in m/N), magnitude (in m/N), and phase (in deg) of the system frequency response function (FRF) as a function of the frequency ratio, r. Use a range of 0 to 2 for r (note that $r = 1$ is the resonant frequency). [Hint: for the phase plot, try using the MATLAB® atan2 function. It considers the quadrant dependence of the tan^{-1} function.]

6. A single degree of freedom spring–mass–damper system which is initially at rest at its equilibrium position is excited by an impulsive force over a time interval of 1.5 ms; see Fig. P3.6. If the mass is 2 kg, the stiffness is 1×10^6 N/m, and the viscous damping coefficient is 10 N-s/m, complete the following.

Fig. P3.6 Spring–mass–
damper system excited by an
impulsive force

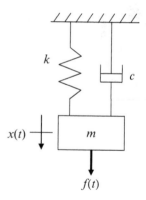

(a) Determine the maximum allowable force magnitude if the maximum deflection is to be 1 mm.

(b) Plot the impulse response function, $h(t)$, for this system. Use a time step size of 0.001 s.

(c) Calculate the impulse of the force (N-s).

7. For a single degree of freedom spring–mass–damper system subject to forced harmonic vibration, the following FRF was measured (two figures are provided with different frequency ranges). Using the "peak picking" fitting method, determine m (in kg), k (in N/m), and c (in N-s/m).

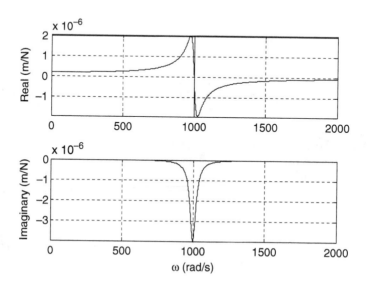

Fig. P3.7a Measured FRF

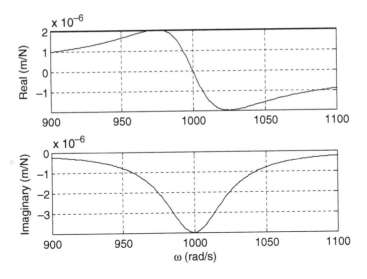

Fig. P3.7b Measured FRF (smaller frequency scale)

8. For a single degree of freedom spring–mass–damper system with $m = 2$ kg, $k = 1 \times 10^7$ N/m, and $c = 200$ N-s/m, complete the following for the case of forced harmonic vibration.

   (a) Calculate the natural frequency (in rad/s) and damping ratio.
   (b) Plot the Argand diagram (real part vs imaginary part of the system FRF).
   (c) Identify the point on the Argand diagram that corresponds to resonance.
   (d) Determine the magnitude of vibration (in m) for this system at a forcing frequency of 2,000 rad/s if the harmonic force magnitude is 100 N.

9. In a crank-slider setup, it is desired to maintain a constant rotational speed for driving the crank. Therefore, a flywheel was added to increase the spindle inertia and reduce the speed sensitivity to the driven load. See Fig. P3.9. If the

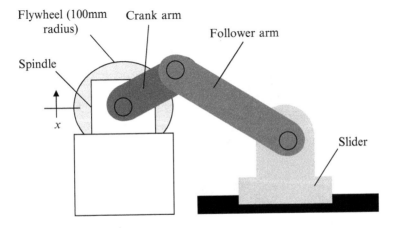

Fig. P3.9 Crank-slider with flywheel

spindle rotating speed is 120 rpm, determine the maximum allowable eccentricity-mass product, *me* (in kg-m), for the flywheel if the spindle vibration magnitude is to be less than 25 μm. The total spindle/flywheel mass is 10 kg, the effective spring stiffness (for the spindle and its support) is 1×10^6 N/m, and the corresponding damping ratio is 0.05 (5%).

Given your *me* result, comment on the accuracy requirements for the flywheel manufacture (you may assume no rotating unbalance in the spindle).

10. A single degree of freedom spring–mass–damper system with $m = 1.2$ kg, $k = 1 \times 10^7$ N/m, and $c = 364.4$ N-s/m is subjected to a forcing function $f(t) = 15e^{i\omega_n t}$ N, where ω_n is the system's natural frequency. Determine the steady-state magnitude (in μm) and phase (in deg) of the vibration due to this harmonic force.

References

http://en.wikipedia.org/wiki/Slinky

Kamen E (1990) Introduction to signals and systems, 2nd edn. Macmillan, New York

Smith S (2000) Flexures: elements of elastic mechanisms. CRC Press, Boca Raton

Xu Z, Naughton J, Lindberg W (2009) 2-D and 3-D numerical modelling of a dynamic resonant shear stress sensor. Comput Fluids 38:340–346

Chapter 4
Two Degree of Freedom Free Vibration

Since we cannot know all that there is to be known about anything, we ought to know a little about everything.

– Blaise Pascal

4.1 Equations of Motion

Let's extend our free vibration analysis from Chap. 2 to include two degrees of freedom in the model. This would make sense, for example, if we completed a measurement to determine the frequency response function (FRF) for a system and saw that there were obviously two modes of vibration within the frequency range of interest; see Fig. 4.1.

The two degree of freedom lumped parameter, chain-type model is shown in Fig. 4.2, where damping is neglected for now. The free body diagrams for the upper and lower masses give two equations of motion. Summing the forces in the x_1 direction for the top mass gives:

$$m_1\ddot{x}_1 + k_1 x_1 - k_2(x_2 - x_1) = 0,$$

which can be rewritten (by grouping terms) to obtain:

$$m_1\ddot{x}_1 + (k_1 + k_2)x_1 - k_2 x_2 = 0. \tag{4.1}$$

Similarly, for the bottom mass we have:

$$m_2\ddot{x}_2 + k_2(x_2 - x_1) = 0$$

T.L. Schmitz and K.S. Smith, *Mechanical Vibrations: Modeling and Measurement*,
DOI 10.1007/978-1-4614-0460-6_4, © Springer Science+Business Media, LLC 2012

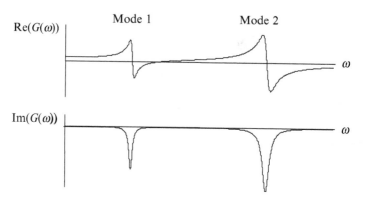

Fig. 4.1 Two-mode FRF measurement that justifies a two degree of freedom model

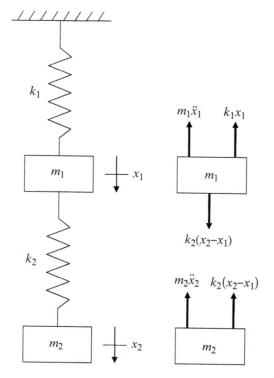

Fig. 4.2 Lumped parameter, chain-type model for an undamped two degree of freedom system

or

$$m_2\ddot{x}_2 + k_2x_2 - k_2x_1 = 0. \tag{4.2}$$

The force in the spring k_2 deserves further discussion. For the lower-mass free body diagram in Fig. 4.2, the force is directed up (opposite the x_2 direction) and is written as $k_2(x_2 - x_1)$. This is because the motion of mass m_2 depends on the motion of mass m_1 and vice versa for the chain-type, lumped parameter model. If the position of m_1 was held fixed and m_2 was displaced down by x_2, then the force in the spring k_2 would oppose this motion and its value would be $k_2(x_2 - 0) = k_2 x_2$. However, if both m_1 and m_2 were displaced down by the same amount, Δx, the force in the spring would be $k_2(\Delta x - \Delta x) = 0$. An equal and opposite force is applied to the top mass.

IN A NUTSHELL   It is often useful to consider the effect of the displacement (or velocity or acceleration) of one coordinate while holding the other coordinate(s) motionless.

4.2   Eigensolution for the Equations of Motion

Let's now organize Eqs. 4.1 and 4.2 into matrix form. Equation 4.3 includes two rows in the matrix expressions. The top row describes the behavior of the top coordinate and the bottom row describes the motion of the bottom coordinate. This equation can be more compactly written, as shown in Eq. 4.4, where the mass matrix is $m = \begin{bmatrix} m_1 & 0 \\ 0 & m_2 \end{bmatrix}$, the acceleration vector is $\{\ddot{x}\} = \begin{Bmatrix} \ddot{x}_1 \\ \ddot{x}_2 \end{Bmatrix}$, the stiffness matrix is $k = \begin{bmatrix} k_1 + k_2 & -k_2 \\ -k_2 & k_2 \end{bmatrix}$, and the displacement vector is $\{\vec{x}\} = \begin{Bmatrix} x_1 \\ x_2 \end{Bmatrix}$.

$$\begin{bmatrix} m_1 & 0 \\ 0 & m_2 \end{bmatrix} \begin{Bmatrix} \ddot{x}_1 \\ \ddot{x}_2 \end{Bmatrix} + \begin{bmatrix} k_1 + k_2 & -k_2 \\ -k_2 & k_2 \end{bmatrix} \begin{Bmatrix} x_1 \\ x_2 \end{Bmatrix} = \begin{Bmatrix} 0 \\ 0 \end{Bmatrix} \tag{4.3}$$

$$[m]\{\ddot{x}\} + [k]\{\vec{x}\} = \{0\} \tag{4.4}$$

We will treat Eq. 4.4 as an *eigenvalue problem* to determine the:

1. *eigenvalues*, which lead to the system's natural frequencies; and
2. *eigenvectors*, or *mode shapes*, which describe the characteristic relative motion of the individual degrees of freedom (typically normalized to one of the degrees of freedom). Each mode shape is associated with a particular natural frequency.

IN A NUTSHELL   Eigenvalue problems are found in many areas of engineering. Such problems are called eigenvalue problems because the German word "eigen," which can be translated as "own," "characteristic," or "peculiar to," emphasizes that these values belong to the system model and do not depend on external perturbations (http://en.wiktionary.org/wiki/eigen). Given the eigensolution, we can then use this information to determine the time-domain response of the system.

Let's begin by assuming harmonic vibration so that we can write a Laplace-domain ($s = i\omega$) form for the solution to the differential equations of motion. The displacement is $x(t) = Xe^{st}$ and the corresponding acceleration is $\ddot{x}(t) = s^2 Xe^{st}$. Substitution into Eq. 4.4 gives:

$$[m]s^2\{\vec{X}\}e^{st} + [k]\{\vec{X}\}e^{st} = [[m]s^2 + [k]]\{\vec{X}\}e^{st} = \{0\}, \qquad (4.5)$$

where $\{\vec{X}\} = \begin{Bmatrix} X_1 \\ X_2 \end{Bmatrix}$. There are two possibilities for satisfying Eq. 4.5. The first is that $\{\vec{X}\} = \{0\}$. This means that there is no motion and, while this is a valid result, it is not very useful to us. As in the single degree of freedom case, it is referred to as the *trivial solution*. The second possibility is that $[[m]s^2 + [k]] = \{0\}$ and it is the option that we will use to find the eigenvalues. In order for this equation to have nontrivial solutions, it is required that the determinant of the left-hand side be equal to zero. This is shown in Eq. 4.6 and is referred to as the *characteristic equation*. The roots of this equation are the eigenvalues.

$$\left| [m]s^2 + [k] \right| = 0 \qquad (4.6)$$

For the two degree of freedom model displayed in Fig. 4.2, the corresponding characteristic equation is:

$$\left| \begin{bmatrix} m_1 & 0 \\ 0 & m_2 \end{bmatrix} s^2 + \begin{bmatrix} k_1 + k_2 & -k_2 \\ -k_2 & k_2 \end{bmatrix} \right| = 0. \qquad (4.7)$$

As we saw in Sect. 2.4.5, the determinant of a 2×2 matrix is the difference between the product of the on-diagonal terms and the product of the off-diagonal terms, i.e., $(1,1)(2,2) - (1,2)(2,1)$, where these indices identify the (row, column). Rewriting Eq. 4.7 in standard 2×2 form yields:

$$\begin{vmatrix} m_1 s^2 + k_1 + k_2 & -k_2 \\ -k_2 & m_2 s^2 + k_2 \end{vmatrix} = 0. \qquad (4.8)$$

The characteristic equation (determinant) is:

$$\left(m_1 s^2 + k_1 + k_2 \right)\left(m_2 s^2 + k_2 \right) - (-k_2)(-k_2) = 0. \qquad (4.9)$$

This gives an equation of the form $as^4 + bs^2 + c = 0$, which is quadratic in s^2. Using the quadratic equation, the two roots, s_1^2 and s_2^2, are given by $s_{1,2}^2 = \frac{-b \pm \sqrt{b^2 - 4ac}}{2a}$. These roots, or eigenvalues, are used to determine the system natural frequencies:

$$s_1^2 = -\omega_{n1}^2$$
$$s_2^2 = -\omega_{n2}^2, \qquad (4.10)$$

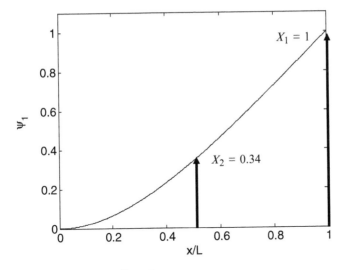

Fig. 4.3 First mode shape for a cantilever beam

where $\omega_{n1} < \omega_{n2}$ by convention (in other words, the roots are ordered such that the first root gives the lowest natural frequency). Using the eigenvalues, we can next determine the eigenvectors, or mode shapes. These represent the relative magnitude and direction for the model's degrees of freedom (coordinates) during vibration.

IN A NUTSHELL   We cannot solve for X_1 and X_2 in Eq. 4.5 once s_1^2 and s_2^2 have been determined. It might look like two equations with two unknowns, but it is not. The s^2 values were found by requiring the determinant to be equal to zero, which means that the two equations are not independent.

Suppose the model in Fig. 4.2 was used to describe the vibrating motion of a cantilever beam. As we discussed in Sect. 1.4, a continuous beam has an infinite number of degrees of freedom. However, in this case, let's assume it is adequate to describe the motion at two locations only: at the midpoint of the beam and at its free end. We will let x_2 represent the midpoint and x_1 the tip. The first mode shape, ψ_1, for a cantilever beam is provided in Fig. 4.3. The vibration in this mode shape occurs at the first natural frequency, ω_{n1}. The relative magnitudes of vibration at x_1 and x_2 are also shown in Fig. 4.3. We see that the magnitude of vibration at the free end is larger than at the beam's midpoint by a ratio of approximately 1:0.34 for vibration at the first natural frequency, ω_{n1}.

The second mode shape, ψ_2, is displayed in Fig. 4.4. In this case, the motion at x_1 is again larger, but it is out of phase with the motion at x_2. The corresponding ratio is approximately 1:−0.71. Motion in this mode shape occurs at the second natural frequency, ω_{n2}. For the model in Fig. 4.2, we only have two coordinates and therefore would not have any way of knowing the high-resolution mode shapes

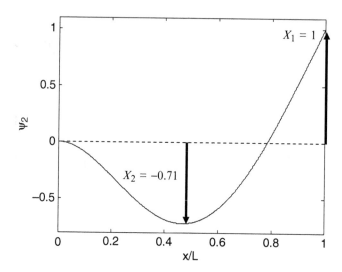

Fig. 4.4 Second mode shape for a cantilever beam

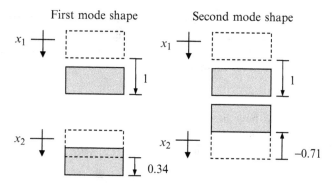

Fig. 4.5 Mode shapes for a two degree of freedom representation of a cantilever beam

given in Figs. 4.3 and 4.4. We can represent the two mode shapes for our two degree of freedom model, as shown in Fig. 4.5. As with Figs. 4.3 and 4.4, we will normalize the motion at x_1 to 1 and plot the corresponding magnitude (and direction) for the motion at x_2. The choice to set the vibration magnitude at x_1 to 1 is for convenience; the mode shape only gives the relative motion between coordinates, so the scaling is arbitrary. In some cases, it could make more sense to normalize to x_2. We will explore this issue in more detail as we move forward. From Fig. 4.5, we see that the motion at x_2 has a smaller magnitude (the ratio is 1:0.34) but the same direction when vibrating at ω_{n1} in the first mode shape. For the second mode shape, the motion at x_2 is again smaller than at x_1 (the ratio is 1:−0.71), but now in the opposite direction. This matches the behavior we observed in Figs. 4.3 and 4.4. Note that we have two natural frequencies and mode shapes because our model has two degrees of freedom. For a three degree of freedom model, we'd have three natural frequencies and mode shapes.

IN A NUTSHELL   There are many ways to normalize mode shapes. They can be normalized to a position: scaling so that the mode shape component at a particular coordinate is 1. They may be length normalized: scaling so that the length of the vector is 1. Mode shapes can be scaled in any convenient way because they show relative motion between coordinates, but not the absolute size of motion.

The functions used to plot the mode shapes for beams with various *boundary conditions* (such as free-free, pinned-pinned, and fixed-free) have been tabulated by Blevins (Blevins 2001). The cantilever, or fixed-free, mode shape function is:

$$\Psi_i = \frac{1}{2}\left(\cosh\left(\frac{\lambda_i x}{L}\right) - \cos\left(\frac{\lambda_i x}{L}\right) - \sigma_i\left(\sinh\left(\frac{\lambda_i x}{L}\right) - \sin\left(\frac{\lambda_i x}{L}\right)\right)\right), \quad (4.11)$$

where i is the mode shape number, x is the distance along the beam of length L, and the constants λ_i and σ_i are provided in Table 4.1. The code used to produce Figs. 4.3 and 4.4 is provided in MATLAB® MOJO 4.1. Note that Eq. 4.11 is normalized to have a value of one at the free end of the cantilever.

MATLAB® MOJO 4.1

```
% matlab_mojo_4_1.m

clc
clear all
close all

% define variables
L = 1;                    % m
x = 0:0.001:L;

% define first mode shape
lambda = 1.87510407;
sigma = 0.734095514;
psi = (cosh(lambda*x/L) - cos(lambda*x/L) - sigma*(sinh(lambda*x/L) -
sin(lambda*x/L)))/2;

figure(1)
plot(x, psi, 'k-')
set(gca,'FontSize', 14)
axis([0 1 0 1.1])
xlabel('x/L')
ylabel('\psi_1')

index = find(x == 0.5);
psi(index)

% define second mode shape
lambda = 4.69409113;
sigma = 1.018467319;
psi = -(cosh(lambda*x/L) - cos(lambda*x/L) - sigma*(sinh(lambda*x/L) -
sin(lambda*x/L)))/2;

figure(2)
plot(x, psi, 'k-')
set(gca,'FontSize', 14)
axis([0 1 -0.8 1.1])
xlabel('x/L')
ylabel('\psi_2')

index = find(x == 0.5);
psi(index)
```

Table 4.1 Constants for
cantilever beam mode shape
calculation in Eq. 4.11
(Blevins 2001)

Mode i	λ_i	σ_i
1	1.87510407	0.734095514
2	4.69409113	1.018467319

In order to determine the eigenvalues of a system model, we set the determinant
of the matrix form for the equations of motion equal to zero; see Eq. 4.6. This means
that the two equations (one for each of the two degrees of freedom) are *linearly
dependent*. We can think about linear dependence in the following way. If you were
given directions from the grocery store to the library that said "go north three blocks
and then go west four blocks," this would provide all the necessary information to
reach the library. The two statements for the north and west travel cannot be
described in terms of the other; they are linearly independent. However, if the
directions were augmented to be "go north three blocks and then go west four
blocks; the library is five blocks northwest of the grocery store," then the final
statement is not independent of the first two. These three statements are linearly
dependent because they give redundant information; one of the three is not required
(http://en.wikipedia.org/wiki/Linear_independence).

Because the two equations in Eq. 4.6 are linearly dependent, we can select either
one to determine the two eigenvectors. They will both give the same result. For the
two-degree-of-freedom model in Fig. 4.2, we have:

$$\left[\begin{bmatrix} m_1 & 0 \\ 0 & m_2 \end{bmatrix} s^2 + \begin{bmatrix} k_1 + k_2 & -k_2 \\ -k_2 & k_2 \end{bmatrix} \right] \begin{Bmatrix} X_1 \\ X_2 \end{Bmatrix} = \begin{Bmatrix} 0 \\ 0 \end{Bmatrix}. \tag{4.12}$$

The top row equation, which corresponds to motion of the top mass, is:

$$\left(m_1 s^2 + k_1 + k_2 \right) X_1 + (-k_2) X_2 = 0. \tag{4.13}$$

The bottom row equation, which corresponds to motion of the bottom mass, is:

$$(-k_2) X_1 + \left(m_2 s^2 + k_2 \right) X_2 = 0. \tag{4.14}$$

We can pick either Eq. 4.13 or 4.14 to determine the two eigenvectors – one
eigenvector for s_1^2 and one for s_2^2. We can also choose to normalize our eigenvectors
to either coordinate x_1 or x_2. We generally select the coordinate that is of most
interest for the particular application (such as the end of the cantilever beam as
in the previous example). To normalize to x_1, we need the ratio $\frac{X_2}{X_1}$. Using Eq. 4.13,
we have:

$$\frac{X_2}{X_1} = \frac{m_1 s^2 + k_1 + k_2}{k_2}. \tag{4.15}$$

The corresponding eigenvector expression is:

$$\psi_{1,2} = \left\{ \begin{array}{c} \dfrac{X_1}{X_1} \\ \dfrac{X_2}{X_1} \end{array} \right\} = \left\{ \begin{array}{c} 1 \\ \dfrac{m_1 s_{1,2}^2 + k_1 + k_2}{k_2} \end{array} \right\}, \tag{4.16}$$

where we substitute s_1^2 to find the first eigenvector, Ψ_1, and s_2^2 to find the second eigenvector, Ψ_2. As described previously, Ψ_1 describes the relative magnitude of the coordinates for vibration at the first natural frequency and Ψ_2 describes the relative magnitude of the coordinates for vibration at the second natural frequency. We would obtain exactly the same results using Eq. 4.14. In this case, the ratio is:

$$\frac{X_2}{X_1} = \frac{k_2}{m_2 s^2 + k_2}. \tag{4.17}$$

In order to normalize to x_2, the required ratio is $\frac{X_1}{X_2}$. Using Eq. 4.13, we find that:

$$\frac{X_1}{X_2} = \frac{k_2}{m_1 s^2 + k_1 + k_2}. \tag{4.18}$$

The eigenvector expression for normalization to x_2 is:

$$\Psi_{1,2} = \left\{ \begin{array}{c} \dfrac{X_1}{X_2} \\ \dfrac{X_2}{X_2} \end{array} \right\} = \left\{ \begin{array}{c} \dfrac{k_2}{m_1 s_{1,2}^2 + k_1 + k_2} \\ 1 \end{array} \right\}. \tag{4.19}$$

Again, substituting s_1^2 gives the first eigenvector, Ψ_1, and substituting s_2^2 gives the second eigenvector, Ψ_2. We would obtain the same eigenvectors using Eq. 4.14, where the ratio is:

$$\frac{X_1}{X_2} = \frac{m_2 s^2 + k_2}{k_2}. \tag{4.20}$$

For a two degree of freedom system, there are two eigenvectors. The first corresponds to vibration at ω_{n1} and the second to vibration at ω_{n2}. The system vibration (due to some set of initial conditions) occurs in: (1) the first mode shape; (2) the second mode shape; or (3) a linear combination of the two. The latter is the general result, but the final behavior depends on the initial conditions.

Fig. 4.6 *By the Numbers 4.1* – Example two degree of freedom model with parameters and initial conditions

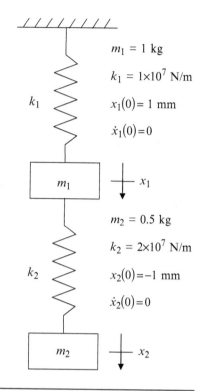

$m_1 = 1$ kg

$k_1 = 1 \times 10^7$ N/m

$x_1(0) = 1$ mm

$\dot{x}_1(0) = 0$

$m_2 = 0.5$ kg

$k_2 = 2 \times 10^7$ N/m

$x_2(0) = -1$ mm

$\dot{x}_2(0) = 0$

IN A NUTSHELL We have seen that free vibration for a single degree of freedom system always occurs at the system's natural frequency. In a two degree of freedom system, the motion may occur at the first natural frequency with the first mode shape, at the second natural frequency with the second mode shape, or in a linear combination of the two modes simultaneously, depending on the initial conditions. No other motions are possible in free vibration.

By the Numbers 4.1

Consider the two degree of freedom system shown in Fig. 4.6. This is the same model we used in Fig. 4.2, so we can write the equations of motion in matrix form directly.

$$\begin{bmatrix} 1 & 0 \\ 0 & 0.5 \end{bmatrix} \begin{Bmatrix} \ddot{x}_1 \\ \ddot{x}_2 \end{Bmatrix} + \begin{bmatrix} 1 \times 10^7 + 2 \times 10^7 & -2 \times 10^7 \\ -2 \times 10^7 & 2 \times 10^7 \end{bmatrix} \begin{Bmatrix} x_1 \\ x_2 \end{Bmatrix} = \begin{Bmatrix} 0 \\ 0 \end{Bmatrix}$$

We should note here that as long as the coordinates are measured with respect to ground, the mass and stiffness matrices are always *symmetric*. In other words, the off-diagonal terms are equal: $m(1,2) = m(2,1)$ and $k(1,2) = k(2,1)$. Alternately, we can write: $m_{12} = m_{21}$ and $k_{12} = k_{21}$ to denote the symmetry. Validating symmetry for the mass and stiffness matrices provides a good check for the equations of motion.

Using the Laplace notion for our (assumed) harmonic solution and substituting gives the characteristic equation:

$$\begin{vmatrix} 1s^2 + 3 \times 10^7 & -2 \times 10^7 \\ -2 \times 10^7 & 0.5s^2 + 2 \times 10^7 \end{vmatrix} = 0.$$

Calculating the determinant gives:

$$\left(1s^2 + 3 \times 10^7\right)\left(0.5s^2 + 2 \times 10^7\right) - \left(-2 \times 10^7\right)\left(-2 \times 10^7\right) = 0.$$

This equation can be expanded and grouped to obtain:

$$0.5s^4 + 3.5 \times 10^7 s^2 + 2 \times 10^{14} = 0.$$

Using the quadratic equation, the roots (eigenvalues) are:

$$s_{1,2}^2 = \frac{-3.5 \times 10^7 \pm \sqrt{\left(3.5 \times 10^7\right)^2 - 4(0.5)2 \times 10^{14}}}{2(0.5)}$$

$$= -3.5 \times 10^7 \pm 2.872 \times 10^7.$$

This gives $s_1^2 = (-3.5 + 2.872) \times 10^7 = -6.28 \times 10^6 = -\omega_{n1}^2$ and the first natural frequency is $\omega_{n1} = 2,506$ rad/s. The second eigenvalue is $s_2^2 = (-3.5 - 2.872) \times 10^7 = -6.37 \times 10^7 = -\omega_{n2}^2$ and the second natural frequency is $\omega_{n2} = 7,981$ rad/s. Note that $\omega_{n1} < \omega_{n2}$. We can also express these natural frequencies in units of Hz: $f_{n1} = \frac{\omega_{n1}}{2\pi} = 398.8$ Hz and $f_{n2} = \frac{\omega_{n2}}{2\pi} = 1,270$ Hz.

To determine the eigenvectors, let's arbitrarily choose the top equation of motion and normalize to coordinate x_2. The eigenvector equation is:

$$\left(1s^2 + 3 \times 10^7\right)X_1 + \left(-2 \times 10^7\right)X_2 = 0,$$

and the required ratio is:

$$\frac{X_1}{X_2} = \frac{2 \times 10^7}{1s^2 + 3 \times 10^7}.$$

Substituting $s_1^2 = -6.28 \times 10^6$ gives:

$$\left.\frac{X_1}{X_2}\right|_{s_1^2} = \frac{X_{11}}{X_{21}} = \frac{2 \times 10^7}{1\left(-6.28 \times 10^6\right) + 3 \times 10^7} = 0.843,$$

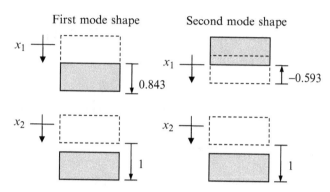

Fig. 4.7 *By the Numbers 4.1* – Mode shapes for the example two-degree-of-freedom system

where the first subscript in $\frac{X_{11}}{X_{21}}$ identifies the coordinate number and the second subscript gives the mode number. Because we normalized to coordinate x_2, the first eigenvector is therefore:

$$\psi_1 = \left\{ \begin{array}{c} \dfrac{X_{11}}{X_{21}} \\ 1 \end{array} \right\} = \left\{ \begin{array}{c} 0.843 \\ 1 \end{array} \right\}.$$

The ratio for the second eigenvector is calculated by substituting $s_2^2 = -6.37 \times 10^7$.

$$\left.\frac{X_1}{X_2}\right|_{s_2^2} = \frac{X_{12}}{X_{22}} = \frac{2 \times 10^7}{1\left(-6.37 \times 10^7\right) + 3 \times 10^7} = -0.593$$

The second eigenvector is:

$$\psi_2 = \left\{ \begin{array}{c} \dfrac{X_{12}}{X_{22}} \\ 1 \end{array} \right\} = \left\{ \begin{array}{c} -0.593 \\ 1 \end{array} \right\}.$$

These mode shapes are demonstrated graphically in Fig. 4.7. For the model type pictured in Fig. 4.6 with the eigenvalues ordered such that $\omega_{n1} < \omega_{n2}$, the vibration of coordinates x_1 and x_2 are *always in phase* for the first mode shape and *always out of phase* for the second mode shape.

How would the mode shapes have changed if we had normalized to coordinate x_1? In this case, we require the ratio:

$$\frac{X_2}{X_1} = \frac{1s^2 + 3 \times 10^7}{2 \times 10^7}.$$

Using the first eigenvalue, $s_1^2 = -6.28 \times 10^6$, we obtain:

$$\frac{X_{21}}{X_{11}} = \frac{1\left(-6.28 \times 10^6\right) + 3 \times 10^7}{2 \times 10^7} = 1.186,$$

and the first eigenvector is:

$$\psi_1 = \left\{\begin{array}{c} 1 \\ \dfrac{X_{21}}{X_{11}} \end{array}\right\} = \left\{\begin{array}{c} 1 \\ 1.186 \end{array}\right\}.$$

Note that 1.186 is simply the reciprocal of 0.843, which we calculated when normalizing to x_2. Similarly, if we use the second eigenvalue, $s_2^2 = -6.37 \times 10^7$, we obtain:

$$\frac{X_{22}}{X_{12}} = \frac{1\left(-6.37 \times 10^7\right) + 3 \times 10^7}{2 \times 10^7} = -1.686,$$

and the second eigenvector is:

$$\psi_2 = \left\{\begin{array}{c} 1 \\ \dfrac{X_{22}}{X_{12}} \end{array}\right\} = \left\{\begin{array}{c} 1 \\ -1.686 \end{array}\right\}.$$

Again, -1.686 is the reciprocal of -0.593, which we determined by normalizing to x_2.

4.3 Time-Domain Solution

Let's now find the time-domain responses, $x_1(t)$ and $x_2(t)$, for the system in Fig. 4.6 (*By the Numbers 4.1*). As we saw in Eq. 2.12 for the free vibration of single degree of freedom systems, the total response is the sum of all possible solutions. For each of the eigenvalues, there are two roots and, therefore, two solutions. For the first eigenvalue, we have $s_{1a} = +i\omega_{n1}$ and $s_{1b} = -i\omega_{n1}$. For the second eigenvalue, we have $s_{2a} = +i\omega_{n2}$ and $s_{2b} = -i\omega_{n2}$. For $x_1(t)$, the sum of the harmonic solutions that correspond to these four roots is:

$$x_1(t) = X_{11}e^{i\omega_{n1}t} + X_{11}^*e^{-i\omega_{n1}t} + X_{12}e^{i\omega_{n2}t} + X_{12}^*e^{-i\omega_{n2}t}, \qquad (4.21)$$

where the first two terms represent motion of the top coordinate from Fig. 4.6 in the first natural frequency, the second two identify motion in the second natural frequency, and X_{11} and X_{11}^*, as well as X_{12} and X_{12}^*, are complex conjugates (they are identical except for the sign of their imaginary parts). Similarly, the time response $x_2(t)$, which describes the motion of the bottom coordinate, has four terms:

$$x_2(t) = X_{21}e^{i\omega_{n1}t} + X_{21}^*e^{-i\omega_{n1}t} + X_{22}e^{i\omega_{n2}t} + X_{22}^*e^{-i\omega_{n2}t}. \tag{4.22}$$

From *By the Numbers 4.1*, the natural frequencies are $\omega_{n1} = 2,506$ rad/s and $\omega_{n2} = 7,981$ rad/s. Substituting in Eqs. 4.20 and 4.21 yields:

$$x_1(t) = X_{11}e^{i2,506t} + X_{11}^*e^{-i2,506t} + X_{12}e^{i7,981t} + X_{12}^*e^{-i7,981t}$$

and

$$x_2(t) = X_{21}e^{i2,506t} + X_{21}^*e^{-i2,506t} + X_{22}e^{i7,981t} + X_{22}^*e^{-i7,981t}.$$

The velocities are determined by calculating the time derivatives.

$$\dot{x}_1(t) = i2{,}506\left(X_{11}e^{i2,506t} - X_{11}^*e^{-i2,506t}\right) + i7{,}981\left(X_{12}e^{i7,981t} - X_{12}^*e^{-i7,981t}\right)$$

$$\dot{x}_2(t) = i2{,}506\left(X_{21}e^{i2,506t} - X_{21}^*e^{-i2,506t}\right) + i7{,}981\left(X_{22}e^{i7,981t} - X_{22}^*e^{-i7,981t}\right)$$

We can now apply the initial conditions specified in Fig. 4.6.

$$x_1(0) = X_{11} + X_{11}^* + X_{12} + X_{12}^* = 1 \text{ mm}$$

$$x_2(0) = X_{21} + X_{21}^* + X_{22} + X_{22}^* = -1 \text{ mm}$$

$$\dot{x}_1(0) = i2{,}506\left(X_{11} - X_{11}^*\right) + i7{,}981\left(X_{12} - X_{12}^*\right) = 0$$

$$\dot{x}_2(0) = i2{,}506\left(X_{21} - X_{21}^*\right) + i7{,}981\left(X_{22} - X_{22}^*\right) = 0$$

We have a problem, though. There are four equations, but eight unknowns. We can remove this obstacle, however, by applying the eigenvector relationships. From our previous analysis (normalizing to x_2), we found that $\frac{X_{11}}{X_{21}} = 0.843$ and $\frac{X_{12}}{X_{22}} = -0.593$. Rearranging gives $X_{11} = 0.843X_{21}$ and $X_{12} = -0.593X_{22}$. Substituting these relationships gives a system of four equations with four unknowns.

$$0.843X_{21} + 0.843X_{21}^* - 0.593X_{22} - 0.593X_{22}^* = 1 \text{ mm}$$

$$X_{21} + X_{21}^* + X_{22} + X_{22}^* = -1 \text{ mm}$$

$$i2{,}506(0.843)\left(X_{21} - X_{21}^*\right) + i7{,}981(-0.593)\left(X_{22} - X_{22}^*\right) = 0$$

$$i2{,}506\left(X_{21} - X_{21}^*\right) + i7{,}981\left(X_{22} - X_{22}^*\right) = 0$$

We can write these four equations in matrix form.

$$\begin{bmatrix} 0.843 & 0.843 & -0.593 & -0.593 \\ 1 & 1 & 1 & 1 \\ i2,112 & -i2,112 & -i4,733 & i4,733 \\ i2,506 & -i2,506 & i7,981 & -i7,981 \end{bmatrix} \begin{Bmatrix} X_{21} \\ X_{21}^* \\ X_{22} \\ X_{22}^* \end{Bmatrix} = \begin{Bmatrix} 1 \\ -1 \\ 0 \\ 0 \end{Bmatrix}$$

We need to solve for the magnitude vector composed of X_{21}, X_{21}^*, X_{22}, and X_{22}^*. Like we saw in Sect. 2.4.5, we can invert the A matrix in the $AX = B$ equation to determine $X = A^{-1}B$. We can complete this operation at the MATLAB® command prompt ($\gg$) using the following statements.

```
>> A = [0.843 0.843 -0.593 -0.593;1 1 1 1;i*2112 -i*2112 -i*4733 i*4733;
i*2506 -i*2506 i*7981 -i*7981]

A =

  1.0e+003 *

  0.0008              0.0008             -0.0006             -0.0006
  0.0010              0.0010              0.0010              0.0010
       0 + 2.1120i         0 - 2.1120i         0 - 4.7330i         0 + 4.7330i
       0 + 2.5060i         0 - 2.5060i         0 + 7.9810i         0 - 7.9810i

>> B = [1 -1 0 0]'

B =

   1
  -1
   0
   0
```

The $'$ operator here indicates that the *transpose* operation is to be performed on the B vector. The transpose operator switches the rows and columns for a matrix. In this case, with only one row, the row becomes the only column. Alternately, B could have been defined using B = [1; -1; 0; 0].

```
>> X = inv(A)*B

X =

   0.1417
   0.1417
  -0.6417
  -0.6417
```

Now that we know X_{21}, X_{21}^*, X_{22}, and X_{22}^*, we can again use the eigenvector relationships to determine X_{11}, X_{11}^*, X_{12}, and X_{12}^*.

$$X_{11} = 0.843X_{21} = 0.843(0.1417) = 0.1195 = X_{11}^*$$

$$X_{12} = -0.593X_{22} = -0.593(-0.6417) = 0.3805 = X_{12}^*$$

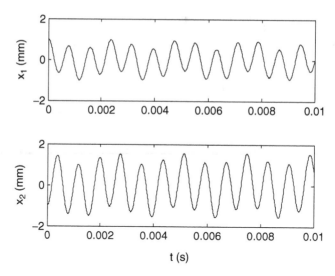

Fig. 4.8 Time-domain responses for the two degree of freedom system described in Fig 4.6

The complex conjugates are real-valued and equal in this case because there is no damping in the model. Substituting the coefficients into the original x_1 equation gives:

$$x_1(t) = 0.1195e^{i2,506t} + 0.1195e^{-i2,506t} + 0.3805e^{i7,981t} + 0.3805e^{-i7,981t}.$$

Using Eq. 1.12 ($2\cos(\theta) = e^{i\theta} + e^{-i\theta}$), we can rewrite the x_1 time response as:

$$x_1(t) = 0.2390\cos(2,506t) + 0.761\cos(7,981t),$$

where the first term describes the portion of the motion oscillating at ω_{n1} and the second describes the portion oscillating at ω_{n2}. With a two degree of freedom system, we can have free oscillation in both natural frequencies.

Similarly, for x_2 we have:

$$x_2(t) = 0.1417e^{i2506t} + 0.1417e^{-i2506t} - 0.6417e^{i7981t} - 0.6417e^{-i7981t}.$$

Again applying Eq. 1.12, we can alternately express this result as:

$$x_2(t) = 0.2834\cos(2,506t) - 1.2834\cos(7,981t).$$

Viewing the time-domain responses for coordinates x_1 and x_2, we see that the two degree of freedom system vibrates in a linear combination of the two mode shapes at the two corresponding natural frequencies. These results are displayed in Fig. 4.8, which was generated using the code in MATLAB® Mojo 4.2.

MATLAB® MOJO 4.2
```
% matlab_mojo_4_2.m

clc
clear all
close all

% define variables
t = 0:1e-5:0.01;   % s

% define functions
x1 = 0.2390*cos(2506*t) + 0.761*cos(7981*t);     % mm
x2 = 0.2834*cos(2506*t) - 1.2834*cos(7981*t);

figure(1)
subplot(211)
plot(t, x1, 'k-')
set(gca,'FontSize', 14)
axis([0 0.01 -2 2])
ylabel('x_1 (mm)')
subplot(212)
plot(t, x2, 'k-')
set(gca,'FontSize', 14)
axis([0 0.01 -2 2])
xlabel('t (s)')
ylabel('x_2 (mm)')
```

Now let's consider a three degree of freedom system; see Fig. 4.9. The equations of motion are:

$$m_1\ddot{x}_1 + (k_1 + k_2)x_1 - k_2x_2 = 0$$
$$m_2\ddot{x}_2 + (k_2 + k_3)x_2 - k_2x_1 - k_3x_3 = 0$$
$$m_3\ddot{x}_3 + k_3x_3 - k_3x_2 = 0 \qquad (4.23)$$

for the three masses from top to bottom, respectively. In matrix form, these equations are written as:

$$\begin{bmatrix} m_1 & 0 & 0 \\ 0 & m_2 & 0 \\ 0 & 0 & m_3 \end{bmatrix} \begin{Bmatrix} \ddot{x}_1 \\ \ddot{x}_2 \\ \ddot{x}_3 \end{Bmatrix} + \begin{bmatrix} k_1 + k_2 & -k_2 & 0 \\ -k_2 & k_2 + k_3 & -k_3 \\ 0 & -k_3 & k_3 \end{bmatrix} \begin{Bmatrix} x_1 \\ x_2 \\ x_3 \end{Bmatrix} = \begin{Bmatrix} 0 \\ 0 \\ 0 \end{Bmatrix}, \qquad (4.24)$$

where the mass and stiffness matrices are symmetric as we discussed previously. For harmonic vibration, we can assume the solution form $x = Xe^{st}$. Substituting for displacement and acceleration, $\ddot{x} = s^2Xe^{st}$, gives:

$$\left[\begin{bmatrix} m_1 & 0 & 0 \\ 0 & m_2 & 0 \\ 0 & 0 & m_3 \end{bmatrix} s^2 + \begin{bmatrix} k_1 + k_2 & -k_2 & 0 \\ -k_2 & k_2 + k_3 & -k_3 \\ 0 & -k_3 & k_3 \end{bmatrix} \right] \begin{Bmatrix} X_1 \\ X_2 \\ X_3 \end{Bmatrix} e^{st} = \begin{Bmatrix} 0 \\ 0 \\ 0 \end{Bmatrix}, \qquad (4.25)$$

Fig. 4.9 Three degree of
freedom system with
associated free body diagrams

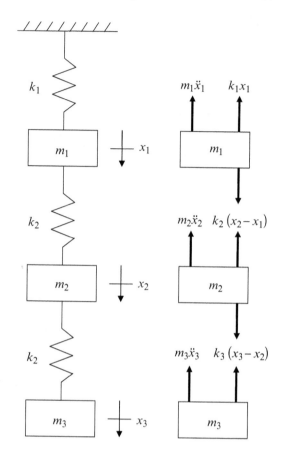

and the characteristic equation is:

$$\left| \begin{bmatrix} m_1 & 0 & 0 \\ 0 & m_2 & 0 \\ 0 & 0 & m_3 \end{bmatrix} s^2 + \begin{bmatrix} k_1 + k_2 & -k_2 & 0 \\ -k_2 & k_2 + k_3 & -k_3 \\ 0 & -k_3 & k_3 \end{bmatrix} \right| = 0. \qquad (4.26)$$

Calculating the determinant, we obtain an equation that is now *cubic* in s^2 so that we obtain three eigenvalues (i.e., the roots of the characteristic equation): s_1^2, s_2^2, and s_3^2. We use these eigenvalues to determine the three associated eigenvectors. To find the time-domain responses, we can follow the same steps as for the two degree of freedom system. Recall that in order to find X_{11}, X_{12}, X_{21}, and X_{22} (and their complex conjugates), it was necessary to invert a $2^2 \times 2^2 = 4 \times 4$ matrix. For the three degree of freedom system, we would need to invert a $3^2 \times 3^2 = 9 \times 9$ matrix. For large structures, 50 degrees of freedom or more may be necessary to fully describe the complicated system behavior. In this case, we'd have to invert a $50^2 \times 50^2 = 2,500 \times 2,500$ matrix. The "squared" scaling on our matrix size is clearly computationally unfriendly. Fortunately, there is an alternative to this approach. It is called *modal analysis* and we will discuss it next.

IN A NUTSHELL   Using this approach, we also have a problem computing the determinant and finding the roots of the characteristic equation. For a three degree of freedom system, the characteristic equation is cubic and explicit solutions exist for the roots, similar to the quadratic equation. If we had a four degree of freedom system however, we'd have to resort to numerical techniques to find the roots. If we had 50 degrees of freedom, then the characteristic equation would have the form $As^{100} + Bs^{98} + \cdots = 0$. This would be difficult to formulate and solve indeed.

4.4   Modal Analysis

In this approach, the local (i.e., the model or physical) coordinates are transformed into *modal coordinates*. While the modal coordinate system does not have a physical basis, it does provide a coordinate frame where the individual degrees of freedom are uncoupled.

IN A NUTSHELL   There are many choices of coordinates that can be used to describe a physical system. Some of them are "easier" than others to derive and some are more convenient mathematically. The modal coordinate system is mathematically easy because it "decouples" a multiple degree of freedom system into separate single degree of freedom systems.

Figure 4.10 demonstrates the modal coordinate concept, where the local coordinate system gives a stiffness matrix that is coupled. This coupling, or dependence of the response of x_1 on x_2 and vice versa, is manifested by the stiffness matrix with nonzero off-diagonal terms. Alternately, the modal coordinate system yields an uncoupled modal stiffness matrix and two separate single degree of freedom systems; note the new modal coordinates q_1 and q_2 and modal mass and stiffness values identified by the q subscripts in Fig. 4.10. The single degree of freedom free and forced vibration solutions we discussed in Chap. 2 and 3 can therefore be applied individually to each single degree of freedom system. Eureka[1]! Once the solutions are determined in modal coordinates, they are transformed back

---

[1] The Greek scholar *Archimedes* is historically credited with this interjection. As the story goes, he noticed that the water level rose in proportion to his body's volume when he stepped into a bath. The account continues that he was so excited by this discovery that he ran through the streets of Syracuse naked (http://en.wikipedia.org/wiki/Eureka_(word)). Typical engineer!

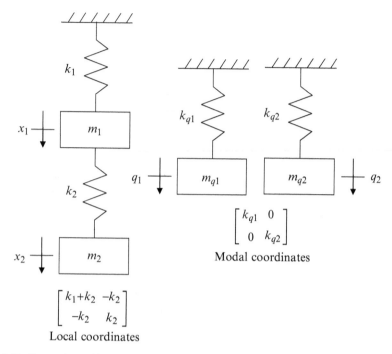

Fig. 4.10 Comparison of local and modal coordinates for a two degree of freedom system

into local coordinates for the final result. In order to convert between the two coordinate systems, the modal matrix, P, is applied. Its columns are the ordered system eigenvectors. The first column is the first eigenvector (that corresponds to the first, lowest natural frequency). The second column is the second eigenvector, and so on. The modal matrix is always a *square matrix*; the number or rows is equal to the number of columns. See Eq. 4.27.

$$P = [\psi_1 \quad \psi_2 \quad \cdots] \tag{4.27}$$

Let's complete an example to demonstrate the modal analysis procedure. We will use a two degree of freedom system that can be described by the two equations of motion:

$$0.05\ddot{x}_1 - 0.05\ddot{x}_2 + 10\dot{x}_1 + 1 \times 10^5 x_1 = 0$$
$$0.25\ddot{x}_2 - 0.05\ddot{x}_1 + 20\dot{x}_2 + 2 \times 10^5 x_2 = 0,$$

which are written in matrix form as:

$$\begin{bmatrix} 0.05 & -0.05 \\ -0.05 & 0.25 \end{bmatrix} \begin{Bmatrix} \ddot{x}_1 \\ \ddot{x}_2 \end{Bmatrix} + \begin{bmatrix} 10 & 0 \\ 0 & 20 \end{bmatrix} \begin{Bmatrix} \dot{x}_1 \\ \dot{x}_2 \end{Bmatrix} + \begin{bmatrix} 1 \times 10^5 & 0 \\ 0 & 2 \times 10^5 \end{bmatrix} \begin{Bmatrix} x_1 \\ x_2 \end{Bmatrix} = \begin{Bmatrix} 0 \\ 0 \end{Bmatrix}.$$

These equations of motion do not represent a chain-type system, as shown in Fig. 4.2. They are coupled in the mass matrix, not the stiffness or damping matrices. However, the modal analysis procedure still works in the same way. The mass, damping, and stiffness matrices are $m = \begin{bmatrix} 0.05 & -0.05 \\ -0.05 & 0.25 \end{bmatrix}$ kg, $c = \begin{bmatrix} 10 & 0 \\ 0 & 20 \end{bmatrix}$ N-s/m, and $k = \begin{bmatrix} 1 \times 10^5 & 0 \\ 0 & 2 \times 10^5 \end{bmatrix}$ N/m. The initial conditions are $x_1(0) = x_2(0) = 1$ mm and $\dot{x}_1(0) = \dot{x}_2(0) = 0$ is missing its unit. $\dot{x}_1(0) = \dot{x}_2(0) = 0$.

In order to carry out the modal analysis approach, *proportional damping* is required. Mathematically, proportional damping exists if the damping matrix can be written as a linear combination of the mass and stiffness matrices: $[c] = \alpha[m] + \beta[k]$, where α and β are real numbers. Physically, proportional damping means that the individual modes reach their maximum values at the same time. They are either exactly in phase or exactly out of phase. While this is not true for all systems, it is a good approximation for those with low damping (as we typically observe for mechanical structures).

For the example we are considering here, $[c] = \alpha[m] + \beta[k]$ is true when $\alpha = 0$ and $\beta = \frac{1}{1 \times 10^4}$. Given that the proportional damping requirement is satisfied, we can neglect damping to find the eigensolution. For the eigenvalues, we need the characteristic equation. In this case, it is:

$$\left| [m]s^2 + [k] \right| = \left| \begin{bmatrix} 0.05 & -0.05 \\ -0.05 & 0.25 \end{bmatrix} s^2 + \begin{bmatrix} 1 \times 10^5 & 0 \\ 0 & 2 \times 10^5 \end{bmatrix} \right| = 0.$$

Calculating the determinant gives $0.01s^4 + 3.5 \times 10^4 s^2 + 2 \times 10^{10} = 0$. Using the quadratic equation, the roots are determined using:

$$s_{1,2}^2 = \frac{-3.5 \times 10^4 \pm \sqrt{(3.5 \times 10^4)^2 - 4(0.01)2 \times 10^{10}}}{2(0.01)}$$

$$= -1.75 \times 10^6 \pm 1.031 \times 10^6.$$

The first eigenvalue is $s_1^2 = -1.75 \times 10^6 + 1.031 \times 10^6 = -7.19 \times 10^5 = -\omega_{n1}^2$ and the second eigenvalue is $s_2^2 = -1.75 \times 10^6 - 1.031 \times 10^6 = -2.781 \times 10^6 = -\omega_{n2}^2$. The corresponding natural frequencies are $\omega_{n1} = 847.9$ rad/s and $\omega_{n2} = 1667.6$ rad/s. Note that $\omega_{n1} < \omega_{n2}$. In units of Hz, the natural frequencies are $f_{n1} = \frac{\omega_{n1}}{2\pi} = 134.95$ Hz and $f_{n2} = \frac{\omega_{n2}}{2\pi} = 265.41$ Hz.

To find the eigenvectors, we can choose either equation of motion and again neglect damping. The Laplace-domain representation for the equations of motion is:

$$[[m]s^2 + [k]]\{\vec{X}\} = \left[\begin{bmatrix} 0.05 & -0.05 \\ -0.05 & 0.25 \end{bmatrix} s^2 + \begin{bmatrix} 1 \times 10^5 & 0 \\ 0 & 2 \times 10^5 \end{bmatrix} \right] \begin{Bmatrix} X_1 \\ X_2 \end{Bmatrix} = \begin{Bmatrix} 0 \\ 0 \end{Bmatrix}.$$

Arbitrarily selecting the top equation gives:

$$(0.05s^2 + 1 \times 10^5)X_1 - 0.05s^2 X_2 = 0.$$

If we wish to normalize to coordinate x_2, we require the ratio:

$$\frac{X_1}{X_2} = \frac{0.05s^2}{0.05s^2 + 1 \times 10^5}.$$

To find the first eigenvector, ψ_1, substitute $s_1^2 = -7.19 \times 10^5$ to obtain:

$$\psi_1 = \left\{ \begin{array}{c} X_{11} \\ X_{21} \\ 1 \end{array} \right\} = \left\{ \begin{array}{c} \dfrac{0.05s_1^2}{0.05s_1^2 + 1 \times 10^5} \\ 1 \end{array} \right\} = \left\{ \begin{array}{c} -0.561 \\ 1 \end{array} \right\}.$$

This eigenvector represents the relative magnitudes of X_1 and X_2 for vibration at ω_{n1}. The first eigenvector gives motion that is out of phase in this case because the model is not a chain-type system. We find the second eigenvector, ψ_2, by substituting $s_2^2 = -2.781 \times 10^6$ to obtain:

$$\psi_2 = \left\{ \begin{array}{c} X_{12} \\ X_{22} \\ 1 \end{array} \right\} = \left\{ \begin{array}{c} \dfrac{0.05s_2^2}{0.05s_2^2 + 1 \times 10^5} \\ 1 \end{array} \right\} = \left\{ \begin{array}{c} 3.56 \\ 1 \end{array} \right\}.$$

We see that the relative motion for vibration at ω_{n2} is in phase and the magnitude of motion at X_1 is 3.56 times larger than the motion at X_2. We can now write the modal matrix, P.

$$P = \begin{bmatrix} -0.561 & 3.56 \\ 1 & 1 \end{bmatrix}$$

The relationship between the local and modal coordinates depends on P.

$$\{\vec{x}\} = [P]\{\vec{q}\}$$
$$\{\dot{\vec{x}}\} = [P]\{\dot{\vec{q}}\}$$
$$\{\ddot{\vec{x}}\} = [P]\{\ddot{\vec{q}}\} \tag{4.28}$$

To transform from local coordinates to modal coordinates, where the degrees of freedom are uncoupled, we substitute for x and its time derivatives in the system equation of motion using Eq. 4.28.

$$[m]\{\ddot{\vec{x}}\} + [c]\{\dot{\vec{x}}\} + [k]\{\vec{x}\} = \{0\}$$
$$[m][P]\{\ddot{\vec{q}}\} + [c][P]\{\dot{\vec{q}}\} + [k][P]\{\vec{q}\} = \{0\} \tag{4.29}$$

If we now premultiply[2] each term in Eq. 4.29 by the transpose of the modal matrix, $P^T = \begin{bmatrix} -0.561 & 3.56 \\ 1 & 1 \end{bmatrix}^T = \begin{bmatrix} -0.561 & 1 \\ 3.56 & 1 \end{bmatrix}$, we obtain the mass, damping, and stiffness matrices in modal coordinates. This process is referred to as *diagonalization*.

$$[P]^T[m][P]\{\vec{\ddot{q}}\} + [P]^T[c][P]\{\vec{\dot{q}}\} + [P]^T[k][P]\{\vec{q}\} = \{0\}$$
$$[m_q]\{\vec{\ddot{q}}\} + [c_q]\{\vec{\dot{q}}\} + [k_q]\{\vec{q}\} = \{0\} \qquad (4.30)$$

The modal matrices, $[m_q]$, $[c_q]$, and $[k_q]$, are diagonal matrices – their off-diagonal terms are zero. This yields uncoupled equations of motion.

$$[m_q] = [P]^T[m][P] = \begin{bmatrix} m_{q1} & 0 \\ 0 & m_{q2} \end{bmatrix} \qquad (4.31)$$

$$[c_q] = [P]^T[c][P] = \begin{bmatrix} c_{q1} & 0 \\ 0 & c_{q2} \end{bmatrix} \qquad (4.32)$$

$$[k_q] = [P]^T[k][P] = \begin{bmatrix} k_{q1} & 0 \\ 0 & k_{q2} \end{bmatrix} \qquad (4.33)$$

The two new equations of motion in modal coordinates are:

$$m_{q1}\ddot{q}_1 + c_{q1}\dot{q}_1 + k_{q1}q_1 = 0$$
$$m_{q2}\ddot{q}_2 + c_{q2}\dot{q}_2 + k_{q2}q_2 = 0. \qquad (4.34)$$

Like the single degree of freedom free vibration systems we studied in Chap. 2, we require the initial conditions in order to solve these differential equations for the time responses $q_1(t)$ and $q_2(t)$. However, we now require these initial conditions to be in modal coordinates. We determine the required initial conditions by rearranging Eq. 4.28. See Eq. 4.35, where the zero subscript denotes the value when the time is zero.

$$\{\vec{q}_0\} = [P]^{-1}\{\vec{x}_0\}$$
$$\{\vec{\dot{q}}_0\} = [P]^{-1}\{\vec{\dot{x}}_0\} \qquad (4.35)$$

---

[2] Matrix multiplication is not commutative, in general, so the order of multiplication matters. The term premultiply means the term appears on the left of the product. The term postmultiply means that the term appears on the right.

As we saw in Chap. 2, we can express the damped free vibration response in various forms. One possible form is:

$$q_1(t) = e^{-\zeta_{q1}\omega_{n1}t}(A_1\cos(\omega_{d1}t) + B_1\sin(\omega_{d1}t)), \tag{4.36}$$

which describes motion in the first natural frequency, but note that this motion is not associated with any physical coordinate. In Eq. 4.36, the *modal damping ratio* is $\zeta_{q1} = \frac{c_{q1}}{2\sqrt{m_{q1}k_{q1}}}$ and the modal damped natural frequency is $\omega_{d1} = \omega_{n1}\sqrt{1 - \zeta_{q1}^2}$.

For vibration in the second natural frequency, the response is:

$$q_2(t) = e^{-\zeta_{q2}\omega_{n2}t}(A_2\cos(\omega_{d2}t) + B_2\sin(\omega_{d2}t)), \tag{4.37}$$

where $\zeta_{q2} = \frac{c_{q2}}{2\sqrt{m_{q2}k_{q2}}}$ and $\omega_{d2} = \omega_{n2}\sqrt{1 - \zeta_{q2}^2}$. To complete the solution, we must transform back into local coordinates.

$$\begin{Bmatrix} x_1 \\ x_2 \end{Bmatrix} = [P]\begin{Bmatrix} q_1 \\ q_2 \end{Bmatrix} \tag{4.38}$$

For our example, we have that $P = \begin{bmatrix} -0.561 & 3.56 \\ 1 & 1 \end{bmatrix}$ and $P^T = \begin{bmatrix} -0.561 & 1 \\ 3.56 & 1 \end{bmatrix}$. The modal mass matrix is then:

$$[m_q] = [P]^T[m][P] = \begin{bmatrix} -0.561 & 1 \\ 3.56 & 1 \end{bmatrix}\begin{bmatrix} 0.05 & -0.05 \\ -0.05 & 0.25 \end{bmatrix}\begin{bmatrix} -0.561 & 3.56 \\ 1 & 1 \end{bmatrix}.$$

Here, we have to multiply three matrices. *Matrix multiplication* for 2×2 matrices can be completed, as shown in Eq. 4.39. For example, the (1,1) term for the product is the first row of the left matrix multiplied in a term-by-term fashion by the first column of the right matrix.

$$\begin{bmatrix} a & b \\ c & d \end{bmatrix}\begin{bmatrix} e & f \\ g & h \end{bmatrix} = \begin{bmatrix} ae + bg & af + bh \\ ce + dg & cf + dh \end{bmatrix} \tag{4.39}$$

Let's now calculate the modal mass matrix. First, we will multiply the left and middle matrices. Then, we multiply the two remaining matrices.

$$[m_q] = \begin{bmatrix} -0.0781 & 0.2781 \\ 0.128 & 0.072 \end{bmatrix}\begin{bmatrix} -0.561 & 3.56 \\ 1 & 1 \end{bmatrix} = \begin{bmatrix} 0.322 & 0 \\ 0 & 0.528 \end{bmatrix} \text{kg}$$

Similarly, the modal stiffness matrix is:

$$[k_q] = [P]^T[k][P] = \begin{bmatrix} -0.561 & 1 \\ 3.56 & 1 \end{bmatrix}\begin{bmatrix} 1 \times 10^5 & 0 \\ 0 & 2 \times 10^5 \end{bmatrix}\begin{bmatrix} -0.561 & 3.56 \\ 1 & 1 \end{bmatrix}.$$

Performing the matrix multiplications yields[3]:

$$[k_q] = \begin{bmatrix} 2.32 \times 10^5 & 0 \\ 0 & 1.47 \times 10^6 \end{bmatrix} \text{N/m}.$$

Because the undamped natural frequencies are the same in local and modal coordinates, we can check our results so far. For the first natural frequency, substitution gives:

$$\omega_{n1} = \sqrt{\frac{k_{q1}}{m_{q1}}} = \sqrt{\frac{2.32 \times 10^5}{0.322}} = 848.8 \,\text{rad/s}.$$

The value we obtained from the (local coordinates) eigenvalue was 847.9 rad/s. The difference is due to round-off error, but the results match well enough to validate our modal values. For the second natural frequency, we have:

$$\omega_{n2} = \sqrt{\frac{k_{q2}}{m_{q2}}} = \sqrt{\frac{1.46 \times 10^6}{0.528}} = 1,668.6 \,\text{rad/s},$$

where the result from the second eigenvalue (determined from local coordinates) was 1667.6 rad/s.

We can calculate the modal damping matrix in two ways. First, we can simply perform the matrix multiplications:

$$[c_q] = [P]^T[c][P] = \begin{bmatrix} -0.561 & 1 \\ 3.56 & 1 \end{bmatrix} \begin{bmatrix} 10 & 0 \\ 0 & 20 \end{bmatrix} \begin{bmatrix} -0.561 & 3.56 \\ 1 & 1 \end{bmatrix}$$

$$= \begin{bmatrix} 23.15 & 0 \\ 0 & 146.7 \end{bmatrix} \text{N-s/m}.$$

Second, we can use the proportional damping relationship:

$$[c_q] = \alpha[m_q] + \beta[k_q] = 0 \cdot [m_q] + \frac{1}{1 \times 10^4}[k_q] = \begin{bmatrix} 23.15 & 0 \\ 0 & 146.7 \end{bmatrix} \text{N-s/m}.$$

We can now calculate the modal damping ratios and corresponding damped natural frequencies.

$$\zeta_{q1} = \frac{c_{q1}}{2\sqrt{m_{q1}k_{q1}}} = \frac{23.5}{2\sqrt{0.322(2.32 \times 10^5)}} = 0.042$$

---

[3] Due to round-off error, the off-diagonal terms in the modal matrices may not be identically zero. However, they will be significantly smaller than the on-diagonal terms.

$$\zeta_{q2} = \frac{c_{q2}}{2\sqrt{m_{q2}k_{q2}}} = \frac{146.7}{2\sqrt{0.528(1.47 \times 10^6)}} = 0.083$$

$$\omega_{d1} = \omega_{n1}\sqrt{1 - \zeta_{q1}^2} = 847.2 \, \text{rad/s}$$

$$\omega_{d2} = \omega_{n2}\sqrt{1 - \zeta_{q2}^2} = 1661.5 \, \text{rad/s}$$

As shown in Eqs. 4.36 and 4.37, one form for the underdamped single degree of freedom vibration solution is:

$$q(t) = e^{-\zeta_q \omega_n t}(A \cos(\omega_d t) + B \sin(\omega_d t)). \tag{4.40}$$

To determine the coefficients, we first calculate the velocity and then apply the initial conditions $q(0) = q_0$ and $\dot{q}(0) = \dot{q}_0$. The velocity is:

$$\dot{q}(t) = \zeta_q \omega_n e^{-\zeta_q \omega_n t}(A \cos(\omega_d t) + B \sin(\omega_d t))$$
$$+ e^{-\zeta_q \omega_n t}(-\omega_d A \sin(\omega_d t) + \omega_d B \cos(\omega_d t)). \tag{4.41}$$

Substituting $t = 0$ in Eqs. 4.40 and 4.41 and using the initial conditions, we obtain Eqs. 4.42 and 4.43.

$$q_1(t) = e^{-\zeta_{q1}\omega_{n1}t}\left(q_{01} \cos(\omega_{d1}t) + \frac{(\dot{q}_{01} + \zeta_{q1}\omega_{n1}q_{01})}{\omega_{d1}} \sin(\omega_{d1}t) \right) \tag{4.42}$$

$$q_2(t) = e^{-\zeta_{q2}\omega_{n2}t}\left(q_{02} \cos(\omega_{d2}t) + \frac{(\dot{q}_{02} + \zeta_{q2}\omega_{n2}q_{02})}{\omega_{d2}} \sin(\omega_{d2}t) \right) \tag{4.43}$$

In order to calculate the initial displacement and velocities in modal coordinates, we use Eq. 4.35.

$$\begin{Bmatrix} q_{01} \\ q_{02} \end{Bmatrix} = [P]^{-1}\begin{Bmatrix} x_{01} \\ x_{02} \end{Bmatrix} = \begin{bmatrix} -0.561 & 3.56 \\ 1 & 1 \end{bmatrix}^{-1}\begin{Bmatrix} 1 \\ 1 \end{Bmatrix}$$

$$\begin{Bmatrix} \dot{q}_{01} \\ \dot{q}_{02} \end{Bmatrix} = [P]^{-1}\begin{Bmatrix} \dot{x}_{01} \\ \dot{x}_{02} \end{Bmatrix} = \begin{bmatrix} -0.561 & 3.56 \\ 1 & 1 \end{bmatrix}^{-1}\begin{Bmatrix} 0 \\ 0 \end{Bmatrix}$$

To invert the 2×2 modal matrix, we complete three steps: (1) switch the on-diagonal terms; (2) change the signs of the off-diagonal terms; and (3) divide the resulting matrix by the determinant of the original modal matrix.

$$[P]^{-1} = \begin{bmatrix} -0.561 & 3.56 \\ 1 & 1 \end{bmatrix}^{-1} = \frac{\begin{bmatrix} 1 & -3.56 \\ -1 & -0.561 \end{bmatrix}}{-0.561(1) - 3.56(1)} = \begin{bmatrix} -0.243 & 0.864 \\ 0.243 & 0.136 \end{bmatrix}$$

Substitution   gives $\begin{Bmatrix} q_{01} \\ q_{02} \end{Bmatrix} = \begin{bmatrix} -0.243 & 0.864 \\ 0.243 & 0.136 \end{bmatrix} \begin{Bmatrix} 1 \\ 1 \end{Bmatrix} = \begin{Bmatrix} 0.621 \\ 0.379 \end{Bmatrix}$ mm.

Because the initial velocities in local coordinates are zero, the initial velocities in modal coordinates are also zero. The modal displacements from Eqs. 4.42 and 4.43 can now be determined.

$$q_1(t) = e^{-35.95t}(0.621 \cos(847.2t) + 0.264 \sin(847.2t))$$

$$q_2(t) = e^{-139.05t}(0.379 \cos(1661.5t) + 0.0317 \sin(1661.5t))$$

Finally, we must transform back into local coordinates using Eq. 4.38.

$$\begin{Bmatrix} x_1 \\ x_2 \end{Bmatrix} = \begin{bmatrix} -0.561 & 3.56 \\ 1 & 1 \end{bmatrix} \begin{Bmatrix} q_1 \\ q_2 \end{Bmatrix}$$

Performing the matrix multiplication, we obtain the expressions for x_1 and x_2.

$$x_1 = -0.561q_1 + 3.56q_2$$
$$x_2 = q_1 + q_2$$

We see that x_2 is the sum of the modal contributions q_1 and q_2. This result is obtained because we normalized to coordinate x_2 when we determined the eigenvectors. Substituting for q_1 and q_2 gives the following results. Each response is a linear combination of motion in both damped natural frequencies.

$$x_1(t) = -0.561e^{-35.95t}(0.621 \cos(847.2t) + 0.264 \sin(847.2t))$$
$$+ 3.56e^{-139.05t}(0.379 \cos(1,661.5t) + 0.0317 \sin(1,661.5t))$$

$$x_2(t) = e^{-35.95t}(0.621 \cos(847.2t) + 0.264 \sin(847.2t))$$
$$+ e^{-139.05t}(0.379 \cos(1,661.5t) + 0.0317 \sin(1,661.5t))$$

IN A NUTSHELL   We derive the equations of motion in local coordinates, which make physical sense to us. We transform them into modal coordinates, which are mathematically easy, since we already know how to solve single degree of freedom problems. We then transform back to local coordinates to express the final solution. While this may not seem like a big time-saver for two degree of freedom systems, the procedure does not get more complicated as the number of degrees of freedom increases and its benefit becomes clearly apparent.

Let's review the modal analysis steps.

1. Write the equations of motion in matrix form.
2. Verify that proportional damping exists.
3. Neglect damping and write the characteristic equation $\left|[m]s^2 + [k]\right| = 0$.
4. Calculate the eigenvalues (i.e., the roots of the characteristic equation). Using the eigenvalues, determine the undamped natural frequencies $s_i^2 = -\omega_{ni}^2$.
5. Select any one of the linearly dependent equations of motion to find the eigenvectors (mode shapes). Normalize to the coordinate of interest (e.g., this may be the location where it is desired to minimize the vibration magnitude).
6. Using the eigenvectors, assemble the modal matrix $P = [\psi_1 \quad \psi_2 \quad \dots]$.
7. Transform the equations of motion into (uncoupled) modal coordinates. The diagonal modal mass, damping, and stiffness matrices are $[m_q] = [P]^T[m][P]$, $[c_q] = [P]^T[c][P]$, and $[k_q] = [P]^T[k][P]$.
8. Write the solutions to the uncoupled (single degree of freedom) equations of motion in modal coordinates. An example solution form is:

$$q(t) = e^{-\zeta_q \omega_n t}\left(q_0 \cos(\omega_d t) + \frac{(\dot{q}_0 + \zeta_q \omega_n q_0)}{\omega_d} \sin(\omega_d t)\right).$$

Note that the initial conditions must be transformed into modal coordinates to solve the equations of motion.
9. Transform back into local coordinates.

The reason that this approach works is that the eigenvectors possess a very unique property. They are orthogonal with respect to the mass, damping, and stiffness matrices. This is a generalization of the concept of perpendicularity. More information is available in Appendix B.

By the Numbers 4.2

In *By the Numbers 4.1*, we determined the time responses for the two degree of freedom system shown in Fig. 4.6. Let's repeat this example, but now use modal analysis. We will apply the modal analysis steps we just reviewed.

1. Write the equations of motion in matrix form.

$$\begin{bmatrix} 1 & 0 \\ 0 & 0.5 \end{bmatrix}\begin{Bmatrix} \ddot{x}_1 \\ \ddot{x}_2 \end{Bmatrix} + \begin{bmatrix} 1 \times 10^7 + 2 \times 10^7 & -2 \times 10^7 \\ -2 \times 10^7 & 2 \times 10^7 \end{bmatrix}\begin{Bmatrix} x_1 \\ x_2 \end{Bmatrix} = \begin{Bmatrix} 0 \\ 0 \end{Bmatrix}$$

2. Verify that proportional damping exists.
   There is no damping in this example.
3. Neglect damping and write the characteristic equation.

$$\begin{vmatrix} 1s^2 + 3 \times 10^7 & -2 \times 10^7 \\ -2 \times 10^7 & 0.5s^2 + 2 \times 10^7 \end{vmatrix} = 0$$

Calculating and simplifying the determinant gives:

$$0.5s^4 + 3.5 \times 10^7 s^2 + 2 \times 10^{14} = 0.$$

4. Calculate the eigenvalues and determine the undamped natural frequencies.

$$s_{1,2}^2 = \frac{-3.5 \times 10^7 \pm \sqrt{(3.5 \times 10^7)^2 - 4(0.5)2 \times 10^{14}}}{2(0.5)}$$

$$= -3.5 \times 10^7 \pm 2.872 \times 10^7$$

The first eigenvalue is:

$$s_1^2 = (-3.5 + 2.872) \times 10^7 = -6.28 \times 10^6 = -\omega_{n1}^2,$$

and the first natural frequency is $\omega_{n1} = 2{,}506$ rad/s. The second eigenvalue is:

$$s_2^2 = (-3.5 - 2.872) \times 10^7 = -6.37 \times 10^7 = -\omega_{n2}^2,$$

and the second natural frequency is $\omega_{n2} = 7{,}981$ rad/s.
5. Select one of the linearly dependent equations of motion to find the eigenvectors. We will arbitrarily use the top equation and normalize to x_2. The corresponding ratio is:

$$\frac{X_1}{X_2} = \frac{2 \times 10^7}{1s^2 + 3 \times 10^7}.$$

Substituting the first eigenvalue gives the first eigenvector.

$$\psi_1 = \left\{ \begin{matrix} \frac{X_{11}}{X_{21}} \\ 1 \end{matrix} \right\} = \left\{ \begin{matrix} 0.843 \\ 1 \end{matrix} \right\}$$

Substituting the second eigenvalue gives the second eigenvector.

$$\psi_2 = \left\{ \begin{matrix} \frac{X_{12}}{X_{22}} \\ 1 \end{matrix} \right\} = \left\{ \begin{matrix} -0.593 \\ 1 \end{matrix} \right\}$$

6. Assemble the modal matrix.

$$P = \begin{bmatrix} 0.843 & -0.593 \\ 1 & 1 \end{bmatrix}$$

7. Transform the equations of motion into (uncoupled) modal coordinates. The diagonal modal mass, damping, and stiffness matrices are $[m_q] = [P]^T[m][P]$, $[c_q] = [P]^T[c][P]$, and $[k_q] = [P]^T[k][P]$.

$$[m_q] = \begin{bmatrix} 0.843 & -0.593 \\ 1 & 1 \end{bmatrix}^T \begin{bmatrix} 1 & 0 \\ 0 & 0.5 \end{bmatrix} \begin{bmatrix} 0.843 & -0.593 \\ 1 & 1 \end{bmatrix} = \begin{bmatrix} 1.211 & 0 \\ 0 & 0.852 \end{bmatrix} \text{kg}$$

$$[k_q] = \begin{bmatrix} 0.843 & -0.593 \\ 1 & 1 \end{bmatrix}^T \begin{bmatrix} 3 \times 10^7 & -2 \times 10^7 \\ -2 \times 10^7 & 2 \times 10^7 \end{bmatrix} \begin{bmatrix} 0.843 & -0.593 \\ 1 & 1 \end{bmatrix}$$

$$= \begin{bmatrix} 7.6 \times 10^6 & 0 \\ 0 & 5.43 \times 10^7 \end{bmatrix} \text{N/m}$$

Actually, due to round-off error, the modal stiffness matrix off-diagonal terms were $k_{q12} = 3 \times 10^3$ N/m and $k_{q21} = 3.5 \times 10^4$ N/m after the previous matrix multiplications were completed. However, these values are two to four orders of magnitude smaller than the on-diagonal terms and were neglected.

8. Write the solutions to the uncoupled (single degree of freedom) equations of motion in modal coordinates. The equations of motion are:

$$1.211\ddot{q}_1 + 7.6 \times 10^6 q_1 = 0$$
$$0.852\ddot{q}_2 + 5.43 \times 10^7 q_2 = 0.$$

These equations are now uncoupled and represent two separate single degree of freedom systems. As a check, let's calculate the natural frequencies.

$$\omega_{n1} = \sqrt{\frac{7.6 \times 10^6}{1.211}} = 2,505 \text{ rad/s}$$

$$\omega_{n2} = \sqrt{\frac{5.43 \times 10^7}{0.852}} = 7,983 \text{ rad/s}$$

These results match the natural frequencies obtained from the eigenvalues, as they should. Natural frequencies are system properties that do not depend on choice of coordinates. To solve the equations of motion, we require the initial conditions in modal coordinates.

$$q(t) = e^{-\zeta_q \omega_n t} \left(q_0 \cos(\omega_d t) + \frac{(\dot{q}_0 + \zeta_q \omega_n q_0)}{\omega_d} \sin(\omega_d t) \right)$$

Note that the initial conditions must be transformed into modal coordinates to solve the equations of motion.

$$\left\{ \begin{array}{c} q_{01} \\ q_{02} \end{array} \right\} = [P]^{-1} \left\{ \begin{array}{c} x_{01} \\ x_{02} \end{array} \right\} = \left[\begin{array}{cc} 0.843 & -0.593 \\ 1 & 1 \end{array} \right]^{-1} \left\{ \begin{array}{c} 1 \\ -1 \end{array} \right\}$$

The inverse of the modal matrix is:

$$[P]^{-1} = \frac{\left[\begin{array}{cc} 1 & 0.593 \\ -1 & 0.843 \end{array} \right]}{0.843(1) - (-0.593)1} = \left[\begin{array}{cc} 0.696 & 0.413 \\ -0.696 & 0.587 \end{array} \right].$$

The initial displacements are therefore:

$$\left\{ \begin{array}{c} q_{01} \\ q_{02} \end{array} \right\} = \left[\begin{array}{cc} 0.696 & 0.413 \\ -0.696 & 0.587 \end{array} \right] \left\{ \begin{array}{c} 1 \\ -1 \end{array} \right\} = \left\{ \begin{array}{c} 0.283 \\ -1.283 \end{array} \right\} \text{ mm.}$$

The zero initial velocities in modal coordinates give zero initial velocities in modal coordinates.

$$\left\{ \begin{array}{c} \dot{q}_{01} \\ \dot{q}_{02} \end{array} \right\} = [P]^{-1} \left\{ \begin{array}{c} 0 \\ 0 \end{array} \right\} = \left\{ \begin{array}{c} 0 \\ 0 \end{array} \right\}$$

For nondamping, one single degree of freedom vibration solution form is:

$$q(t) = q_0 \cos(\omega_n t) + \frac{\dot{q}_0}{\omega_n} \sin(\omega_n t).$$

Substituting the appropriate natural frequencies and initial conditions gives:

$$q_1(t) = 0.283 \cos(2{,}506t)$$

and

$$q_2(t) = -1.283 \cos(7{,}981t).$$

9. Transform back into local coordinates using $\left\{ \begin{array}{c} x_1 \\ x_2 \end{array} \right\} = \left[\begin{array}{cc} 0.843 & -0.593 \\ 1 & 1 \end{array} \right] \left\{ \begin{array}{c} q_1 \\ q_2 \end{array} \right\}.$
The results are:

$$x_1(t) = 0.239 \cos(2{,}506t) + 0.761 \cos(7{,}981t) \text{ mm}$$

and

$$x_2(t) = 0.283 \cos(2{,}506t) - 1.283 \cos(7{,}981t) \text{ mm.}$$

Naturally, this is the same result that we obtained from the previous analysis. However, we avoided the need to invert the 4×4 matrix. This is particularly beneficial for models with many more degrees of freedom.

 IN A NUTSHELL Note that the motion of the coordinate to which the mode shapes are normalized is just the sum of the motions expressed in modal coordinates. If you are concerned about the motion of a particular point, simply normalize the mode shapes so that their value is 1 at that coordinate.

Let's conclude the chapter by applying new initial conditions to this example. For the initial conditions provided in Fig. 4.6, we obtained time-domain responses for the two degrees of freedom that were linear combinations of vibration in the two natural frequencies. Could we select initial conditions that would yield vibration in only one natural frequency? The answer is yes. If we choose initial displacements (and zero initial velocities) that match one of the eigenvectors (i.e., the ratios of the magnitude of vibrations between the individual local coordinates), then we will obtain vibration only in the natural frequency that corresponds to the selected eigenvector.

Let's use local coordinate initial displacements of $x_1(0) = -1.686$ mm and $x_2(0) = -2$ mm. Note that these initial displacements match the ratio provided by the first eigenvector. The initial displacements in modal coordinates are:

$$\left\{ \begin{array}{c} q_{01} \\ q_{02} \end{array} \right\} = \left[\begin{array}{cc} 0.696 & 0.413 \\ -0.696 & 0.587 \end{array} \right] \left\{ \begin{array}{c} -1.686 \\ -2 \end{array} \right\} = \left\{ \begin{array}{c} -2 \\ 0 \end{array} \right\} \text{ mm,}$$

and the initial velocities are zero.

The modal coordinate, time-domain solutions in the form $q(t) = q_0 \cos(\omega_n t) + \frac{\dot{q}_0}{\omega_n} \sin(\omega_n t)$ are:

$$q_1(t) = -2 \cos(2,506t) \text{ mm}$$

and

$$q_2(t) = 0.$$

Transforming back into local coordinates using $\left\{ \begin{array}{c} x_1 \\ x_2 \end{array} \right\} = \left[\begin{array}{cc} 0.843 & -0.593 \\ 1 & 1 \end{array} \right] \left\{ \begin{array}{c} q_1 \\ q_2 \end{array} \right\}$ gives:

$$x_1(t) = -1.686 \cos(2,506t) \text{ mm}$$

and

$$x_2(t) = -2 \cos(2,506t) \text{ mm.}$$

As expected, the time-domain solutions in local coordinates include vibration in only the first natural frequency and the ratio of displacement magnitudes between coordinates x_1 and x_2 matches the first eigenvector.

Chapter Summary

- The eigensolution gives the eigenvalues, which identify the system's natural frequencies, and the eigenvectors, or mode shapes, which describe the relative motion of the individual degrees of freedom.
- The number of eigenvalue/eigenvector pairs is equal to the number of degrees of freedom in the system model.
- The mode shapes are typically normalized to one of the degrees of freedom for the system model since they only provide the ratio of vibration magnitude between coordinates.
- The roots of the characteristic equation are the eigenvalues. The characteristic equation is determined from the equations of motion.
- The eigenvalues are used to determine the eigenvectors. The eigenvalues, and corresponding eigenvectors, are ordered in ascending natural frequency values.
- In modal analysis, the local (model) coordinates are transformed into modal coordinates. The equations of motion are uncoupled in modal coordinates and, therefore, they can each be treated as a single degree of freedom system.
- The modal matrix is used to transform between local and modal coordinates. It is also used to diagonalize the mass, stiffness, and damping matrices. Its columns are the system's eigenvectors (ordered from left to right in the matrix).
- Modal analysis requires proportional damping. Mathematically, proportional damping exists if the damping matrix can be written as a linear combination of the mass and stiffness matrices. Physically, proportional damping means that the individual modes reach their maximum values at the same time.
- The undamped natural frequencies are the same in both local and modal coordinates.
- The selection of initial conditions determines whether the system's time-domain free vibration responses will oscillate in the first natural frequency, second natural frequency, or a linear combination of the two. If the initial displacements (with zero initial velocities) match the ratio provided by one of the eigenvectors, then the system will vibrate only in the natural frequency that corresponds to that eigenvector.

Exercises

1. Given the eigenvalues and eigenvectors for the two degree of freedom system shown in Fig. P4.1, determine the modal matrices m_q (kg), c_q (N-s/m), and k_q (N/m).

$$s_1^2 = -1 \times 10^6 \mathrm{rad/s^2}$$

$$s_2^2 = -7 \times 10^6 \mathrm{rad/s^2}$$

$$\psi_1 = \begin{Bmatrix} 0.5 \\ 1 \end{Bmatrix} \quad \psi_2 = \begin{Bmatrix} -2.5 \\ 1 \end{Bmatrix}$$

Fig. P4.1 Two degree of freedom spring-mass-damper system

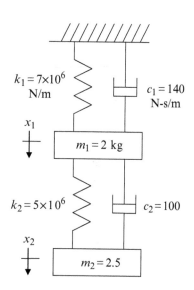

$k_1 = 7 \times 10^6$ N/m

$c_1 = 140$ N-s/m

x_1

$m_1 = 2$ kg

$k_2 = 5 \times 10^6$

$c_2 = 100$

x_2

$m_2 = 2.5$

2. Given the two degree of freedom system in Fig. P4.2, complete the following.

   (a) Write the equations of motion in matrix form.
   (b) Write the system characteristic equation using Laplace notation. Your solution should be a polynomial that is quadratic in s^2 with appropriate numerical coefficients.
   (c) Calculate the natural frequencies (in Hz).
   (d) Determine the two mode shapes (normalize to coordinate x_1).

Fig. P4.2 Two degree of
freedom spring-mass system

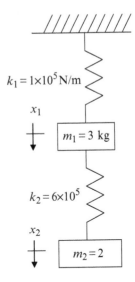

$k_1 = 1 \times 10^5$ N/m

x_1

$m_1 = 3$ kg

$k_2 = 6 \times 10^5$

x_2

$m_2 = 2$

Fig. P4.3 Two degree of
freedom spring-mass system

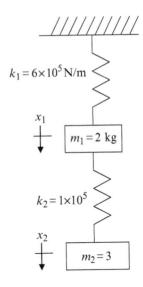

$k_1 = 6 \times 10^5$ N/m

x_1

$m_1 = 2$ kg

$k_2 = 1 \times 10^5$

x_2

$m_2 = 3$

3. Given the two degree of freedom system shown in Fig. P4.3, complete the
   following.

   (a) Write the equations of motion in matrix form.
   (b) Write the system characteristic equation using Laplace notation. Your
       solution should be a polynomial that is quadratic in s^2 with appropriate
       numerical coefficients.
   (c) Determine the natural frequencies (in rad/s).
   (d) Determine the mode shapes (normalize to coordinate x_2).

Fig. P4.4 Two degree of
freedom spring-mass system

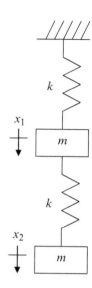

4. A two degree of freedom spring-mass system is shown in Fig. P4.4. For harmonic
   free vibration, complete the following if $k = 5 \times 10^6$ N/m and $m = 2$ kg.

   (a) Draw the free body diagram showing the forces on the two masses during
       vibration.
   (b) Write the two equations of motion in matrix form. First show the equations
       symbolically and then substitute the numerical values for m and k.
   (c) Write the characteristic equation for this system. First show the equation
       symbolically and then substitute the numerical values for m and k.
   (d) Determine the numerical roots of the characteristic equation (which is
       quadratic in s^2). What do these two roots represent?
   (e) Determine the two mode shapes for this system. Normalize the mode
       shapes to coordinate x_1.

5. A two degree of freedom spring-mass system is displayed in Fig. P4.5. For
   harmonic free vibration, complete the following if $k_1 = 2 \times 10^6$ N/m,
   $m_1 = 0.8$ kg, $k_2 = 1 \times 10^6$ N/m, and $m_2 = 1.4$ kg. The initial displacements
   for the system's free vibration are $x_1(0) = 2$ mm and $x_2(0) = 1$ mm and
   the initial velocities are $\dot{x}_1(0) = 0$ mm/s and $\dot{x}_2(0) = 5$ mm/s.

   (a) Calculate the two natural frequencies and mode shapes. Normalize the
       mode shapes (eigenvectors) to coordinate x_2.
   (b) Define the modal matrix and determine the modal mass and stiffness
       matrices.
   (c) Write the uncoupled single degree of freedom time responses for the modal
       coordinates q_1 and q_2. Use the following form: $q_{1,2}(t) = A_{1,2} \cos(\omega_{n_{1,2}} t) +$
       $B_{1,2} \sin(\omega_{n_{1,2}} t)$ with units of mm.

Fig. P4.5 Two degree of
freedom spring-mass system

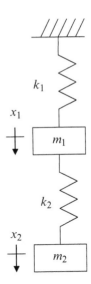

(d) Write the time responses for the local coordinates x_1 and x_2 (in mm).

(e) Plot the time responses for x_1 and x_2 (in mm). Define the time vector as: $t = 0:0.0001:0.2$; (in seconds).

6. For the same system as described in problem 5, complete the following.

   (a) The initial displacements for the system's free vibration are $x_1(0) = 0.378$ mm and $x_2(0) = 1$ mm and the initial velocities are $\dot{x}_1(0) = 0$ mm/s and $\dot{x}_2(0) = 0$ mm/s. Plot the time responses for x_1 and x_2 (in mm). Define the time vector as: $t = 0:0.0001:0.2$; (in seconds). What is the vibrating frequency for both $x_1(t)$ and $x_2(t)$? What is special about these initial conditions to give this result?

   (b) The initial displacements for the system's free vibration are $x_1(0) = -4.628$ mm and $x_2(0) = 1$ mm and the initial velocities are $\dot{x}_1(0) = 0$ mm/s and $\dot{x}_2(0) = 0$ mm/s. Plot the time responses for x_1 and x_2 (in mm). Define the time vector as: $t = 0:0.0001:0.2$; (in seconds). What is the vibrating frequency for both $x_1(t)$ and $x_2(t)$? What is special about these initial conditions to give this result?

7. A two degree of freedom spring-mass-damper system is shown in Fig. P4.7. For harmonic free vibration, complete the following if $k_1 = 2 \times 10^5$ N/m, $c_1 = 60$ N-s/m, $m_1 = 2.5$ kg, $k_2 = 5.5 \times 10^4$ N/m, $c_2 = 16.5$ N-s/m, and $m_2 = 1.2$ kg.

   (a) Verify that proportional damping exists.

   (b) Define the modal matrix and determine the modal mass, stiffness, and damping matrices. Normalize the mode shapes to coordinate x_2.

Fig. P4.7 Two degree of
freedom spring-mass-damper
system under free vibration

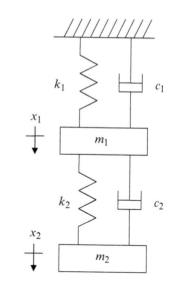

Fig. P4.8 Two degree of
freedom spring-mass system

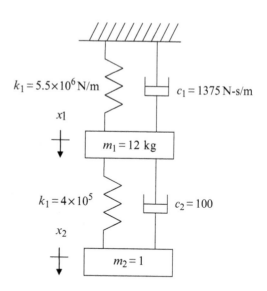

8. Given the two degree of freedom system in Fig. P4.8, complete the following.

(a) Write the equations of motion in matrix form.
(b) Verify that proportional damping exists.
(c) Determine the roots of the characteristic equation. What do these roots represent?
(d) Determine the two mode shapes (normalize to coordinate x_1).

Fig. P4.9 Three degree of freedom spring-mass-damper model

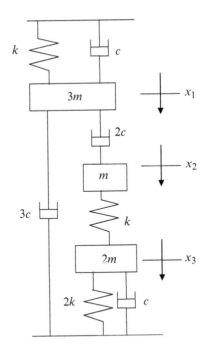

9. Determine the mass, damping, and stiffness matrices in local coordinates for the model shown in Fig. P4.9.
10. Given the mass, damping, and stiffness matrices for the model shown in Fig. P4.9 determined from problem 9, can proportional damping exist for this system? Justify your answer.

References

Blevins RD (2001) Formulas for natural frequency and mode shape. Krieger, Malabar (Table 8–1)
http://en.wiktionary.org/wiki/eigen
http://en.wikipedia.org/wiki/Eureka_(word)
http://en.wikipedia.org/wiki/Linear_independence

Chapter 5
Two Degree of Freedom Forced Vibration

*Our achievements of today are but the sum total of our thoughts
of yesterday.*

— Blaise Pascal

5.1 Equations of Motion

Let's extend the two degree of freedom free vibration analysis from Chap. 4 to
include externally applied forces so that we can analyze two degree of freedom
forced vibration. The general case is that a separate harmonic force is applied at
each coordinate; see Fig. 5.1. However, we are considering only linear systems, so
we can apply *superposition*. This means that we can determine the system response
due to each force separately and then sum the results to find the combined effect.

Using the free body diagrams included in Fig. 5.1, the equations of motion
expressed in matrix form are:

$$\begin{bmatrix} m_1 & 0 \\ 0 & m_2 \end{bmatrix} \begin{Bmatrix} \ddot{x}_1 \\ \ddot{x}_2 \end{Bmatrix} + \begin{bmatrix} c_1 + c_2 & -c_2 \\ -c_2 & c_2 \end{bmatrix} \begin{Bmatrix} \dot{x}_1 \\ \dot{x}_2 \end{Bmatrix} + \begin{bmatrix} k_1 + k_2 & -k_2 \\ -k_2 & k_2 \end{bmatrix} \begin{Bmatrix} x_1 \\ x_2 \end{Bmatrix}$$
$$= \begin{Bmatrix} F_1 e^{i\omega_1 t} \\ F_2 e^{i\omega_2 t} \end{Bmatrix}, \tag{5.1}$$

where the subscripts on the forcing frequencies indicate that they are not necessarily
equal. Because the springs and dampers appear together in the Fig. 5.1 model, the
damping and stiffness matrices have the same format in Eq. 5.1. This is not always
the case, but the mass, damping, and stiffness matrices will be symmetric in all
instances, as long as the coordinates are measured with respect to ground.

T.L. Schmitz and K.S. Smith, *Mechanical Vibrations: Modeling and Measurement*,
DOI 10.1007/978-1-4614-0460-6_5, © Springer Science+Business Media, LLC 2012

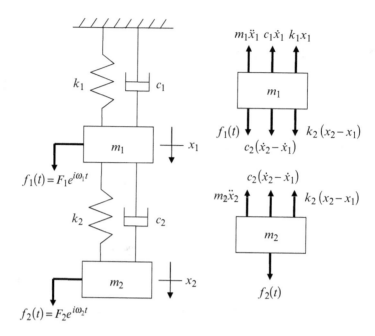

Fig. 5.1 Two degree of freedom chain-type, spring–mass–damper system with harmonic forces $f_1(t)$ and $f_2(t)$ applied at coordinates x_1 and x_2, respectively

 IN A NUTSHELL   If the coordinates are measured with respect to ground and the stiffness matrix (for example) is not symmetric, then we have made a perpetual motion machine. Essentially, the energy required to achieve a specified displacement configuration through one loading path (say moving coordinate 1 and then moving coordinate 2) would not be the same as the energy recovered by returning the coordinates to their original position through a different loading path. By loading and unloading through two different paths, energy could be extracted from the system indefinitely. Simply put, when the coordinates are measured with respect to ground, the mass, stiffness, and damping matrices must be symmetric.

For now let's take advantage of superposition and consider only $f_2(t) = F_2 e^{i\omega t}$ (the frequency subscript is removed because there is just one force). Similar to the single degree of freedom forced vibration analysis in Chap. 3, we can assume a harmonic form for the solution which mimics the forcing function $x(t) = X e^{i\omega t}$. This gives $\dot{x}(t) = i\omega X e^{i\omega t}$ and $\ddot{x}(t) = (i\omega)^2 X e^{i\omega t} = -\omega^2 X e^{i\omega t}$. Substitution in Eq. 5.1 yields:

$$\left[-\omega^2 \begin{bmatrix} m_1 & 0 \\ 0 & m_2 \end{bmatrix} + i\omega \begin{bmatrix} c_1 + c_2 & -c_2 \\ -c_2 & c_2 \end{bmatrix} + \begin{bmatrix} k_1 + k_2 & -k_2 \\ -k_2 & k_2 \end{bmatrix} \right] \begin{Bmatrix} X_1 \\ X_2 \end{Bmatrix} e^{i\omega t}$$

$$= \begin{Bmatrix} 0 \\ F_2 e^{i\omega t} \end{Bmatrix}. \tag{5.2}$$

In the generic case, we can express Eq. 5.2 as:

$$\left[-\omega^2[m] + i\omega[c] + [k]\right]\{\vec{X}\}e^{i\omega t} = \{\vec{F}\}e^{i\omega t}. \tag{5.3}$$

We will apply two methods to solve the system of coupled differential equations represented by Eq. 5.3. In both cases, we will determine the system frequency response functions (FRFs). These FRFs identify the frequency-dependent, steady-state vibration behavior and, because this is our desired result, we will neglect the transients which rapidly decay in general. The first solution method is referred to as *complex matrix inversion*. While this approach is more computationally expensive, it does not require proportional damping. The second method is *modal analysis*, which we introduced in Chap. 4. This method is applicable to systems with any number of degrees of freedom, but proportional damping must be satisfied (or assumed).

5.2 Complex Matrix Inversion

To implement this approach, it is helpful to rewrite Eq. 5.3 in a more compact form; see Eq. 5.4. In this form, the sum of the mass, damping, and stiffness matrices, with the appropriate frequency multipliers on the mass and damping matrices, is represented by the matrix A. To determine the system FRFs, we simply need to invert the complex matrix A. This is demonstrated in Eq. 5.5.

$$\left[-\omega^2[m] + i\omega[c] + [k]\right]\{\vec{X}\}e^{i\omega t} = [A]\{\vec{X}\}e^{i\omega t} = \{\vec{F}\}e^{i\omega t} \tag{5.4}$$

$$\{\vec{X}\} = [A]^{-1}\{\vec{F}\} \tag{5.5}$$

For the two degree of freedom system displayed in Fig. 5.1, A is a 2×2 matrix and can be expressed as shown in Eq. 5.6. This matrix is frequency dependent. One way to visualize A is to consider it as a book where each page provides the four a_{ij} values at a particular frequency. The beginning of the book gives the low frequency values and the end gives the high frequency values.

$$[A] = \begin{bmatrix} a_{11} & a_{12} \\ a_{21} & a_{22} \end{bmatrix} = \begin{bmatrix} -\omega^2 m_1 + i\omega(c_1 + c_2) + (k_1 + k_2) & -i\omega c_2 - k_2 \\ -i\omega c_2 - k_2 & -\omega^2 m_2 + i\omega c_2 + k_2 \end{bmatrix} \tag{5.6}$$

As we saw in Sect. 2.4.5, we determine the inverse of the A matrix by switching the on-diagonals, changing the sign of the off-diagonals, and dividing each term by the determinant of A; see Eq. 5.7. In this case, however, we have to repeat the inversion for each frequency value within our range (or *bandwidth*) of interest. For example, we might be interested in a system's response between 0 and 5,000 Hz. With a frequency resolution of 1 Hz, we would need to invert A 5,001 times.

$$[A]^{-1} = \frac{1}{|A|} \begin{bmatrix} a_{22} & -a_{12} \\ -a_{21} & a_{11} \end{bmatrix} = \frac{1}{(a_{11}a_{22} - a_{12}a_{21})} \begin{bmatrix} a_{22} & -a_{12} \\ -a_{21} & a_{11} \end{bmatrix} = \begin{bmatrix} \alpha_{11} & \alpha_{12} \\ \alpha_{21} & \alpha_{22} \end{bmatrix}$$

(5.7)

The four entries in the inverted A matrix are provided in Eqs. 5.8–5.11.

$$\alpha_{11} = \frac{-\omega^2 m_2 + i\omega c_2 + k_2}{|A|}$$

(5.8)

$$\alpha_{12} = \frac{i\omega c_2 + k_2}{|A|}$$

(5.9)

$$\alpha_{21} = \frac{i\omega c_2 + k_2}{|A|}$$

(5.10)

$$\alpha_{22} = \frac{-\omega^2 m_1 + i\omega(c_1 + c_2) + (k_1 + k_2)}{|A|}$$

(5.11)

Let's assume that the external force is applied to coordinates x_1 only in Fig. 5.1. We can therefore write:

$$\begin{Bmatrix} X_1 \\ X_2 \end{Bmatrix} = \begin{bmatrix} \alpha_{11} & \alpha_{12} \\ \alpha_{21} & \alpha_{22} \end{bmatrix} \begin{Bmatrix} F_1 \\ 0 \end{Bmatrix}.$$

(5.12)

Using Eq. 5.12, we can solve for X_1.

$$X_1 = \alpha_{11} F_1$$

(5.13)

From Eq. 5.13, we see that $\alpha_{11} = X_1/F_1$ gives X_1 for a force F_1. This is called a *direct FRF* because the response is measured at the same location where the force is applied (the α_{ij} subscripts match). From the bottom row in Eq. 5.12, we have $\alpha_{21} = X_2/F_1$, which relates X_2 and F_1. This is a *cross FRF* because the response is measured at a different location than where the force is applied (the α_{ij} subscripts do not match). If the force is applied to coordinate x_2 only in Fig. 5.1, then $\begin{Bmatrix} X_1 \\ X_2 \end{Bmatrix} = \begin{bmatrix} \alpha_{11} & \alpha_{12} \\ \alpha_{21} & \alpha_{22} \end{bmatrix} \begin{Bmatrix} 0 \\ F_2 \end{Bmatrix}$ and we have:

$$X_2 = \alpha_{22} F_2.$$

(5.14)

From Eq. 5.14, $\alpha_{22} = X_2/F_2$ is a direct FRF that relates X_2 and F_2. Also, $\alpha_{12} = X_1/F_2$ is a cross FRF that relates X_1 and F_2. The FRFs X_2/F_2 and X_2/F_1 are depicted in Fig. 5.2. All together, our two degree of freedom system has four direct and cross FRFs (2^2). A three degree of freedom system has nine (3^2).

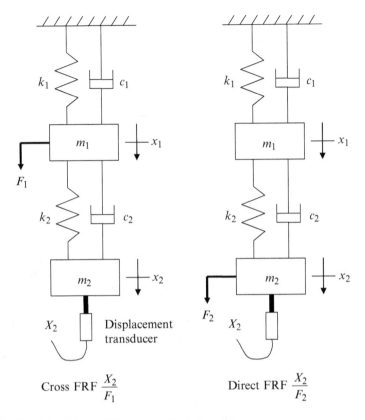

Fig. 5.2 The direct and cross FRFs for coordinate x_2 are shown

From Eqs. 5.9 and 5.10, we see that $\alpha_{12} = \alpha_{21}$ because $[A]^{-1}$ is symmetric. This attribute is referred to as *reciprocity* and can be observed by comparing cross FRFs measured on actual systems. Let's consider the cantilever, or fixed-free, beam shown in Fig. 5.3 for two cases. First, a harmonic force is applied at coordinate 1 (the free end) and the response is measured at coordinate 2 somewhere along the beam, let's say the midpoint. Second, the same force is applied at coordinate 2, but the response is measured at coordinate 1. In the first case, if we measured the response and converted both the force and response into the frequency domain (using the Fourier transform), we would obtain the cross FRF X_2/F_1. In the second case, we would determine the cross FRF X_1/F_2 from our measurements. Due to reciprocity, these two cross FRFs are equal.

IN A NUTSHELL   Reciprocity will be a handy tool later when we explore the physical measurement of FRFs. Sometimes it is convenient to switch the excitation and measurement locations; we are free to do so because of this property.

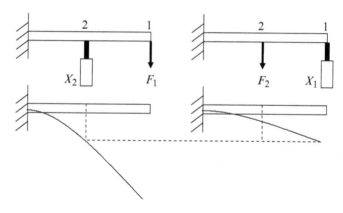

Fig. 5.3 Reciprocity demonstration for cantilever beam cross FRF measurements

To aid in understanding this concept, Fig. 5.3 depicts the two cases where the forcing frequency is equal to the beam's first (lowest) natural frequency so that it vibrates in the corresponding mode shape (see Sect. 4.2). When the force is applied at the free end, the magnitude of the response is largest everywhere and the corresponding X_2 is obtained. When the force is instead applied at coordinate 2, the response is not so big, but it is largest at the free end where we measure X_1. As shown in the figure, the magnitudes X_1 and X_2 are equal and, therefore, the cross FRFs are equal.

To consider a limiting case, assume that the force is applied at the free end and the response is measured at the fixed end, now labeled as coordinate 2. Because the boundary condition is fixed, no matter what force is applied at the free end (1), the measured response will be zero at the fixed end (2). Therefore, the cross FRF X_2/F_1 will be zero for all frequencies. If, on the other hand, the force is applied at the fixed end and the response is measured at the free end, the cross FRF X_1/F_2 will still be zero because a force at the base will not serve to excite vibration in the beam. Reciprocity again holds.

To determine the system FRFs by complex matrix inversion, we solve Eq. 5.7 for each frequency within the range of interest. At each frequency, we extract the desired direct or cross FRF from the inverted matrix and then plot the results on a frequency-by-frequency basis. An example is provided in *By the Numbers 5.1*.

By the Numbers 5.1

Consider the system shown in Fig. 5.1 with $m_1 = 2$ kg, $c_1 = 150$ N-s/m, $k_1 = 1 \times 10^6$ N/m, $m_2 = 4$ kg, $c_2 = 50$ N-s/m, and $k_2 = 2 \times 10^6$ N/m. Note that proportional damping is not satisfied for this system ($[c] \neq \alpha[m] + \beta[k]$ for any combination of real α and β values). Substitution in Eq. 5.6 gives:

$$[A] = \begin{bmatrix} -\omega^2 \cdot 2 + i\omega \cdot 200 + 3 \times 10^6 & -i\omega \cdot 50 - 2 \times 10^6 \\ -i\omega \cdot 50 - 2 \times 10^6 & -\omega^2 \cdot 4 + i\omega \cdot 50 + 2 \times 10^6 \end{bmatrix}.$$

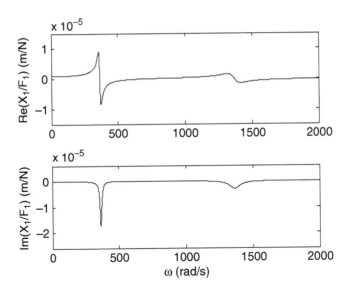

Fig. 5.4 *By the Numbers 5.1* – Direct FRF X_1/F_1 for the example system

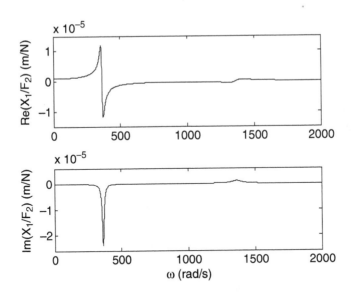

Fig. 5.5 *By the Numbers 5.1* – Cross FRF X_1/F_2 for the example system

Computing the inverse of A for a frequency range between 0 and 2000 rad/s gives the four system FRFs. The direct FRF X_1/F_1 is shown in Fig. 5.4. Vibration modes are observed at 366 and 1,365.5 rad/s for the two degree of freedom system; these are the natural frequencies. The cross FRF X_1/F_2 is displayed in Fig. 5.5. We see that the higher frequency mode is inverted; its motion is out of phase with the lower frequency mode. We also see that its magnitude is quite small relative to

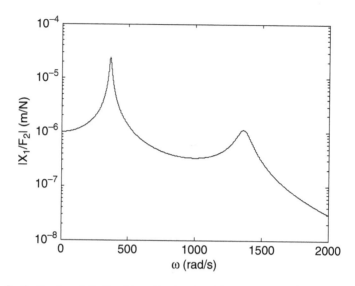

Fig. 5.6 *By the Numbers 5.1* – Semi-logarithmic plot of the cross FRF X_1/F_2

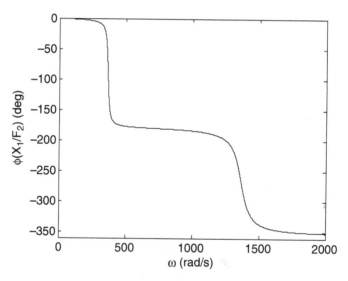

Fig. 5.7 *By the Numbers 5.1* – Phase plot of the cross FRF X_1/F_2

the lower frequency mode. To better view modes with very different magnitudes in a single plot, a semi-logarithmic representation is often used. In this case, the logarithmic vertical (magnitude) axis is plotted against the linear horizontal (frequency) axis. Figure 5.6 was produced using the MATLAB® command `semilogy`. To complete the story, the corresponding phase is provided in Fig. 5.7. Figures 5.4–5.7 were generated using MATLAB® MOJO 5.1.

MATLAB® MOJO 5.1

```
% matlab_mojo_5_1.m

clc
clear all
close all

% Define variables
omega = 0:0.5:2000;          % frequency, rad/s

% Define function
for cnt = 1:length(omega)
    w = omega(cnt);
    a11 = -w^2*2 + i*w*200 + 3e6;
    a12 = -i*w*50 - 2e6;
    a21 = a12;
    a22 = -w^2*4 + i*w*50 + 2e6;
    A = [a11 a12; a21 a22];
    inverted_A = inv(A);
    X1_F1(cnt) = inverted_A(1,1);
    X1_F2(cnt) = inverted_A(1,2);
    X2_F1(cnt) = inverted_A(2,1);
    X2_F2(cnt) = inverted_A(2,2);
end

figure(1)
subplot(211)
plot(omega, real(X1_F1), 'k-')
set(gca,'FontSize', 14)
ylabel('Re(X_1/F_1) (m/N)')
axis([0 2000 -1.5e-5 1.5e-5])
subplot(212)
plot(omega, imag(X1_F1), 'k-')
set(gca,'FontSize', 14)
xlabel('\omega (rad/s)')
ylabel('Im(X_1/F_1) (m/N)')
axis([0 2000 -2.6e-5 6e-6])

figure(2)
subplot(211)
plot(omega, real(X1_F2), 'k-')
set(gca,'FontSize', 14)
ylabel('Re(X_1/F_2) (m/N)')
axis([0 2000 -1.5e-5 1.5e-5])
subplot(212)
plot(omega, imag(X1_F2), 'k-')
set(gca,'FontSize', 14)
xlabel('\omega (rad/s)')
ylabel('Im(X_1/F_2) (m/N)')
axis([0 2000 -2.6e-5 6e-6])

figure(3)
semilogy(omega, abs(X1_F2), 'k-')
set(gca,'FontSize', 14)
xlabel('\omega (rad/s)')
ylabel('Mag(X_1/F_2) (m/N)')

figure(4)
plot(omega, unwrap(angle(X1_F2))*180/pi, 'k-')
set(gca,'FontSize', 14)
xlabel('\omega (rad/s)')
ylabel('\phi(X_1/F_2) (deg)')
ylim([-360 0])
```

5.3   Modal Analysis

Let's next solve the two degree of freedom forced vibration problem using the modal analysis approach. Recall that this method requires that proportional damping exists (or can be assumed). The analysis steps are similar to those we discussed for two degree of freedom free vibration in Sect. 4.4. For the chain-type, spring–mass–damper system shown in Fig. 5.8, we will consider only $f_2(t)$ for now. We could also determine the response to $f_1(t)$ and add the results using linear superposition. The system equations of motion in matrix form are:

$$[m]\{\ddot{\vec{x}}\} + [c]\{\dot{\vec{x}}\} + [k]\{\vec{x}\} = \{\vec{F}\}e^{i\omega t}, \tag{5.15}$$

where $[m] = \begin{bmatrix} m_1 & 0 \\ 0 & m_2 \end{bmatrix}$, $[c] = \begin{bmatrix} c_1 + c_2 & -c_2 \\ -c_2 & c_2 \end{bmatrix}$, $[k] = \begin{bmatrix} k_1 + k_2 & -k_2 \\ -k_2 & k_2 \end{bmatrix}$, and $\{\vec{F}\} = \begin{Bmatrix} 0 \\ F_2 \end{Bmatrix}$. To determine the forced response, we complete the following steps.

IN A NUTSHELL   Because the modal analysis technique is so powerful, we often assume that the damping is proportional even when it is not. In other situations, when we do not know the nature of the damping because a physical incarnation of our design does not exist, we often assume proportional damping during the analysis.

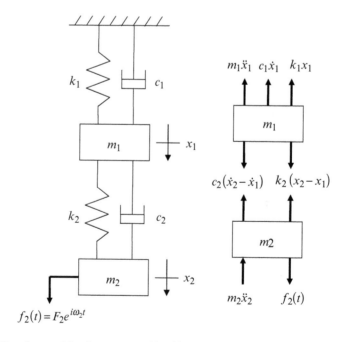

Fig. 5.8 Two degree of freedom system with $f_2(t)$ applied at x_2

1. Verify proportional damping using $[c] = \alpha[m] + \beta[k]$.
2. Ignore damping and the force to find the eigenvalues and eigenvectors.

$$[[m]s^2 + [k]]\left\{\begin{array}{c} X_1 \\ X_2 \end{array}\right\} = \left\{\begin{array}{c} 0 \\ 0 \end{array}\right\}, \text{ where } \left\{\begin{array}{c} x_1 \\ x_2 \end{array}\right\} = \left\{\begin{array}{c} X_1 \\ X_2 \end{array}\right\}e^{st} \tag{5.16}$$

Obtain the eigenvalues from the characteristic equation:

$$\left|[m]s^2 + [k]\right| = 0. \tag{5.17}$$

For the two degree of freedom system, the two roots, s_1^2 and s_2^2, are the eigenvalues. These give the natural frequencies $s_1^2 = -\omega_{n1}^2$ and $s_2^2 = -\omega_{n2}^2$, where $\omega_{n1} < \omega_{n2}$. Use either equation of motion and normalize to X_2 (because the force is applied at this location). The eigenvectors are:

$$\psi_1 = \left\{\begin{array}{c} \left(\frac{X_1}{X_2}\right)_1 \\ 1 \end{array}\right\} = \left\{\begin{array}{c} p_1 \\ 1 \end{array}\right\} \text{ and } \psi_2 = \left\{\begin{array}{c} \left(\frac{X_1}{X_2}\right)_2 \\ 1 \end{array}\right\} = \left\{\begin{array}{c} p_2 \\ 1 \end{array}\right\}, \tag{5.18}$$

where ψ_1 is evaluated using $s^2 = s_1^2$ and ψ_2 is obtained using $s^2 = s_2^2$.
3. Construct the modal matrix using the eigenvectors.

$$[P] = [\psi_1 \quad \psi_2] \tag{5.19}$$

Use the modal matrix to transform into modal coordinates and uncouple the equations of motion.

$$[m_q] = \begin{bmatrix} m_{q1} & 0 \\ 0 & m_{q2} \end{bmatrix} = [P]^T[m][P] \tag{5.20}$$

$$[c_q] = \begin{bmatrix} c_{q1} & 0 \\ 0 & c_{q2} \end{bmatrix} = [P]^T[c][P] \tag{5.21}$$

$$[k_q] = \begin{bmatrix} k_{q1} & 0 \\ 0 & k_{q2} \end{bmatrix} = [P]^T[k][P] \tag{5.22}$$

Transform the force vector from local to modal coordinates.

$$\{\bar{R}\} = \left\{\begin{array}{c} R_1 \\ R_2 \end{array}\right\} = [P]^T[\bar{F}] = \begin{bmatrix} p_1 & p_2 \\ 1 & 1 \end{bmatrix}\left\{\begin{array}{c} 0 \\ F_2 \end{array}\right\} = \left\{\begin{array}{c} F_2 \\ F_2 \end{array}\right\} \tag{5.23}$$

In modal coordinates, the same force is applied to both single degree of freedom systems. Recall that modal coordinates may not make physical sense to us.

Fig. 5.9 Real and imaginary parts of Q_1/R_1 FRF (modal coordinates)

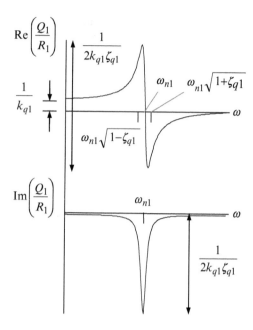

4. Write the FRFs for the two single degree of freedom systems in modal coordinates, Q_1 and Q_2. Note that Q_1 and Q_2 are the frequency-domain representations of the time-domain modal coordinates, q_1 and q_2, that we introduced in Sect. 4.4. For the first natural frequency, the FRF is:

$$\frac{Q_1}{R_1} = \frac{1}{k_{q1}} \left(\frac{\left(1 - r_1^2\right) - i\left(2\zeta_{q1}r_1\right)}{\left(1 - r_1^2\right)^2 + \left(2\zeta_{q1}r_1\right)^2} \right), \tag{5.24}$$

where $r_1 = \omega/\omega_{n1}$ and $\zeta_{q1} = c_{q1}/2\sqrt{k_{q1}m_{q1}}$. Plot the real and imaginary parts as shown in Fig. 5.9 and the magnitude and phase as displayed in Fig. 5.10. For the second natural frequency, the FRF is:

$$\frac{Q_2}{R_2} = \frac{1}{k_{q2}} \left(\frac{\left(1 - r_2^2\right) - i\left(2\zeta_{q2}r_2\right)}{\left(1 - r_2^2\right)^2 + \left(2\zeta_{q2}r_2\right)^2} \right), \tag{5.25}$$

where $r_2 = \omega/\omega_{n2}$ and $\zeta_{q2} = c_{q2}/2\sqrt{k_{q2}m_{q2}}$. The plots are similar to Figs. 5.9 and 5.10.

5. Transform back to local coordinates using:

$$\{\vec{X}\} = [P]\{\vec{Q}\} = \begin{bmatrix} p_1 & p_2 \\ 1 & 1 \end{bmatrix} \begin{Bmatrix} Q_1 \\ Q_2 \end{Bmatrix}. \tag{5.26}$$

Fig. 5.10 Magnitude and
phase of Q_1/R_1 FRF (modal
coordinates)

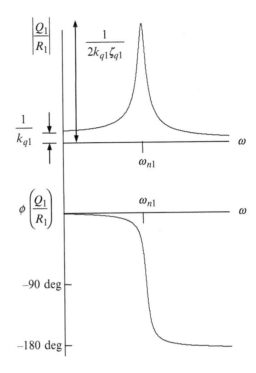

Using Eq. 5.26, we solve for X_1 and X_2.

$$X_1 = p_1 Q_1 + p_2 Q_2 \text{ and } X_2 = Q_1 + Q_2. \tag{5.27}$$

The direct FRF is:

$$\frac{X_2}{F_2} = \frac{Q_1 + Q_2}{F_2} = \frac{Q_1}{R_1} + \frac{Q_2}{R_2}. \tag{5.28}$$

Because $R_1 = R_2 = F_2$. The direct FRF at the location where the mode shapes
were normalized is the *sum of the modal contributions*. See Fig. 5.11. The cross
FRF is:

$$\frac{X_1}{F_2} = \frac{p_1 Q_1 + p_2 Q_2}{F_2} = p_1 \frac{Q_1}{R_1} + p_2 \frac{Q_2}{R_2}. \tag{5.29}$$

The cross FRF is the *sum of the modal contributions scaled by the eigenvectors*.
See Fig. 5.12.

The five fundamental steps for modal analysis are summarized in Table 5.1.

Fig. 5.11 Direct FRF X_2/F_2
(local coordinates)

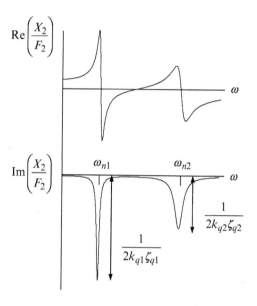

Fig. 5.12 Cross FRF X_1/F_2
(local coordinates)

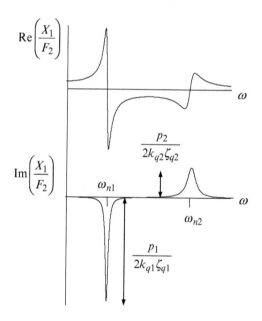

By the Numbers 5.2

Let's complete an example to demonstrate the steps we have just discussed. For the two degree of freedom model in Fig. 5.8, the parameters are $k_1 = 4 \times 10^5$ N/m,

Table 5.1 Modal analysis steps

Step	Action
1	Verify proportional damping.
2	Ignore damping and external force to find the eigenvalues and eigenvectors.
3	Construct the modal matrix using the eigenvectors.
4	Write the FRFs for the single degree of freedom systems in modal coordinates.
5	Transform back to local coordinates.

$c_1 = 80$ N-s/m, $m_1 = 2$ kg, $k_2 = 6 \times 10^5$ N/m, $c_2 = 120$ N-s/m, $m_2 = 1$ kg, and $f_2(t) = 100e^{i\omega t}$ N. The mass, damping, and stiffness matrices are:

$$[m] = \begin{bmatrix} m_1 & 0 \\ 0 & m_2 \end{bmatrix} = \begin{bmatrix} 2 & 0 \\ 0 & 1 \end{bmatrix} \text{kg},$$

$$[c] = \begin{bmatrix} c_1 + c_2 & -c_2 \\ -c_2 & c_2 \end{bmatrix} = \begin{bmatrix} 200 & -120 \\ -120 & 120 \end{bmatrix} \text{N-s/m} ,$$

and

$$[k] = \begin{bmatrix} k_1 + k_2 & -k_2 \\ -k_2 & k_2 \end{bmatrix} = \begin{bmatrix} 1 \times 10^6 & -6 \times 10^5 \\ -6 \times 10^5 & 6 \times 10^5 \end{bmatrix} \text{N/m}.$$

1. Verify proportional damping. The equality is true when $\alpha = 0$ and $\beta = 1/5000$, so proportional damping holds.

$$\begin{bmatrix} 200 & -120 \\ -120 & 120 \end{bmatrix} = \alpha \begin{bmatrix} 2 & 0 \\ 0 & 1 \end{bmatrix} + \beta \begin{bmatrix} 1 \times 10^6 & -6 \times 10^5 \\ -6 \times 10^5 & 6 \times 10^5 \end{bmatrix}$$

2. Ignore damping and the force to find the eigenvalues and eigenvectors.

$$\left[\begin{bmatrix} 2 & 0 \\ 0 & 1 \end{bmatrix} s^2 + \begin{bmatrix} 1 \times 10^6 & -6 \times 10^5 \\ -6 \times 10^5 & 6 \times 10^5 \end{bmatrix} \right] \begin{Bmatrix} X_1 \\ X_2 \end{Bmatrix} = \begin{Bmatrix} 0 \\ 0 \end{Bmatrix}$$

Obtain the eigenvalues from the characteristic equation:

$$\begin{vmatrix} 2s^2 + 1 \times 10^6 & -6 \times 10^5 \\ -6 \times 10^5 & 1s^2 + 6 \times 10^5 \end{vmatrix} = 0,$$

or

$$(2s^2 + 1 \times 10^6)(1s^2 + 6 \times 10^5) - (-6 \times 10^5)^2$$
$$= 2s^4 + 2.2 \times 10^6 s^2 + 2.4 \times 10^{11} = 0.$$

The roots are $s_1^2 = -122,799.81 = -\omega_{n1}^2$ and $s_2^2 = -977,200.19 = -\omega_{n2}^2$. The natural frequencies are $\omega_{n1} = 350.43$ rad/s and $\omega_{n2} = 988.53$ rad/s. Alternately, $f_{n1} = 55.77$ Hz and $f_{n2} = 157.33$ Hz. The top equation of motion from the matrix format is $(2s^2 + 1 \times 10^6)X_1 - 6 \times 10^5 X_2 = 0$. In order to normalize the eigenvectors to x_2 (the force location), the required ratio is:

$$\frac{X_1}{X_2} = \frac{6 \times 10^5}{2s^2 + 1 \times 10^6}.$$

Substituting $s_1^2 = -122,799.81$ gives the first eigenvector.

$$\psi_1 = \left\{ \begin{array}{c} \left(\frac{6 \times 10^5}{2(-122,799.81)+1 \times 10^6} \right) \\ 1 \end{array} \right\} = \left\{ \begin{array}{c} 0.795 \\ 1 \end{array} \right\}$$

Substituting $s_2^2 = -977,200.19$ gives the second eigenvector.

$$\psi_2 = \left\{ \begin{array}{c} \left(\frac{6 \times 10^5}{2(-977,200.19)+1 \times 10^6} \right) \\ 1 \end{array} \right\} = \left\{ \begin{array}{c} -0.629 \\ 1 \end{array} \right\}$$

3. Construct the modal matrix using the eigenvectors.

$$[P] = \begin{bmatrix} 0.795 & -0.629 \\ 1 & 1 \end{bmatrix}$$

Use the modal matrix to transform into modal coordinates and uncouple the equations of motion.

$$[m_q] = [P]^T [m][P] = \begin{bmatrix} 2.264 & 0 \\ 0 & 1.791 \end{bmatrix} \text{ kg}$$

$$[c_q] = [P]^T [c][P] = \begin{bmatrix} 55.605 & 0 \\ 0 & 350.088 \end{bmatrix} \text{ N-s/m}$$

$$[k_q] = [P]^T [k][P] = \begin{bmatrix} 2.780 \times 10^5 & 0 \\ 0 & 1.750 \times 10^6 \end{bmatrix} \text{ N/m}$$

Transform the force vector from local to modal coordinates.

$$\{\bar{R}\} = \left\{ \begin{array}{c} R_1 \\ R_2 \end{array} \right\} = [P]^T [\bar{F}] = \begin{bmatrix} 0.795 & -0.629 \\ 1 & 1 \end{bmatrix} \left\{ \begin{array}{c} 0 \\ 100 \end{array} \right\} = \left\{ \begin{array}{c} 100 \\ 100 \end{array} \right\} \text{ N}$$

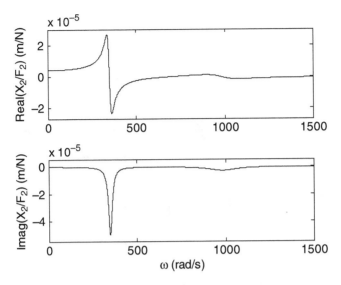

Fig. 5.13 *By the Numbers 5.2* – Direct FRF X_2/F_2 (local coordinates)

4. Write the FRFs for the two single degree of freedom systems in modal coordinates. For the first natural frequency, the FRF is:

$$\frac{Q_1}{R_1} = \frac{1}{2.780 \times 10^5} \left(\frac{\left(1 - r_1^2\right) - i(2(0.0358)r_1)}{\left(1 - r_1^2\right)^2 + (2(0.0358)r_1)^2} \right),$$

where $r_1 = \omega/350.43$ and the damping ratio is $\zeta_{q1} = \frac{55.605}{2\sqrt{2.780 \times 10^5 (2.264)}} = 0.0358$.

For the second natural frequency, the FRF is:

$$\frac{Q_2}{R_2} = \frac{1}{1.750 \times 10^6} \left(\frac{\left(1 - r_2^2\right) - i(2(0.0989)r_2)}{\left(1 - r_2^2\right)^2 + (2(0.0989)r_2)^2} \right),$$

where $r_2 = \omega/988.53$ and $\zeta_{q2} = \frac{350.088}{2\sqrt{1.750 \times 10^6 (1.791)}} = 0.0989$.

5. Transform back to local coordinates. The direct FRF, $\frac{X_2}{F_2} = \frac{Q_1}{R_1} + \frac{Q_2}{R_2}$, is plotted in Fig. 5.13. The cross FRF, $\frac{X_1}{F_2} = p_1 \frac{Q_1}{R_1} + p_2 \frac{Q_2}{R_2}$, is plotted in Fig. 5.14.

In Figs. 5.13 and 5.14, the value of the real part at zero frequency ($\omega = 0$) represents the DC compliance (i.e., the inverse of stiffness). For the direct FRF in Fig. 5.13, the value is:

$$\operatorname{Re}\left(\frac{X_2}{F_2}\right)\Bigg|_{\omega=0} = \frac{1}{k_{q1}} + \frac{1}{k_{q2}} = 4.169 \times 10^{-6} \text{ m/N}. \tag{5.30}$$

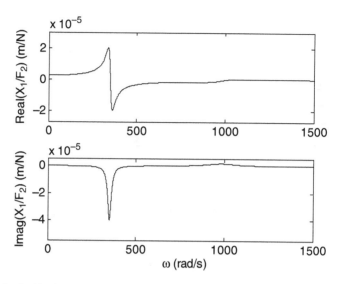

Fig. 5.14 *By the Numbers 5.2* – Cross FRF X_1/F_2 (local coordinates)

This indicates that the modal stiffness for each mode is added in series to give the local stiffness for the model. For the cross FRF in Fig. 5.14, the DC compliance is:

$$\mathrm{Re}\left(\frac{X_1}{F_2}\right)\bigg|_{\omega=0} = \frac{p_1}{k_{q1}} + \frac{p_2}{k_{q2}} = 2.500 \times 10^{-6} \text{ m/N}. \tag{5.31}$$

Again, the modal stiffness values are added in series, but each is scaled by the appropriate eigenvector. Given the force magnitude F_2, we could use Eqs. 5.30 and 5.31 to determine the real-valued deflections, X_1 and X_2, due to the DC (non-oscillating) force. At any nonzero forcing frequency, the responses are complex-valued and describe the steady-state forced vibration; see Figs. 5.13 and 5.14.

5.4 Dynamic Absorber

Let's now investigate a special application of complex matrix inversion. Consider the undamped single degree of freedom system subject to forced vibration shown in Fig. 5.15. Let's assume that the magnitude X_1 due to the harmonic force $f_1(t) = F_1 e^{i\omega_f t}$ is too large and causes, for example, mechanical failure, passenger discomfort, manufacturing errors, etc. In order to eliminate this problem, we would like X_1 to ideally be zero at the forcing frequency ω_f. In other words, the magnitude

Fig. 5.15 Single degree of freedom, undamped system with the harmonic force $f_1(t) = F_1 e^{i\omega_f t}$ which causes excessive vibration at coordinate x_1

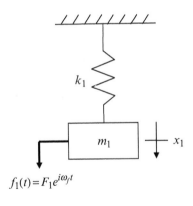

Fig. 5.16 The addition of the dynamic absorber gives a new two degree of freedom system

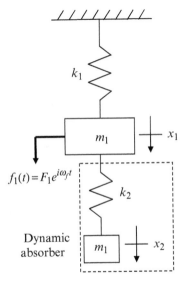

of motion at coordinate x_1 is zero when the system is forced at the frequency ω_f. This result is expressed in Eq. 5.32.

$$\frac{X_1}{F_1} = 0 \text{ at } \omega = \omega_f \qquad (5.32)$$

Let's add a second degree of freedom to the system and see what happens. See Fig. 5.16, where the system model is now:

$$\left[-\omega^2 \begin{bmatrix} m_1 & 0 \\ 0 & m_2 \end{bmatrix} + \begin{bmatrix} k_1 + k_2 & -k_2 \\ -k_2 & k_2 \end{bmatrix} \right] \begin{Bmatrix} X_1 \\ X_2 \end{Bmatrix} e^{i\omega_f t} = \begin{Bmatrix} F_1 \\ 0 \end{Bmatrix} e^{i\omega_f t}. \qquad (5.33)$$

Equation 5.33 can be more compactly expressed as $[A]\{\vec{X}\} = \{\vec{F}\}$. As we saw in Sect. 5.2, we can use complex matrix inversion to determine the system FRFs.

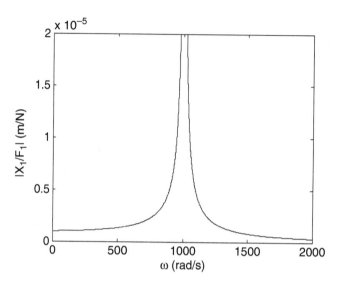

Fig. 5.17   Direct FRF for the original single degree of freedom system

Similar to Eq. 5.8, the direct FRF at coordinate x_1 for the new two degree of freedom system is:

$$\alpha_{11} = \frac{X_1}{F_1} = \frac{-\omega_f^2 m_2 + k_2}{\left(-\omega_f^2 m_1 + k_1 + k_2\right)\left(-\omega_f^2 m_2 + k_2\right) - \left(-k_2\right)^2}. \tag{5.34}$$

We want this FRF to be zero, so it is required that $-\omega_f^2 m_2 + k_2 = 0$ or:

$$\omega_f = \sqrt{\frac{k_2}{m_2}}. \tag{5.35}$$

When the k_2 and m_2 values for the added spring and mass, together known as a *dynamic absorber*, are selected by applying this design rule, the response at x_1 is zero, even though $f_1(t) = F_1 e^{i\omega_f t}$ remains. Let's sketch the magnitude plots for the original and new systems. Figure 5.17 shows the original single degree of freedom FRF magnitude with $m_1 = 1$ kg and $k_1 = 1 \times 10^6$ N/m. We see that the response is infinite at resonance in the absence of damping. For this example, we will assume that the forcing frequency is equal to the natural frequency, $\omega_f = \sqrt{\frac{k_1}{m_1}} = 1{,}000$ rad/s. Clearly, the response would be too large in this case! In Fig. 5.18, the new direct FRF X_1/F_1 is shown, where m_2 was selected to be 0.1 kg and the corresponding spring stiffness was $k_2 = \omega_f^2 \cdot m_2 = 1 \times 10^5$ N/m according to Eq. 5.35. The new two degree of freedom system naturally has two modes (and natural frequencies), but the response is zero at $\omega = \omega_f$.

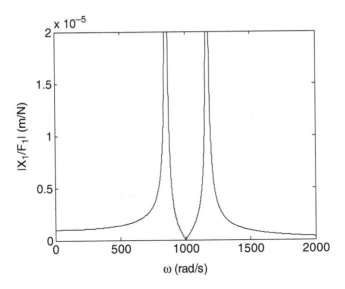

Fig. 5.18 Direct FRF X_1/F_1 for the new two degree of freedom system

We have now eliminated the vibration at x_1 due to the force f_1 with a frequency of ω_f. What about the motion of the added mass? To answer this question, we need the cross FRF X_2/F_1. Similar to Eq. 5.10 we have:

$$\alpha_{21} = \frac{X_2}{F_1} = \frac{k_2}{\left(-\omega_f^2 m_1 + k_1 + k_2\right)\left(-\omega_f^2 m_2 + k_2\right) - \left(-k_2\right)^2}. \qquad (5.36)$$

Expanding the denominator gives:

$$\alpha_{21} = \frac{X_2}{F_1} = \frac{k_2}{m_1 m_2 \omega_f^4 - k_2 m_1 \omega_f^2 - (k_1 + k_2) m_2 \omega_f^2 + k_1 k_2 + k_2^2 - k_2^2}. \qquad (5.37)$$

According to the dynamic absorber design rule, $\omega_f^2 = k_2/m_2$. Substituting yields:

$$\alpha_{21} = \frac{X_2}{F_1} = \frac{k_2}{m_1 m_2 \dfrac{k_2^2}{m_2^2} - k_2 m_1 \dfrac{k_2}{m_2} - k_1 m_2 \dfrac{k_2}{m_2} - k_2 m_2 \dfrac{k_2}{m_2} + k_1 k_2}. \qquad (5.38)$$

Simplifying gives:

$$\alpha_{21} = \frac{X_2}{F_1} = \frac{k_2}{\dfrac{m_1}{m_2} k_2^2 - \dfrac{m_1}{m_2} k_2^2 - k_1 k_2 - k_2^2 + k_1 k_2} = \frac{k_2}{-k_2^2} = -\frac{1}{k_2}. \qquad (5.39)$$

Equation 5.39 can be rewritten as $X_2 = -F_1/k_2$. This tells us that the motion of x_2 is 180° out of phase with x_1. The role of the added spring and mass, therefore, is for its inertial force to counteract the external force at x_1 and cause the vibration to be zero – hence the name, dynamic *absorber*. It effectively absorbs the external force's energy at ω_f. Equation 5.22 also shows that a stiffer dynamic absorber spring decreases the magnitude of the x_2 vibration. While we may want k_2 to be larger to keep X_2/F_1 smaller, we must also satisfy $\omega_f^2 = k_2/m_2$. A larger k_2 means a proportionally larger m_2 and there is typically a practical limit on how much mass can be added as the dynamic absorber.

 IN A NUTSHELL   The dynamic absorber is a valuable tool. If we encounter a forced vibration that is too large, we can eliminate it by adding a new spring and mass, which alone would have a natural frequency that matches the forcing frequency. The added mass will move, but the motion of the attachment point will be dramatically reduced. Dynamic absorbers are found, for example, in automobile transmissions, at the top of skyscrapers, on power lines, and in machine tools.

To conclude this section, we should recognize that all systems include some level of damping. We can conveniently analyze the damped system response using Eqs. 5.8 and 5.10. Given the base system (Fig. 5.15) description, we could tune the dynamic absorber stiffness (assuming its mass was preselected) to minimize the response at the forcing frequency. The design rule provided in Eq. 5.35 still provides a reasonable starting point.

By the Numbers 5.3

Let's again consider the Fig. 5.15 system with $m_1 = 1$ kg and $k_1 = 1 \times 10^6$ N/m, but now add the damper $c_1 = 100$ N - s/m. This gives a single degree of freedom damping ratio of $\zeta_1 = \frac{c_1}{2\sqrt{k_1 m_1}} = \frac{100}{2\sqrt{1 \times 10^6 (1)}} = 0.05 = 5\%$. The FRF is shown in Fig. 5.19, where the damped natural frequency, ω_d, is 998.75 rad/s. We will specify a forcing function of $f_1(t) = 100e^{i\omega_d t}$ N. If the absorber mass, m_2, is 0.1 kg, then an initial guess for the absorber spring stiffness is $k_2 = \omega_d^2 \cdot m_2 = 9.975 \times 10^4$ N/m. We will assume the absorber damping coefficient is $c_2 = 5$ N-s/m. The corresponding two degree of freedom system direct FRF X_1/F_1 is provided in Fig. 5.20. Its value at the forcing frequency, ω_d, is 4.7565×10^{-7} m/N. The vibration magnitude at x_1 is therefore $X_1 = 4.7565 \times 10^{-7} \cdot 100 = 4.7565 \times 10^{-5}$ m ≈ 48 μm for the 100 N magnitude force. This is quite an improvement over the original single degree of freedom response of $X_1 = 1 \times 10^{-5} \cdot 100 = 1 \times 10^{-3}$ m $= 1$ mm. The cross FRF X_2/F_1 for the absorber response is shown in Fig. 5.21. The magnitude at the forcing frequency is $X_2 = 9.5249 \times 10^{-6} \cdot 100 = 9.5249 \times 10^{-4}$ m ≈ 0.952 mm.

While this result is already pretty good, perhaps we can reduce the response at x_1 by adjusting k_2. Figure 5.22 shows the value of X_1/F_1 at $\omega = \omega_f$ for the two degree of freedom system as a function of the modified k_2 value. We see that the minimum

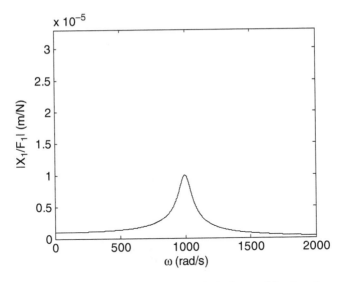

Fig. 5.19 *By the Numbers 5.3* – FRF for the original single degree of freedom damped system

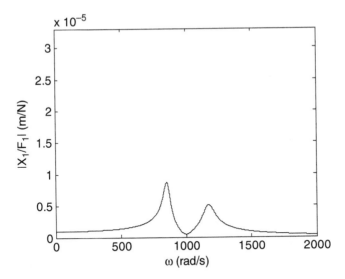

Fig. 5.20 *By the Numbers 5.3* – Direct FRF X_1/F_1 for the two degree of freedom system with the dynamic absorber added

is obtained when $k_2 = 1.0025 \times 10^5$ N/m. The vibration magnitude for this optimum stiffness is $X_1 = 4.7517 \times 10^{-7} \cdot 100 = 4.7517 \times 10^{-5}$ m, which is again approximately 48 μm. The code used to produce Figs. 5.19–5.22 is provided in MATLAB® MOJO 5.2.

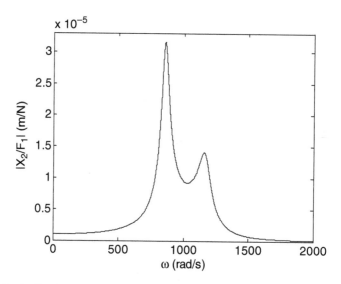

Fig. 5.21 *By the Numbers 5.3* – Cross FRF X_2/F_1 for the two degree of freedom system with the dynamic absorber added

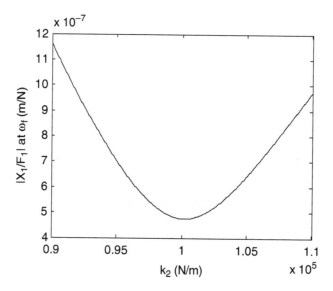

Fig. 5.22 *By the Numbers 5.3* – Variation in X_1/F_1 magnitude at ω_f for a range of k_2 values near the initial guess

MATLAB® MOJO 5.2

```
% matlab_mojo_5_2.m

clc
clear all
close all

% Define variables
omega = 0:0.25:2000;     % rad/s
m1 = 1;                  % kg
k1 = 1e6;                % N/m
c1 = 100;                % N-s/m
wn1 = sqrt(k1/m1);       % rad/s
zeta1 = 0.05;

% Define function
r = omega/wn1;
mag1 = 1/k1*(1./((1-r.^2).^2 + (2*zeta1*r).^2)).^0.5;

figure(1)
plot(omega, mag1, 'k-')
set(gca,'FontSize', 14)
xlabel('\omega (rad/s)')
ylabel('|X_1/F_1| (m/N)')
axis([0 2000 0 3.3e-5])

index = find(omega == wn1);
mag_original = abs(mag1(index))

% Dynamic absorber
m2 = 0.1;
wf = wn1;
k2 = wf^2*m2;
c2 = 5;

for cnt = 1:length(omega)
    w = omega(cnt);
    a11 = -w^2*m1 + i*w*(c1 + c2) + k1 + k2;
    a12 = -i*w*c2 - k2;
    a21 = a12;
    a22 = -w^2*m2 + i*w*c2 + k2;
    A = [a11 a12; a21 a22];
    inverted_A = inv(A);
    X1_F1(cnt) = inverted_A(1,1);
    X1_F2(cnt) = inverted_A(1,2);
    X2_F1(cnt) = inverted_A(2,1);
    X2_F2(cnt) = inverted_A(2,2);
end

figure(2)
plot(omega, abs(X1_F1), 'k-')
set(gca,'FontSize', 14)
xlabel('\omega (rad/s)')
ylabel('|X_1/F_1| (m/N)')
axis([0 2000 0 3.3e-5])

index = find(omega == wf);
mag_direct = abs(X1_F1(index))
mag_cross = abs(X2_F1(index))

figure(3)
plot(omega, abs(X2_F1), 'k-')
set(gca,'FontSize', 14)
xlabel('\omega (rad/s)')
ylabel('|X_2/F_1| (m/N)')
axis([0 2000 0 3.3e-5])
```

```
clear X1_F1

k2_test = 0.9*k2:50:1.1*k2;
for cnt = 1:length(k2_test)
    w = wf;
    k2 = k2_test(cnt);
    a11 = -w^2*m1 + i*w*(c1 + c2) + k1 + k2;
    a12 = -i*w*c2 - k2;
    a21 = a12;
    a22 = -w^2*m2 + i*w*c2 + k2;
    A = [a11 a12; a21 a22];
    inverted_A = inv(A);
    X1_F1(cnt) = inverted_A(1,1);
end

figure(4)
plot(k2_test, abs(X1_F1), 'k-')
set(gca,'FontSize', 14)
xlabel('k_2 (N/m)')
ylabel('|X_1/F_1)| at \omega_f (m/N)')
xlim([min(k2_test) max(k2_test)])

index = find(X1_F1 == min(X1_F1));
k2_test(index)
abs(X1_F1(index))
```

Chapter Summary

- Superposition enables the linear system response to each external force to be calculated individually and the results summed to find the overall response.
- Complex matrix inversion can be used to determine the forced response for systems with two (or more) degrees of freedom. While this approach is more computationally expensive than modal analysis, it does not require proportional damping.
- For a direct FRF, the response is measured at the same location where the force is applied.
- For a cross FRF, the response is measured at a different location than where the force is applied.
- Pairs of cross FRFs with the displacement and force subscripts switched, such as X_1/F_2 and X_2/F_1, are equal. This is referred to as reciprocity.
- A dynamic absorber may be added to a system to attenuate the original system response at a particular frequency. The dynamic absorber may be realized using a simple spring–mass–damper system. The natural frequency of the dynamic absorber is matched to the frequency of interest.
- In modal analysis, the coupled differential equations of motion are uncoupled using the modal matrix, which is composed of the system eigenvectors. Modal analysis requires proportional damping.
- In modal analysis, the direct FRF in local coordinates is the sum of the single degree of freedom FRFs in modal coordinates. The cross FRF is the sum of the modal FRFs scaled by the eigenvectors.

Exercises

1. A two degree of freedom spring–mass–damper system is shown Fig. P5.1. For harmonic forced vibration (due to the external force at coordinate x_2), complete the following if $k_1 = 2 \times 10^5$ N/m, $c_1 = 60$ N-s/m, $m_1 = 2.5$ kg, $k_2 = 5.5 \times 10^4$ N/m, $c_2 = 16.5$ N-s/m, and $m_2 = 1.2$ kg.

Fig. P5.1 Two degree of freedom spring–mass–damper system under forced vibration

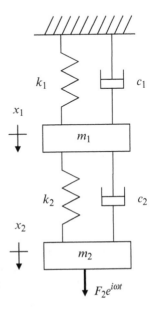

(a) Verify that proportional damping exists.
(b) Define the modal matrix and determine the modal mass, stiffness, and damping matrices. Note that the mode shapes should be normalized to the force location.
(c) Write expressions for the uncoupled single degree of freedom FRFs in modal coordinates, Q_1/R_1 and Q_2/R_2, and the direct FRF in local coordinates, X_2/F_2.
(d) Plot the real and imaginary parts of the direct FRF, X_2/F_2. Units should be m/N for the vertical axis and rad/s for the horizontal (frequency) axis. Use a frequency range of omega = 0:0.01:500; (rad/s).

2. A two degree of freedom spring–mass–damper system is shown in Fig. P5.2. For harmonic forced vibration (due to the external force applied at coordinate x_2), complete the following if $k_1 = 8 \times 10^7$ N/m, $c_1 = 1000$ N-s/m, $m_1 = 50$ kg, $k_2 = 5 \times 10^7$ N/m, $c_2 = 500$ N-s/m, and $m_2 = 12$ kg.

(a) Show that proportional damping does not exist.
(b) Write a symbolic expression for the direct FRF X_2/F_2 as a function of the frequency, ω, and mass, stiffness, and damping values, $m_{1,2}$, $k_{1,2}$, and $c_{1,2}$. Use the complex matrix inversion approach.

Fig. P5.2 Two degree of
freedom spring–mass–
damper system under
forced vibration

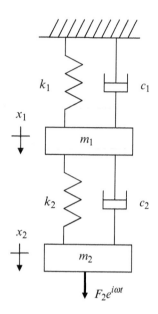

(c) Write a symbolic expression for the cross FRF X_1/F_2 as a function of the
frequency, ω, and mass, stiffness, and damping values, $m_{1,2}$, $k_{1,2}$, and $c_{1,2}$.
Use the complex matrix inversion approach.

(d) Plot the real and imaginary parts of the cross FRF, X_1/F_2. Units should be
m/N for the vertical axis and rad/s for the horizontal (frequency) axis. Use a
frequency range of omega = 0:0.01:3500; (rad/s).

3. Consider the single degree of freedom spring–mass system shown in Fig. P5.3,
where $k = 4 \times 10^5$ N/m and $m = 8$ kg. It is being excited by a harmonic
forcing function, $F_1 e^{i\omega_f t}$, at a frequency, ω_f.

(a) If the excitation frequency is 200 rad/s, design a dynamic absorber to
eliminate the vibration at coordinate x_1. The only available spring for use
in the absorber is identical to the one already used in the system.

Fig. P5.3 Single degree of
freedom system excited by
the harmonic forcing function
$F_1 e^{i\omega_f t}$

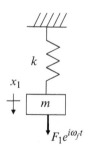

(b) If the 4×10^5 N/m absorber spring is used in conjunction with a 2 kg absorber mass, at what forced excitation frequency (in rad/s) will the steady-state vibration of coordinate x_1 be eliminated?

4. A two degree of freedom spring–mass–damper system is shown in Fig. P5.4. For harmonic forced vibration (due to the external force at coordinate x_1), complete the following if $k_1 = 2 \times 10^5$ N/m, $c_1 = 60$ N-s/m, $m_1 = 2.5$ kg, $k_2 = 5.5 \times 10^4$ N/m, $c_2 = 16.5$ N-s/m, and $m_2 = 1.2$ kg.

(a) Verify that proportional damping exists.
(b) Define the modal matrix and determine the modal mass, stiffness, and damping matrices. Note that the mode shapes should be normalized to the force location.
(c) Write expressions for the uncoupled single degree of freedom FRFs in modal coordinates, Q_1/R_1 and Q_2/R_2, and the direct FRF in local coordinates, X_1/F_1.
(d) Plot the real and imaginary parts of the direct FRF, X_1/F_1. Units should be m/N for the vertical axis and rad/s for the horizontal (frequency) axis. Use a frequency range of omega $= 0:0.1:500;$ (rad/s).

Fig. P5.4 Two degree of freedom spring–mass–damper system under forced vibration

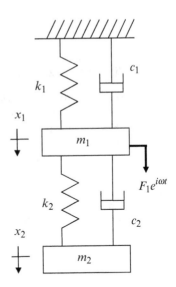

5. For the two degree of freedom spring–mass–damper system shown in Fig. P5.5, complete the following if $k_a = 2 \times 10^5$ N/m, $k_b = 5.5 \times 10^4$ N/m, $c_a = 60$ N-s/m, $c_b = 16.5$ N-s/m, $m_a = 2.5$ kg, and $m_b = 1.2$ kg.

(a) Obtain the equations of motion in matrix form and transform them into modal coordinates q_1 and q_2. Normalize your eigenvectors to the force location, coordinate x_2. Verify that proportional damping exists.

Fig. P5.5 Two degree of
freedom spring–mass–damper
system under forced vibration

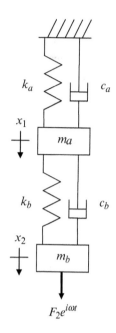

(b) Determine the FRFs Q_1/R_1, Q_2/R_2, and X_2/F_2. Express them in equation
form and then plot the real and imaginary parts (in m/N) versus frequency
(in rad/s). Use a frequency range of $0:0.1:600$; (rad/s).

6. A dynamic absorber is to be designed to eliminate the vibration at coordinate x_1
for the system shown in Fig. P5.6, where the excitation frequency is 400 rad/s
and the force magnitude is 100 N. For the given system constants, determine the
values of the mass and spring constant for the dynamic absorber if the magnitude
of vibration for the absorber mass is 5 mm.

Fig. P5.6 Single degree of
freedom system excited by
the harmonic forcing function
$100e^{i400t}$

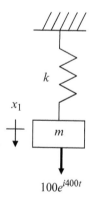

7. Given the modal mass matrix, $m_q = \begin{bmatrix} 2 & 0 \\ 0 & 2 \end{bmatrix}$ kg, the modal stiffness matrix,
$k_q = \begin{bmatrix} 5.858 \times 10^6 & 0 \\ 0 & 3.414 \times 10^7 \end{bmatrix}$ N/m, the modal matrix, $[P] = \begin{bmatrix} 0.707 & -0.707 \\ 1 & 1 \end{bmatrix}$,
and the modal damping ratios, $\zeta_{q1} = 0.04$ and $\zeta_{q2} = 0.02$, complete the following.

(a) Plot the imaginary part (m/N) of the direct FRF X_2/F_2. Use a frequency range of $0:0.1:5000$; (rad/s).

(b) Plot the imaginary part (in m/N) of the cross FRF X_1/F_2. Use a frequency range of $0:0.1:5000$; (rad/s).

8. After installation, it was found that a particular machine exhibited excessive vibration due to a harmonic excitation force with a frequency of 100 Hz. A dynamic absorber was designed and added to the original system to attenuate this vibration. If the resulting vibration magnitude of the absorber mass was 2 mm at 100 Hz and the excitation force magnitude was 25 N, determine the stiffness of the spring (N/m) and mass (kg) used to construct the absorber. You may neglect damping in your analysis.

9. Given the eigenvalues and eigenvectors for the two degree of freedom system shown in Fig. P5.9, complete the following.

$$s_1^2 = -1 \times 10^6 \text{rad/s}^2 \quad s_2^2 = -7 \times 10^6 \text{rad/s}^2$$

$$\Psi_1 = \left\{ \begin{array}{c} 0.5 \\ 1 \end{array} \right\} \qquad \Psi_2 = \left\{ \begin{array}{c} -2.5 \\ 1 \end{array} \right\}$$

(a) Determine the modal matrices m_q (kg), c_q (N-s/m), and k_q (N/m).

(b) Plot the imaginary part (in m/N) of the cross frequency response function, X_1/F_2. Use a frequency range of $0:0.1:3500$; (rad/s).

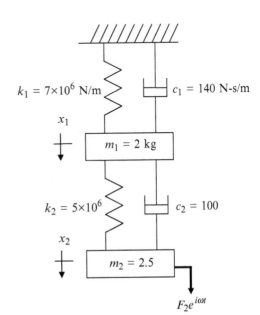

Fig. P5.9 Two degree of freedom spring–mass–damper system under forced vibration

10. Given the eigenvalues and eigenvectors for the two degree of freedom system shown in Fig. P5.9, determine the DC (zero frequency) compliance for the real part of the direct FRF X_2/F_2.

$$s_1{}^2 = -1 \times 10^6 \text{rad/s}^2 \quad s_2{}^2 = -7 \times 10^6 \text{rad/s}^2$$

$$\Psi_1 = \begin{Bmatrix} 0.5 \\ 1 \end{Bmatrix} \qquad \Psi_2 = \begin{Bmatrix} -2.5 \\ 1 \end{Bmatrix}$$

Chapter 6
Model Development by Modal Analysis

Il n'est pas certain que tout soit incertain. (It is not certain that everything is uncertain).

– Blaise Pascal

6.1 The Backward Problem

In Chaps. 1–5, we assumed a model and then used that model to determine the system response in the time or frequency domain (or both). More often, however, we have an actual dynamic system and would like to build a model that we can use to represent its vibratory behavior in response to some external excitation. For example, in milling operations, the flexibility of the cutting tool–holder–spindle–machine structure (and sometimes the workpiece) determines the limiting axial depth of cut to avoid chatter, a self-excited vibration (Schmitz and Smith 2009). In this case, the dynamic response at the free end of the tool (and/or at the cutting location on the workpiece) is measured. Using this measured response, a model in the form of modal parameters can be developed for use in a time-domain simulation[1] of the milling process. How can we work this "backward problem" of starting with a measurement and developing a model? To begin, we need to determine the modal mass, stiffness, and damping values from the measured frequency response function (FRF).

---

[1] In time-domain, or time-marching, simulation, the equations of motion that describe the process behavior are solved at small increments in time using numerical integration (Schmitz and Smith 2009).

T.L. Schmitz and K.S. Smith, *Mechanical Vibrations: Modeling and Measurement*, DOI 10.1007/978-1-4614-0460-6_6, © Springer Science+Business Media, LLC 2012

6.2   Peak Picking

6.2.1   Single Degree of Freedom

After performing a measurement to determine the FRF for a physical structure (see Chap. 7), we can use *peak picking* to estimate the modal parameters from the real and imaginary parts of the FRF. We discussed this previously in Sect. 3.4, but let us review it here and see how it fits into our task of model development.

Figure 6.1 shows a representation of a measured FRF with a single mode within the measurement bandwidth. Therefore, a single degree of freedom model is sufficient to describe this system's dynamic behavior. In the figure, three frequencies, ω_1, ω_2, and ω_3, and one peak value, A, are identified. Frequency ω_1 gives the (undamped) natural frequency, ω_n. Also, ω_2 (from the maximum real part peak) occurs at $\omega_n(1 - \zeta_q)$ and ω_3 (from the minimum real part peak) occurs at $\omega_n(1 + \zeta_q)$. (As we saw in Sect. 3.4, these approximations yield reasonable results when the damping is low).

Differencing ω_3 and ω_2 gives:

$$\omega_3 - \omega_2 = \omega_n(1 + \zeta_q) - \omega_n(1 - \zeta_q) = 2\omega_n\zeta_q. \tag{6.1}$$

Because ω_2, ω_3, and ω_n are known, Eq. 6.1 can be solved for ζ_q. See Eq. 6.2:

$$\zeta_q = \frac{\omega_3 - \omega_2}{2\omega_n} \tag{6.2}$$

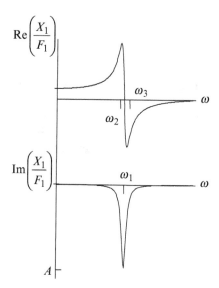

Fig. 6.1 An example FRF measurement. A single mode is included within the measurement bandwidth

Fig. 6.2 Single degree of
freedom model determined
from the measurement
provided in Fig. 6.1

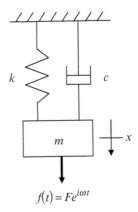

$$f(t) = Fe^{i\omega t}$$

Next, from the imaginary part minimum peak, we have that $A = -1/(2k_q\zeta_q)$.
Rearranging to solve for the modal stiffness, k_q results in:

$$k_q = \frac{-1}{2A\zeta_q}. \tag{6.3}$$

Finally, given that $\omega_n^2 = k_q/m_q$, we can solve for the modal mass m_q using
Eq. 6.4, where compatible SI units are rad/s for the natural frequency, N/m for the
stiffness, and kg for the mass.

$$m_q = \frac{k_q}{\omega_n^2} \tag{6.4}$$

Now we have the natural frequency, stiffness, mass, and damping ratio for the
single degree of freedom system. We can also determine the modal (viscous)
damping coefficient using Eq. 6.5:

$$c_q = 2\zeta_q\sqrt{m_q k_q}. \tag{6.5}$$

The single degree of freedom FRF represents a special situation. In this case,
there is no difference between modal and local coordinates. There is no coordinate
transformation required to uncouple the equations of motion because there is only
one equation of motion. We can therefore define the single degree of freedom
model directly: $k = k_q$, $c = c_q$, and $m = m_q$; see Fig. 6.2. Given this model, we can
describe the transient and steady-state vibration for any harmonic forcing function
or initial conditions.

Fig. 6.3 An FRF measurement is displayed where two modes are contained in the measurement bandwidth

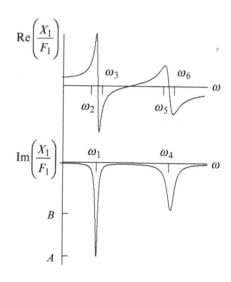

IN A NUTSHELL   Using the model developed by peak picking, it is also possible to inexpensively examine "what if" scenarios. What if we could double the damping? What if a dynamic absorber was added? What if the system was exposed to a given time-varying force? Such questions can be easily and quickly evaluated once we have a model of the physical system. Some parts of a model might be easily estimated, such as the mass. However, other parts are vexingly difficult to predict without a measurement, particularly the damping.

6.2.2   Two Degrees of Freedom

What if there are two modes within the measurement bandwidth for the direct FRF? The simple answer is that we can again use peak picking, but now we must repeat the process for each of the two modes. Of course, a model with two degrees of freedom will now be required as well. Consider the direct FRF measurement data provided in Fig. 6.3 – six frequencies and two peak magnitudes are identified. The three frequencies, ω_1, ω_2, and ω_3, and peak value, A, describe the lower frequency mode with a natural frequency of $\omega_{n1} = \omega_1$. The three frequencies, ω_4, ω_5, and ω_6, and peak value, B, describe the higher frequency mode with a natural frequency of $\omega_{n2} = \omega_4$. Recall from Chap. 4 that the natural frequencies are the same in modal and local coordinates, so we do not need to make any distinction regarding the coordinate system (local or modal) for these frequencies.

For the lower frequency mode, differencing ω_3 and ω_2 gives:

$$\omega_3 - \omega_2 = 2\omega_{n1}\zeta_{q1}.$$

Because ω_2, ω_3, and ω_{n1} are known, this equation can be solved for the modal damping ratio ζ_{q1}.

$$\zeta_{q1} = \frac{\omega_3 - \omega_2}{2\omega_{n1}}$$

From the imaginary part minimum peak, we determine the modal stiffness $k_{q1} = \frac{-1}{2A\zeta_{q1}}$. Next, we can solve for m_{q1} using $m_{q1} = k_{q1}/\omega_{n1}^2$. Finally, the modal damping coefficient is $c_{q1} = 2\zeta_{q1}\sqrt{m_{q1}k_{q1}}$.

For the higher frequency mode, we follow the same steps. Differencing ω_6 and ω_5 gives:

$$\omega_6 - \omega_5 = 2\omega_{n2}\zeta_{q2}.$$

Because ω_5, ω_6, and ω_{n2} are known, we can solve for ζ_{q2}.

$$\zeta_{q2} = \frac{\omega_6 - \omega_5}{2\omega_{n2}}$$

From the imaginary part minimum peak, we have that $k_{q2} = \frac{-1}{2B\zeta_{q2}}$. Next, we determine the modal mass, m_{q2}, using $m_{q2} = \frac{k_{q2}}{\omega_{n2}^2}$. Finally, the modal damping coefficient is $c_{q2} = 2\zeta_{q2}\sqrt{m_{q2}k_{q2}}$.

6.3   Building the Model

The peak picking approach is straightforward to implement, but you may wonder why we can treat the individual modes from a direct FRF measurement separately to find the modal parameters. The answer lies in the "undoing" of the modal analysis steps. Recall that once we uncoupled the equations of motion to determine the single degree of freedom modal coordinate models, we transformed back to local coordinates by: (1) summing the modal FRFs for the direct FRF; and (2) summing the modal FRFs scaled by the eigenvectors for the cross FRF. In this case, we are beginning with the local coordinate direct FRF measurement, X_1/F_1, and determining the modal FRFs, Q_1/R_1 and Q_2/R_2, by peak picking. The two-mode fit of the direct FRF in local coordinates is simply the sum of the individual modal responses:

$$\frac{X_1}{F_1} = \frac{Q_1}{R_1} + \frac{Q_2}{R_2}.$$

In the peak picking technique, we fit Q_1/R_1 and Q_2/R_2 separately. From that fitting exercise, we defined the modal parameters. In matrix form, they are expressed as $[m_q] = \begin{bmatrix} m_{q1} & 0 \\ 0 & m_{q2} \end{bmatrix}$, $[c_q] = \begin{bmatrix} c_{q1} & 0 \\ 0 & c_{q2} \end{bmatrix}$, and $[k_q] = \begin{bmatrix} k_{q1} & 0 \\ 0 & k_{q2} \end{bmatrix}$ for a two degree of freedom system. Note that we have automatically *assumed that proportional damping holds* when using this technique for model identification. However, this is a reasonable assumption for most systems because structural damping is typically low.

In order to complete the local coordinate model, we need to transform the modal mass, damping, and stiffness matrices into local coordinates. We are already familiar with the "forward" version of this transformation from local (or model) to modal coordinates. For the mass matrices, we have:

$$[m_q] = [P]^T[m][P],$$

where $[P]$ is the modal matrix. Its columns are the system eigenvectors (or mode shapes):

$$[P] = [\,\Psi_1 \quad \Psi_2\,].$$

To determine the mass matrix in local coordinates, the "backward" form of the coordinate transformation from modal to local coordinates is:

$$[m] = [P]^{-T}[m_q][P]^{-1},$$

where the "−1" superscript indicates the matrix inverse operation (inv(P) in MATLAB®) and the "−T" represents the inverse of the modal matrix transpose (inv(P') in MATLAB®). Similarly, the damping and stiffness matrices in local coordinates are determined using:

$$[c] = [P]^{-T}[c_q][P]^{-1} \text{ and } [k] = [P]^{-T}[k_q][P]^{-1}.$$

For a two degree of freedom model, the modal matrix can be represented as:

$$[P] = [\,\Psi_1 \quad \Psi_2\,] = \begin{bmatrix} 1 & 1 \\ p_1 & p_2 \end{bmatrix},$$

where the first eigenvector, Ψ_1, corresponds to vibration at ω_{n1} and the second eigenvector, Ψ_2, describes vibration at ω_{n2}. In this case, we have normalized to coordinate x_1, so p_1 and p_2 give the ratio X_2/X_1. This is a cross FRF, so we must measure not only a direct FRF to determine the modal matrices but also a cross FRF to identify the modal matrix. In performing these measurements, the force should be applied to the physical system at the location of most interest in the structure's response.

Given the direct and cross FRF measurements, how do we determine p_1 and p_2? To answer this question, we must again return to the modal analysis steps. As we

Fig. 6.4 Representation of direct and cross FRF measurements on a physical system

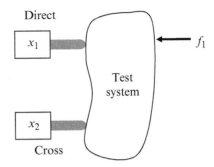

just discussed, the transformation from modal to local coordinates includes a summation of the modal contributions. For the direct FRF, we have:

$$\frac{X_1}{F_1} = \frac{Q_1}{R_1} + \frac{Q_2}{R_2},$$

and the transformation for the cross FRF is:

$$\frac{X_2}{F_1} = p_1 \frac{Q_1}{R_1} + p_2 \frac{Q_2}{R_2}.$$

To determine p_1, we use the first (lowest frequency) mode from both the direct and cross FRFs. For the first mode at ω_{n1}, the cross FRF is given by $p_1(Q_1/R_1)$ and the direct FRF by Q_1/R_1. Taking their ratio enables us to determine p_1; see Eq. 6.6.

$$\left.\frac{\frac{X_2}{F_1}}{\frac{X_1}{F_1}}\right|_{\omega_{n1}} = \frac{p_1 \dfrac{Q_1}{R_1}}{\dfrac{Q_1}{R_1}} = p_1. \tag{6.6}$$

Similarly, we determine p_2 from the second mode (with a natural frequency of ω_{n2}) from the direct and cross FRFs. This is shown in Eq. 6.7.

$$\left.\frac{\frac{X_2}{F_1}}{\frac{X_1}{F_1}}\right|_{\omega_{n2}} = \frac{p_2 \dfrac{Q_1}{R_1}}{\dfrac{Q_1}{R_1}} = p_2. \tag{6.7}$$

Let's graphically examine this identification of the eigenvectors and, subsequently, the modal matrix. Figure 6.4 shows a schematic of direct and cross FRF measurements on a physical system (the "blob" representation). Figures 6.5 and 6.6 display the measurement results. They show the real and imaginary parts of the direct and cross FRFs, respectively. However, the imaginary parts are sufficient to

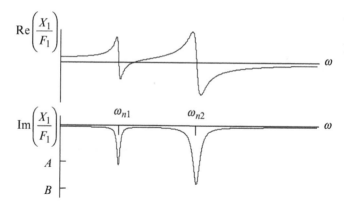

Fig. 6.5 Direct FRF measurement results for the system shown in Fig. 6.4

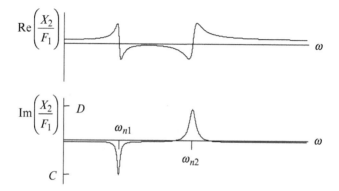

Fig. 6.6 Cross FRF measurement results for the system shown in Fig. 6.4

identify the eigenvectors. In the direct FRF (Fig. 6.5), the local minimum values of the imaginary parts are identified: A for the first mode (at ω_{n1}) and B for the second mode (at ω_{n2}). The peak values C and D are identified in the cross FRF (Fig. 6.6). These peak values can be used in Eqs. 6.6 and 6.7. Using Eq. 6.6, we determine p_1 by:

$$\frac{\mathrm{Im}\left(\dfrac{X_2}{F_1}\right)}{\mathrm{Im}\left(\dfrac{X_1}{F_1}\right)}\Bigg|_{\omega_{n1}} = \frac{C}{A} = p_1. \tag{6.8}$$

The value of p_2 is calculated using Eq. 6.7 and the peak heights B and D; see Eq. 6.9. Note that we must pay attention to the sign of these heights. Once p_1 and p_2 are determined, they are substituted into the modal matrix,

$$[P] = [\Psi_1 \ \Psi_2] = \begin{bmatrix} 1 & 1 \\ p_1 & p_2 \end{bmatrix}.$$

$$\frac{\text{Im}\left(\dfrac{X_2}{F_1}\right)}{\text{Im}\left(\dfrac{X_1}{F_1}\right)}\Bigg|_{\omega_{n2}} = \frac{D}{B} = p_2 \tag{6.9}$$

IN A NUTSHELL   This technique is often used to visualize the mode shape. First, a direct FRF is measured. Then, using a series of measured cross FRFs, the height of the imaginary peak for the mode of interest (relative to the height of the same mode in the direct measurement) provides the mode shape component at that location. With a little practice, it is possible to quickly describe the mode shape.

By the Numbers 6.1

In order to perform a numerical demonstration of the backward problem solution, let's begin with the forward problem solution by modal analysis for a two degree of freedom spring–mass–damper model (as we discussed in Chap. 5). We will then use the local coordinate direct and cross FRFs as the starting points for the backward solution. Commençons![2]

The two degree of freedom chain-type model is shown in Fig. 6.7. A harmonic force, $f_1(t)$, is applied at coordinate x_1. Let's follow the five steps we identified in Sect. 5.4 for the forward solution. The parameters for the model in Fig. 6.7 are $k_1 = 8 \times 10^5$ N/m, $c_1 = 160$ N-s/m, $m_1 = 3$ kg, $k_2 = 4 \times 10^5$ N/m, $c_2 = 80$ N-s/m, $m_2 = 3$ kg, and $f_1(t) = 200e^{i\omega t}$ N. The mass, damping, and stiffness matrices are:

$$left[m] = \begin{bmatrix} m_1 & 0 \\ 0 & m_2 \end{bmatrix} = \begin{bmatrix} 3 & 0 \\ 0 & 3 \end{bmatrix} \text{kg},$$

$$[c] = \begin{bmatrix} c_1 + c_2 & -c_2 \\ -c_2 & c_2 \end{bmatrix} = \begin{bmatrix} 240 & -80 \\ -80 & 80 \end{bmatrix} \text{N-s/m},$$

and

$$[k] = \begin{bmatrix} k_1 + k_2 & -k_2 \\ -k_2 & k_2 \end{bmatrix} = \begin{bmatrix} 1.2 \times 10^6 & -4 \times 10^5 \\ -4 \times 10^5 & 4 \times 10^5 \end{bmatrix} \text{N/m}.$$

Note that the matrices are symmetric (i.e., the off-diagonal terms are equal). We proceed according to the steps identified in Table 5.1.

1. Verify proportional damping. The equality $[c] = \alpha[m] + \beta[k]$ is true when $\alpha = 0$ and $\beta = 1/5000$, so proportional damping holds. We require proportional

---

[2] Let's begin!

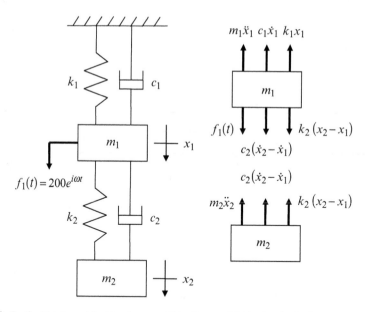

Fig. 6.7 *By the Numbers 6.1* – two degree of freedom model (the free body diagrams for the two masses are also included)

damping for modal analysis, so this check must be completed for each forward problem solution. For the backward problem, however, we *assume* that proportional damping exists, so no check is applied.

$$\begin{bmatrix} 240 & -80 \\ -80 & 80 \end{bmatrix} = \alpha \begin{bmatrix} 3 & 0 \\ 0 & 3 \end{bmatrix} + \beta \begin{bmatrix} 1.2 \times 10^6 & -4 \times 10^5 \\ -4 \times 10^5 & 4 \times 10^5 \end{bmatrix}$$

2. Ignore damping and the force to find the eigenvalues and eigenvectors.

$$\left[\begin{bmatrix} 3 & 0 \\ 0 & 3 \end{bmatrix} s^2 + \begin{bmatrix} 1.2 \times 10^6 & -4 \times 10^5 \\ -4 \times 10^5 & 4 \times 10^5 \end{bmatrix} \right] \begin{Bmatrix} X_1 \\ X_2 \end{Bmatrix} = \begin{Bmatrix} 0 \\ 0 \end{Bmatrix}$$

Obtain the eigenvalues from the characteristic equation:

$$\begin{vmatrix} 3s^2 + 1.2 \times 10^6 & -4 \times 10^5 \\ -4 \times 10^5 & 3s^2 + 4 \times 10^5 \end{vmatrix} = 0$$

or

$$\left(3s^2 + 1.2 \times 10^6 \right) \left(3s^2 + 4 \times 10^5 \right) - \left(-4 \times 10^5 \right)^2$$
$$= 9s^4 + 4.8 \times 10^6 s^2 + 3.2 \times 10^{11} = 0.$$

The roots are $s_1^2 = -78, 104.86 = -\omega_{n1}^2$ and $s_2^2 = -455, 228.47 = -\omega_{n2}^2$. The natural frequencies are $\omega_{n1} = 279.47$ rad/s ($f_{n1} = 44.48$ Hz) and $\omega_{n2} = 674.71$ rad/s ($f_{n2} = 107.38$ Hz). The bottom (second) equation of motion from the matrix format is $-4 \times 10^5 X_1 + (3s^2 + 4 \times 10^5)X_2 = 0$. In order to normalize the eigenvectors to x_1 (the force location), the required ratio is:

$$\frac{X_2}{X_1} = \frac{4 \times 10^5}{3s^2 + 4 \times 10^5}.$$

Substituting $s_1^2 = -78, 104.86$ gives the first eigenvector.

$$\psi_1 = \left\{ \begin{array}{c} 1 \\ \left(\dfrac{4 \times 10^5}{3(-78, 104.86) + 4 \times 10^5} \right) \end{array} \right\} = \left\{ \begin{array}{c} 1 \\ 2.414 \end{array} \right\}$$

Substituting $s_2^2 = -455, 228.47$ gives the second eigenvector.

$$\psi_2 = \left\{ \begin{array}{c} 1 \\ \left(\dfrac{4 \times 10^5}{3(-45, 228.47) + 4 \times 10^5} \right) \end{array} \right\} = \left\{ \begin{array}{c} 1 \\ -0.414 \end{array} \right\}$$

3. Construct the modal matrix using the eigenvectors.

$$[P] = \begin{bmatrix} 1 & 1 \\ 2.414 & -0.414 \end{bmatrix}$$

Use the modal matrix to transform into modal coordinates and uncouple the equations of motion.

$$[m_q] = [P]^T [m][P] = \begin{bmatrix} 20.482 & 0 \\ 0 & 3.514 \end{bmatrix} \text{kg}$$

$$[c_q] = [P]^T [c][P] = \begin{bmatrix} 319.95 & 0 \\ 0 & 319.95 \end{bmatrix} \text{N-s/m}$$

$$[k_q] = [P]^T [k][P] = \begin{bmatrix} 1.600 \times 10^6 & 0 \\ 0 & 1.600 \times 10^6 \end{bmatrix} \text{N/m}$$

Transform the force vector from local to modal coordinates.

$$\{\vec{R}\} = \left\{ \begin{array}{c} R_1 \\ R_2 \end{array} \right\} = [P]^T [\vec{F}] = \begin{bmatrix} 1 & 2.414 \\ 1 & -0.414 \end{bmatrix} \left\{ \begin{array}{c} 200 \\ 0 \end{array} \right\} = \left\{ \begin{array}{c} 200 \\ 200 \end{array} \right\} \text{N}$$

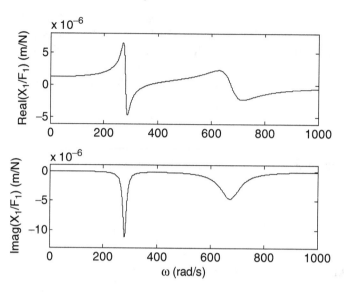

Fig. 6.8 *By the Numbers 6.1* – Direct FRF X_1/F_1 (local coordinates)

As a check, the single degree of freedom natural frequencies can be calculated using the modal parameters; they are $\omega_{n1} = \sqrt{\frac{k_{q1}}{m_{q1}}} = \sqrt{\frac{1.6 \times 10^6}{20.482}} = 279.49$ rad/s and $\omega_{n2} = \sqrt{\frac{k_{q2}}{m_{q2}}} = \sqrt{\frac{1.6 \times 10^6}{3.514}} = 674.78$ rad/s. These values do not match the original natural frequencies (determined from the eigenvalues) exactly due to round-off error, but they do verify that the modal mass and stiffness parameters are correct.

4. Write the FRFs for the two single degree of freedom systems in modal coordinates. For the first natural frequency, the FRF is:

$$\frac{Q_1}{R_1} = \frac{1}{1.600 \times 10^6} \left(\frac{\left(1 - r_1^2\right) - i(2(0.0279)r_1)}{\left(1 - r_1^2\right)^2 + (2(0.0279)r_1)^2} \right),$$

where $r_1 = \omega/279.49$ and the damping ratio is $\zeta_{q1} = \frac{319.95}{2\sqrt{1.600 \times 10^6 (20.482)}} = 0.0279$. For the second natural frequency, the FRF is:

$$\frac{Q_2}{R_2} = \frac{1}{1.600 \times 10^6} \left(\frac{\left(1 - r_2^2\right) - i(2(0.0675)r_2)}{\left(1 - r_2^2\right)^2 + (2(0.0675)r_2)^2} \right),$$

where $r_2 = \omega/674.78$ and $\zeta_{q2} = \frac{319.95}{2\sqrt{1.600 \times 10^6 (3.514)}} = 0.0675$.

5. Transform back to local coordinates. The direct FRF, $X_1/F_1 = (Q_1/R_1) + (Q_2/R_2)$, is plotted in Fig. 6.8. The cross FRF, $X_2/F_1 = p_1(Q_1/R_1) + p_2(Q_2/R_2)$, is displayed in Fig. 6.9.

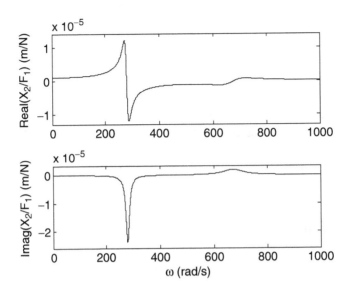

Fig. 6.9 *By the Numbers 6.1* – Cross FRF X_2/F_1 (local coordinates)

We can now treat Figs. 6.8 and 6.9 as measurement results (0.1 rad/s frequency resolution) and identify the two degree of freedom model that can be used to represent this dynamic system; this is the backward solution. We will assume a chain-type model format, as shown in Fig. 6.7. However, this is not necessary in general; we will discuss this further in Sect. 6.6. As shown in Figs. 6.3 and 6.6, the peak picking approach requires that we identify six frequencies and four peak heights; see Figs. 6.10 (direct FRF) and 6.11 (cross FRF). The frequencies and heights are summarized in Table 6.1.

Let's begin with the 279.4 rad/s mode. We determine the modal damping ratio using $\zeta_{q1} = \frac{287.2 - 271.6}{2(279.4)} = 0.0279$. From the imaginary part minimum peak, A, the modal stiffness is $k_{q1} = \frac{-1}{2(-1.125 \times 10^{-5})0.0279} = 1.593 \times 10^6$ N/m. Next, the modal mass is $m_{q1} = \frac{1.593 \times 10^6}{279.4^2} = 20.406$ kg. Finally, the modal damping coefficient is $c_{q1} = 2(0.0279)\sqrt{20.406(1.593 \times 10^6)} = 318.14$ N-s/m.

We complete the same calculations for the 673.2 rad/s mode. The modal damping ratio is $\zeta_{q2} = \frac{718.6 - 628.1}{2(673.2)} = 0.0672$. Using the imaginary part minimum peak, B, the modal stiffness is determined by $k_{q2} = \frac{-1}{2(-4.639 \times 10^{-6})0.0672} = 1.604 \times 10^6$ N/m. The modal mass is $m_{q2} = \frac{1.604 \times 10^6}{673.2^2} = 3.539$ kg and the modal damping coefficient is $c_{q2} = 2(0.0672)\sqrt{3.539(1.604 \times 10^6)} = 320.22$ N-s/m. We now have all the values required to fully populate the modal mass, damping, and stiffness matrices.

$$[m_q] = \begin{bmatrix} 20.406 & 0 \\ 0 & 3.539 \end{bmatrix} \text{kg}$$

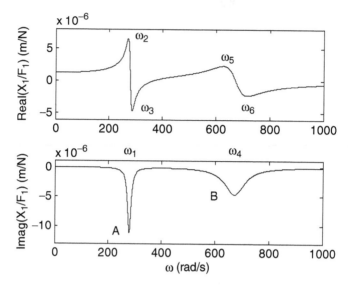

Fig. 6.10 *By the Numbers 6.1* – Peak picking frequencies and heights for the direct FRF X_1/F_1

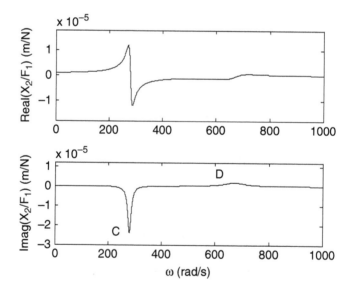

Fig. 6.11 *By the Numbers 6.1* – Peak picking heights for the cross FRF X_2/F_1

$$[c_q] = \begin{bmatrix} 318.14 & 0 \\ 0 & 320.22 \end{bmatrix} \text{N-s/m}$$

$$[k_q] = \begin{bmatrix} 1.593 \times 10^6 & 0 \\ 0 & 1.604 \times 10^6 \end{bmatrix} \text{N/m}$$

Table 6.1 *By the Numbers 6.1* – peak picking frequencies and heights from Figs. 6.10 and 6.11

Frequency (rad/s)	Peak height (m/N)
$\omega_1 = 279.4$	$A = -1.125 \times 10^{-5}$
$\omega_2 = 271.6$	$B = -4.639 \times 10^{-6}$
$\omega_3 = 287.2$	$C = -2.396 \times 10^{-5}$
$\omega_4 = 673.2$	$D = 1.911 \times 10^{-6}$
$\omega_5 = 628.1$	
$\omega_6 = 718.6$	

Next, we determine the eigenvectors using the peak heights A, B, C, and D. For the lower frequency mode, $p_1 = \frac{C}{A} = \frac{-2.396 \times 10^{-5}}{-1.125 \times 10^{-5}} = 2.130$. For the higher frequency mode, $p_2 = \frac{D}{B} = \frac{1.911 \times 10^{-6}}{-4.639 \times 10^{-5}} = -0.412$. The modal matrix is:

$$[P] = \begin{bmatrix} 1 & 1 \\ 2.130 & -0.412 \end{bmatrix}.$$

The mass, damping, and stiffness matrices in local coordinates are:

$$[m] = [P]^{-T} [m_q] [P]^{-1} = \begin{bmatrix} 3.021 & 0 \\ 0 & 3.707 \end{bmatrix} = \begin{bmatrix} m_1 & 0 \\ 0 & m_2 \end{bmatrix} \text{kg},$$

$$[c] = [P]^{-T} [c_q] [P]^{-1} = \begin{bmatrix} 233.19 & -85.27 \\ -85.27 & 98.79 \end{bmatrix} = \begin{bmatrix} c_1 + c_2 & -c_2 \\ -c_2 & c_2 \end{bmatrix} \text{N-s/m},$$

and

$$[k] = [P]^{-T} [k_q] [P]^{-1} = \begin{bmatrix} 1.168 \times 10^6 & -4.272 \times 10^5 \\ -4.272 \times 10^5 & 4.948 \times 10^5 \end{bmatrix}$$

$$= \begin{bmatrix} k_1 + k_2 & -k_2 \\ -k_2 & k_2 \end{bmatrix} \text{N/m}.$$

IN A NUTSHELL   The forms on the right hand side of the local coordinate mass, damping, and stiffness matrices indicate that we had an idea what the model should look like. That is, we assumed a chain-type model. If our idea of the model is right, then the local mass, stiffness, and damping matrices should have this form. We do not always know what the model should be, but the measurements guide us. For example, how many modes do we see in the measured FRF? That is, how many degrees of freedom are required to represent the physical system in the frequency range we measured and in the locations we measured?

We are almost finished, but need to make an engineering decision. For the damping matrix, we have that $c_2 = 98.79$ N-s/m from the (2,2) term. For the (1,2) and (2,1) terms, however, $c_2 = 85.27$ N-s/m. Let's use the average, $c_2 =$

Table 6.2 *By the Numbers 6.1* – results for the backward problem solution

	Original model	Backward solution	Percent error (%)
m_1 (kg)	3	3.021	−0.7
c_1 (N-s/m)	160	141.16	11.8
k_1 (N/m)	8×10^5	7.07×10^5	11.6
m_2	3	3.707	−23.6
c_2	80	92.03	−15.0
k_2	4×10^5	4.61×10^5	−15.3

$(98.79 + 85.27)/2 = 92.03$ N - s/m. From the (1,1) term, then, $c_1 = 233.19 - 92.03 = 141.16$ N-s/m. We complete the same analysis for the stiffness values. Using the (1,2), (2,1), and (2,2) values from the stiffness matrix, $k_2 = (4.948 \times 10^5 + 4.272 \times 10^5)/2 = 4.61 \times 10^5$ N/m. Using the (1,1) term, $k_1 = 1.168 \times 10^6 - 4.610 \times 10^5 = 7.07 \times 10^5$ N/m. The original model parameters, backward solution values, and percent differences are provided in Table 6.2. The disagreement is the result of round-off error in both the forward and backward solutions, assumptions in the peak picking approach, and the engineering decision to use the average c_2 and k_2 values.

6.4 Peak Picking for Multiple Degrees of Freedom

The peak picking method is straightforward to extend to additional degrees of freedom. For the cutting tool–holder–spindle–machine structure mentioned in Sect. 6.1, it is common for many modes to exist in the 0–5,000 Hz frequency range. An example direct FRF measurement for a 47-mm-diameter shell mill[3] which was clamped in a high-speed spindle is provided in Fig. 6.12. The measurement was performed at the free end of the cutting tool. There are five modes which were selected for fitting. The three frequencies and peak height for each mode are included in Table 6.3. The corresponding modal stiffness and damping ratio are also identified.

The five-mode fit and original data are plotted together in Fig. 6.13; the code used to generate the modal fit is provided in MATLAB® MOJO 6.1. The fit to the direct FRF measurement is simply the sum of the modal contributions:

$$\frac{X_1}{F_1} = \sum_{n=1}^{5} \frac{Q_n}{R_n}. \tag{6.10}$$

---
[3] Shell mills are typically used to machine large flat surfaces.

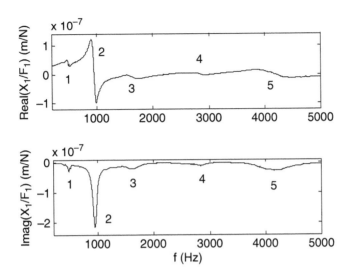

Fig. 6.12 Direct FRF measurement for cutting tool–holder–spindle–machine structure. The modes selected for fitting are identified numerically (1 through 5)

Table 6.3 Modal fitting parameters for the cutting tool–holder–spindle–machine direct FRF

Mode	f_1 (Hz)	f_2 (Hz)	f_3 (Hz)	Imaginary peak (m/N)	f_n (Hz)	k_q (N/m)	ζ_q
1	480	468	493	-2.89×10^{-8}	480	6.65×10^8	0.026
2	950	906	992	-2.14×10^{-7}	950	5.19×10^7	0.045
3	1,630	1,542	1,745	-2.13×10^{-8}	1,630	3.79×10^8	0.062
4	2,842	2,770	2,915	-1.12×10^{-8}	2,842	1.72×10^9	0.026
5	4,150	3,823	4,500	-2.78×10^{-8}	4,150	2.19×10^8	0.082

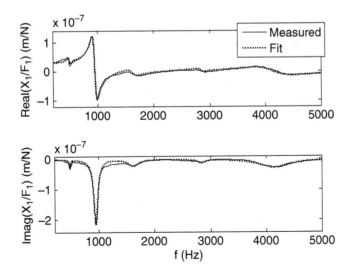

Fig. 6.13 Modal fit to the cutting tool–holder–spindle–machine direct FRF

MATLAB® MOJO 6.1

```
% matlab_mojo_6_1.m

clc
clear all
close all

% Modal fit
f = 200:0.2:5000;    % Hz
kq = 6.65e8;         % N/m
zetaq = 0.026;
fn = 480;            % Hz
r = f/fn;
real_part = 1/kq*(1-r.^2)./((1-r.^2).^2 + (2*zetaq*r).^2);
imag_part = 1/kq*(-2*zetaq*r)./((1-r.^2).^2 + (2*zetaq*r).^2);
Q1_R1 = real_part + 1i*imag_part;    % m/N


kq = 5.19e7;         % N/m
zetaq = 0.045;
fn = 950;            % Hz
r = f/fn;
real_part = 1/kq*(1-r.^2)./((1-r.^2).^2 + (2*zetaq*r).^2);
imag_part = 1/kq*(-2*zetaq*r)./((1-r.^2).^2 + (2*zetaq*r).^2);
Q2_R2 = real_part + 1i*imag_part;    % m/N


kq = 3.79e8;         % N/m
zetaq = 0.062;
fn = 1630;           % Hz
r = f/fn;
real_part = 1/kq*(1-r.^2)./((1-r.^2).^2 + (2*zetaq*r).^2);
imag_part = 1/kq*(-2*zetaq*r)./((1-r.^2).^2 + (2*zetaq*r).^2);
Q3_R3 = real_part + 1i*imag_part;    % m/N


kq = 1.72e9;         % N/m
zetaq = 0.026;
fn = 2842;           % Hz
r = f/fn;
real_part = 1/kq*(1-r.^2)./((1-r.^2).^2 + (2*zetaq*r).^2);
imag_part = 1/kq*(-2*zetaq*r)./((1-r.^2).^2 + (2*zetaq*r).^2);
Q4_R4 = real_part + 1i*imag_part;    % m/N


kq = 2.19e8;         % N/m
zetaq = 0.082;
fn = 4150;           % Hz
r = f/fn;
real_part = 1/kq*(1-r.^2)./((1-r.^2).^2 + (2*zetaq*r).^2);
imag_part = 1/kq*(-2*zetaq*r)./((1-r.^2).^2 + (2*zetaq*r).^2);
Q5_R5 = real_part + 1i*imag_part;    % m/N


X1_F1 = Q1_R1 + Q2_R2 + Q3_R3 + Q4_R4 + Q5_R5;


figure(1)
subplot(211)
plot(f, real(X1_F1), 'k:')
set(gca,'FontSize', 14)
axis([200 5000 -12e-8 14e-8])
ylabel('Real(X_1/F_1)  (m/N)')
subplot(212)
plot(f, imag(X1_F1), 'k:')
axis([200 5000 -24e-8 1e-8])
set(gca,'FontSize', 14)
xlabel('f (Hz)')
ylabel('Imag(X_1/F_1)  (m/N)')
```

IN A NUTSHELL   There are other curve fitting methods, of course. The peak picking technique described here is easy to visualize and works well, provided that the modes are well separated. Naturally, the modal parameters are sensitive to the peaks used in the curve fitting.

6.5   Mode Shape Measurement

It is often of interest to determine the mode shapes (eigenvectors) for a structure, even in the absence of building a model. This can give insight into: (1) the source of a particular natural frequency; and (2) how the structure might be modified to reduce the magnitude of a particular mode.[4]

Consider the steel rod from the *beam experimental platform (BEP)*.[5] What if this beam was not clamped in the platform but was floating in space instead? In this case, the boundary conditions would be free-free; there would be no external restrictions on the beam's motion. Let's discuss the mode shapes for this free-free beam. First, there are two *rigid body modes*. Second, there are an infinite number of *bending modes* for the continuous beam.[6]

One of the rigid body modes is depicted in Fig. 6.14. This mode is a translation of the rigid beam due to a constant (zero frequency or DC) force applied at the center of mass. The corresponding rigid body mode frequency is zero because the beam does not actually oscillate. The other rigid body mode is represented in Fig. 6.15. It is a rotation of the rigid beam due to a constant force that is not applied at the center of mass. The vibration frequency is again zero because there is no relative motion of points along the beam, only the rigid body rotation.

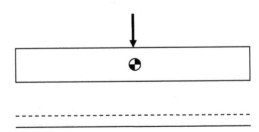

Fig. 6.14 Translational rigid body mode

---

[4] One structural modification technique we have already discussed is the addition of a dynamic absorber (Sect. 5.4).

[5] The BEP was introduced in Sect. 2.6.

[6] There are also other modes of vibration, but we'll discuss these in Chap. 8.

Fig. 6.15 Rotational rigid
body mode

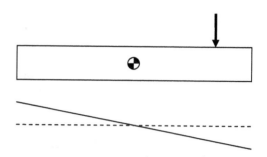

Table 6.4 Constants for the free-free beam mode shape calculation
in Eq. 6.11 (Blevins 2001)

Mode i	λ_i	σ_i
1	4.73004074	0.982502215
2	7.85320462	1.000777312

The bending modes for a free-free, uniform cross-section beam can be described analytically (Blevins 2001). The mode shape function that describes the relative vibration magnitudes at locations, x, along the beam length, l, is:

$$\psi_i = \frac{1}{2}\left(\cosh\left(\frac{\lambda_i x}{l}\right) + \cos\left(\frac{\lambda_i x}{l}\right) - \sigma_i\left(\sinh\left(\frac{\lambda_i x}{l}\right) + \sin\left(\frac{\lambda_i x}{l}\right)\right)\right), \quad (6.11)$$

where $i = 1, 2, 3,\ldots$ is the mode shape number and the constants λ_i and σ_i for the first two modes are provided in Table 6.4. The first two mode shapes, which have been normalized to the end of the beam ($x = 0$), are displayed in Fig. 6.16.

We see in Fig. 6.16 that there are points where the mode shapes change sign and therefore pass through zero. These points are referred to as *nodes* and identify locations along the beam where the vibration response (at the corresponding natural frequency) is zero regardless of the force magnitude. Note, however, that the locations of the two nodes for the first mode, which describes the beam's shape for vibration at the first natural frequency, do not coincide with the locations of the three nodes for the second mode, which corresponds to vibration at the second natural frequency.[7] Because the beam's response to an external force is the superposition of all the modes, there is still motion at all points along the beam, in general, when it is excited.

For the steel rod from the BEP (12.7 mm diameter, 153 mm long), let's assume that we placed the beam on a very flexible support, such as a block of soft foam, to mimic free-free boundary conditions[8] and then measured the direct FRF at location 1 and the cross FRFs at locations 2 through 4. These locations are identified in Fig. 6.17, where 2 (at $x/l = 0.132$) and 4 (at $x/l = 0.5$) are nodes for ψ_2 and 3 (at $x/l = 0.224$)

---

[7] The number of nodes is equal to $i + 1$ for a free-free beam.

[8] Because the foam base is much more flexible than the beam, free-free conditions are approximated. Alternately, we could support the beam using flexible bungee cords. Both techniques are applied in practice.

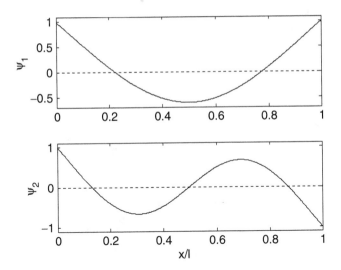

Fig. 6.16 The first two bending mode shapes for a free-free beam

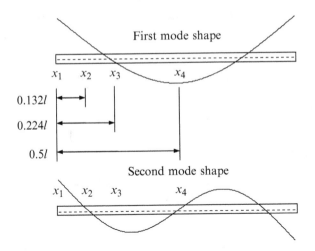

Fig. 6.17 Direct and cross FRF measurement locations for the free-free steel rod

is a node for ψ_1. If we applied the force at location 1 for each measurement, then the four FRFs would be X_1/F_1, X_2/F_1, X_3/F_1, and X_4/F_1. The imaginary parts of these FRFs are displayed in Fig. 6.18, where the measurement bandwidth only includes the first two modes.

For the first mode at a natural frequency of 2,446 Hz, we notice two distinct behaviors. First, the peak height at location 3 is zero. This demonstrates an important consideration for experimental identification of mode shapes – if the transducer is positioned or the force is applied at or near a node, then the response will have a magnitude that is close to (or equal to) zero for the mode shape that includes that node at the selected measurement location. Because the node locations are not known in advance, many measurements are typically required on a given structure to

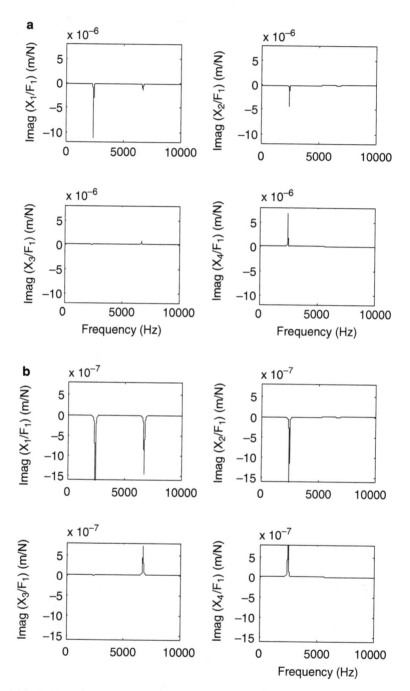

Fig. 6.18 (a) The imaginary parts of the four direct and cross FRFs for the free-free steel rod viewed at full scale. (b) The imaginary parts of the four direct and cross FRFs for the free-free steel rod viewed at a reduced scale to show the second mode

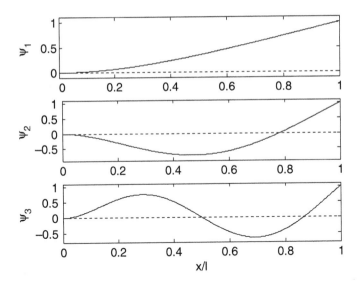

Fig. 6.19 The first three bending mode shapes for a fixed-free beam

Table 6.5 Constants for the fixed-free beam mode shape calculation in Eq. 6.12 (Blevins 2001)

Mode i	λ_i	σ_i
1	1.87510407	0.734095514
2	4.69409113	1.018467319
3	7.85475744	0.999224497

adequately describe all the mode shapes within the measurement bandwidth. Second, the sign of the peak height switches between locations 2 and 4. The motion at these points is out of phase when the beam vibrates in the first mode shape. For the second mode at 6,741 Hz, we see a zero response at both locations 2 and 4. Also, the peak heights switch sign between locations 1 and 3. These sign switches and height variations with measurement location correspond directly to the eigenvector identification approach we detailed in Sect. 6.3.

Next, let's insert the rod in the BEP with an overhang length of 130 mm so that it now has fixed-free boundary conditions (it is a cantilever beam). The first three mode shapes are shown in Fig. 6.19.[9] These are described using the mode shape function in Eq. 6.12. The mode shapes are normalized to the beam's free end and the constants λ_i and σ_i ($i = 1$ to 3) are provided in Table 6.5.

$$\psi_i = \frac{1}{2}\left(\cosh\left(\frac{\lambda_i x}{l}\right) - \cos\left(\frac{\lambda_i x}{l}\right) - \sigma_i\left(\sinh\left(\frac{\lambda_i x}{l}\right) - \sin\left(\frac{\lambda_i x}{l}\right)\right)\right) \qquad (6.12)$$

---

[9] The number of nodes is equal to $i - 1$ for a fixed-free beam.

Fig. 6.20 Direct and cross
FRF measurement locations
for the BEP fixed-free
steel rod

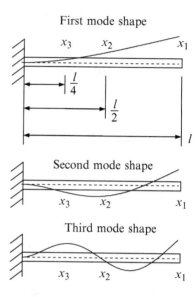

First mode shape

Second mode shape

Third mode shape

We will select three measurement locations along the beam's axis, as shown
in Fig. 6.20: $x_1 = l$, $x_2 = l/2$, and $x_3 = l/4$ (this location is near a node for the
third mode shape); the force is applied at x_1. The imaginary parts of the direct and
cross FRFs are displayed in Fig. 6.21a and b. Because there are three modes within
the measurement bandwidth and we performed measurements at three locations,
we can identify the eigenvectors and modal matrix for the three degree of freedom
system. This is actually a requirement for modal analysis. The number of measurement locations and therefore the number of direct and cross FRFs must always be
equal to the number of degrees of freedom (i.e., the number of modes selected for
modeling). This produces a square modal matrix.[10] Using the peak heights A
through I identified in Fig. 6.21, the modal matrix is:

$$P = \begin{bmatrix} \dfrac{A}{A} = 1 & \dfrac{B}{B} = 1 & \dfrac{C}{C} = 1 \\[2mm] \dfrac{D}{A} & \dfrac{E}{B} & \dfrac{F}{C} \\[2mm] \dfrac{G}{A} & \dfrac{H}{B} & \dfrac{I}{C} \end{bmatrix} = \begin{bmatrix} 1 & 1 & 1 \\ 0.3395 & -0.7137 & 0.0197 \\ 0.0973 & -0.4173 & 0.7245 \end{bmatrix}, \quad (6.13)$$

where the eigenvectors are normalized to the beam's free end.

---

[10] In a square matrix, the number of rows and columns is the same.

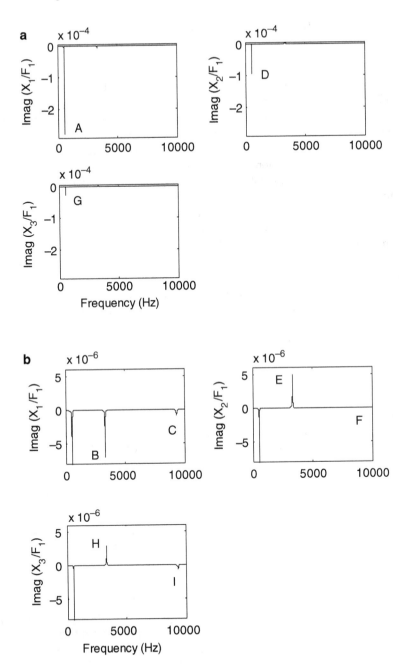

Fig. 6.21 (a) The imaginary parts of the three direct and cross FRFs for the BEP fixed-free steel rod (*full scale*). The peak heights *A*, *D*, and *G* are labeled. (b) The imaginary parts of the three direct and cross FRFs for the BEP fixed-free steel rod (*reduced scale*). The peak heights *B, C, E, F*, *H*, and *I* are labeled

6.6   Shortcut Method for Determining Mass, Stiffness, and Damping Matrices

So far, we have only discussed the chain-type spring–mass–damper model shown in Fig. 6.7 for the backward problem solution. Of course, there may be situations where the structure's design and/or behavior warrant another model type. For example, an automobile's suspension response may be better described using the model shown in Fig. 6.22, which includes a rigid, massless bar, two concentrated masses, two degrees of freedom, and two springs and viscous dampers.

Let's next describe a shortcut method that can be applied to determine the local coordinate mass, damping, and stiffness matrices for lumped-parameter models, such as those shown in Figs. 6.7 and 6.22. We will begin with Fig. 6.7 since we have already identified the matrices using the free body diagrams. In the shortcut method, the force required to give a unit acceleration to the coordinate in question, while holding the other coordinate(s) motionless, is used to determine the on-diagonal terms in the mass matrix. The force that is required to hold the other coordinate(s) motionless defines the off-diagonal terms. Similarly, the damping matrix is determined using the force required to give a unit velocity and the stiffness matrix is identified using the force required to give a unit displacement.

The mass matrix for the two degree of freedom model has four terms: two on-diagonals, m_{11} and m_{22}, and two off-diagonals, m_{12} and m_{21}.

$$m = \begin{bmatrix} m_{11} & m_{12} \\ m_{21} & m_{22} \end{bmatrix}$$

To determine the mass matrix, we neglect the influence of the springs, dampers, and external forces. Let's begin with m_{11} on-diagonal term. We need to find the

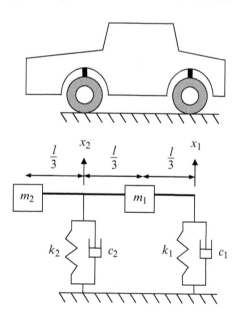

Fig. 6.22 Potential model for describing an automobile's suspension behavior

Fig. 6.23 Force balance for determining m_{11}

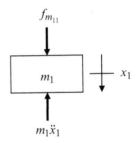

Fig. 6.24 Force balance for determining m_{12}

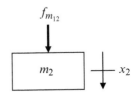

Fig. 6.25 Force balance for determining m_{22}

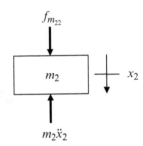

force, $f_{m_{11}}$, required to give coordinate x_1 a unit acceleration, $\ddot{x}_1 = 1$, while holding the other coordinate, x_2, motionless. When we remove the springs and dampers, the two coordinates are not connected, so the analysis is simple. The forces are shown in Fig. 6.23. Summing the forces in the x_1 direction to zero gives:

$$f_{m_{11}} = m_1\ddot{x}_1 = m_1(1) = m_1.$$

The m_{11} on-diagonal term is, therefore, m_1. The m_{12} off-diagonal term is the force, $f_{m_{12}}$, required to hold x_2 stationary while giving x_1 a unit acceleration. In the absence of the springs and dampers, no force is necessary to restrict x_2, as shown in Fig. 6.24. Therefore, the force summation in the x_2 direction gives:

$$f_{m_{12}} = 0.$$

We find the other on-diagonal term, m_{22}, by applying unit acceleration to x_2 while holding x_1 motionless. Using Fig. 6.25, the required force is:

$$f_{m_{22}} = m_2\ddot{x}_2 = m_2(1) = m_2.$$

Fig. 6.26 Force balance for
determining m_{21}

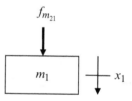

Fig. 6.27 Force balance for
determining k_{11}

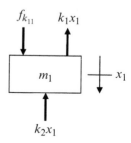

The m_{21} off-diagonal term is the force, $f_{m_{21}}$, required to hold x_1 fixed while giving x_2 a unit acceleration. It is calculated using Fig. 6.26. We find that $f_{m_{21}} = 0$. We did not actually need to calculate $f_{m_{21}}$, however. Because the mass matrix is symmetric, we know that $f_{m_{21}} = f_{m_{12}}$. However, it serves as a good check to ensure that our $f_{m_{12}}$ calculation was correct. We can now populate the mass matrix for Fig. 6.7.

$$m = \begin{bmatrix} m_{11} & m_{12} \\ m_{21} & m_{22} \end{bmatrix} = \begin{bmatrix} m_1 & 0 \\ 0 & m_2 \end{bmatrix}$$

Of course, this is the same result we obtained previously. To determine the stiffness matrix, we neglect the masses, dampers, and external forces.

$$k = \begin{bmatrix} k_{11} & k_{12} \\ k_{21} & k_{22} \end{bmatrix}$$

To find k_{11}, we need the force, $f_{k_{11}}$, required to give coordinate x_1 a unit displacement, $x_1 = 1$, while holding the other coordinate, x_2, motionless. The free body diagram is provided in Fig. 6.27. Summing the forces in the x_1 direction gives:

$$f_{k_{11}} = k_1 x_1 + k_2 x_1 = k_1(1) + k_2(1) = k_1 + k_2.$$

The off-diagonal term, k_{12}, is the force required to hold x_2 stationary while giving x_1 a unit displacement. Using Fig. 6.28, we see that:

$$f_{k_{12}} = -k_2 x_1 = -k_2(1) = -k_2.$$

Fig. 6.28 Force balance for
determining k_{12}

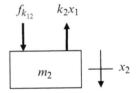

Fig. 6.29 Force balance for
determining k_{22}

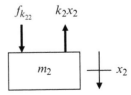

Fig. 6.30 Force balance for
determining k_{21}

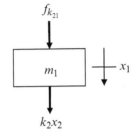

We determine k_{22} from the force, $f_{k_{22}}$, required to give coordinate x_2 a unit displacement while holding x_1 stationary. The free body diagram is displayed in Fig. 6.29. The force summation in the x_2 direction gives:

$$f_{k_{22}} = k_2 x_2 = k_2(1) = k_2.$$

Finally, we find k_{21} from the force required to hold x_1 fixed while giving x_2 a unit displacement. See Fig. 6.30, where the force balance requires that $f_{k_{21}} = -k_2$, which satisfies our symmetry requirement. The system stiffness matrix is:

$$k = \begin{bmatrix} k_{11} & k_{12} \\ k_{21} & k_{22} \end{bmatrix} = \begin{bmatrix} k_1 + k_2 & -k_2 \\ -k_2 & k_2 \end{bmatrix},$$

which agrees with our previous result using the free body diagrams. We could complete the analysis for damping by applying a unit velocity and neglecting the springs, masses, and external forces, but we can see from Fig. 6.7 that the dampers are located at the same physical locations as the springs. Therefore, the damping matrix will have the same form as the stiffness matrix.

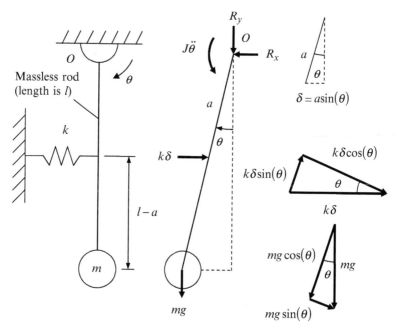

Fig. 6.31 Linearized pendulum and its free body diagram

6.6.1 Linearized Pendulum

Let's next use the shortcut method for the *linearized pendulum*. We will begin with the traditional free body diagram analysis and then follow it with the new shortcut method. The pendulum, relevant geometry, and free body diagram are shown in Fig. 6.31. As seen in Sect. 2.5.2, we can sum the moments, M, about O using the force components that are perpendicular to the massless rod. Because we are summing about O, the reaction components at the pivot may be neglected. Also, the free body diagram includes d'Alembert's inertial moment, $J\ddot{\theta} = ml^2\ddot{\theta}$, with the mass moment of inertia, J, so that $\sum M_O = 0$.

The moment sum about O from Fig. 6.31 is:

$$\sum M_O = J\ddot{\theta} + k\delta \cos(\theta) \cdot a + mg \sin(\theta) \cdot l = 0. \tag{6.14}$$

Substituting $J = ml^2$ and $\delta = a \sin(\theta)$ gives:

$$ml^2\ddot{\theta} + ka^2 \cos(\theta) \sin(\theta) + mgl \sin(\theta) = 0. \tag{6.15}$$

For small angles, we can approximate using $\sin(\theta) \approx \theta$ and $\cos(\theta) \approx 1$. Substitution yields:

$$ml^2\ddot{\theta} + ka^2\theta + mgl\theta = ml^2\ddot{\theta} + (ka^2 + mgl)\theta = 0. \tag{6.16}$$

Fig. 6.32 Relationship between small angles, θ, and horizontal displacements, x, of the pendulum mass

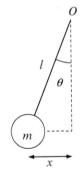

Fig. 6.33 Force balance to determine the mass term

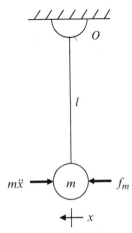

We can convert to small horizontal displacements, x, of the pendulum mass by again applying the small angle approximation. See Fig. 6.32, where $x = l\sin(\theta) \approx l\theta$ and, subsequently, $\ddot{x} \approx l\ddot{\theta}$ for a constant rod length. Rewriting Eq. 6.16 in terms of x gives:

$$ml^2\frac{\ddot{x}}{l} + \left(ka^2 + mgl\right)\frac{x}{l} = ml\ddot{x} + \left(\frac{ka^2}{l} + mg\right)x = m\ddot{x} + \left(\frac{ka^2}{l^2} + \frac{mg}{l}\right)x = 0. \quad (6.17)$$

Now let's use the shortcut method. To find the mass term for the single degree of freedom equation of motion, we need to determine the force, f_m, required to give the pendulum mass a unit (horizontal) acceleration. Figure 6.33 shows the forces, where the spring is neglected. Summing the moments about O gives:

$$\sum M_O = f_m l - m\ddot{x}l = f_m l - m(1)l = f_m l - ml = 0. \quad (6.18)$$

Simplifying Eq. 6.18 gives $f_m = m$. To determine the stiffness term, we find the force, f_k, required to give the pendulum mass a unit displacement. In this case,

Fig. 6.34 Force balance to determine the stiffness term

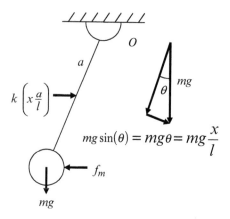

we neglect the inertial influence of the mass in the free body diagram but must include the gravitational effects; see Fig. 6.34. Summing the moments about O gives:

$$\sum M_O = f_k l - k\left(x\frac{a}{l}\right)a - mg\frac{x}{l}l = f_k l - k\left(\frac{a}{l}\right)a - mg = f_k l - k\frac{a^2}{l} - mg. \quad (6.19)$$

Note that the displacement at the spring is not $x = 1$. Rather, it is scaled by the distance from the pivot O and is $x(a/l) = a/l$ for a unit displacement x. Solving for f_k from Eq. 6.19 gives the stiffness term:

$$f_k = \frac{ka^2}{l^2} + \frac{mg}{l}. \quad (6.20)$$

As expected, substitution in the differential equation of motion gives the same result that we obtained from the free body diagram approach; see Eq. 6.17.

6.6.2 Automobile Suspension Model

Let's conclude this section (and chapter) by determining the mass, damping, and stiffness matrices for the automobile suspension model shown in Fig. 6.22. We will begin with the 2×2 mass matrix for the two degree of freedom system. We determine m_{11} from the force, $f_{m_{11}}$, required to apply a unit acceleration to x_1 while holding x_2 stationary. The force balance is shown in Fig. 6.35, where the motionless x_2 is represented as a pivot for the rigid, massless bar. The accelerations at the two masses are both $\ddot{x}_1/2 = 1/2$ because they are half the distance from the pivot relative to x_1. Summing the moments (clockwise moments are taken to be positive) about the x_2 pivot gives:

$$\sum M = -f_{m_{11}} \cdot \frac{2l}{3} + m_1 \frac{\ddot{x}_1}{2} \cdot \frac{l}{3} + m_2 \frac{\ddot{x}_1}{2} \cdot \frac{l}{3}$$

$$= -f_{m_{11}} \cdot \frac{2l}{3} + m_1 \frac{1}{2} \cdot \frac{l}{3} + m_2 \frac{1}{2} \cdot \frac{l}{3} = 0. \quad (6.21)$$

Fig. 6.35 Force balance to determine m_{11}

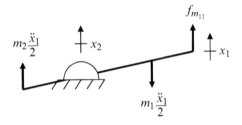

Fig. 6.36 Force balance to determine m_{12}

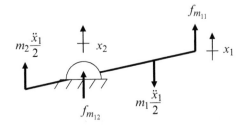

Dividing by $l/3$ and solving for $f_{m_{11}}$ yield:

$$f_{m_{11}} = \frac{m_1 + m_2}{4}. \tag{6.22}$$

The off-diagonal term m_{12} is the force that is necessary to hold x_2 motionless. The forces are displayed in Fig. 6.36. The force summation is:

$$\sum f = f_{m_{11}} + f_{m_{12}} - m_1 \frac{\ddot{x}_1}{2} + m_2 \frac{\ddot{x}_1}{2} = f_{m_{11}} + f_{m_{12}} - m_1 \frac{1}{2} + m_2 \frac{1}{2} = 0. \tag{6.23}$$

Solving Eq. 6.23 for $f_{m_{12}}$ gives:

$$f_{m_{12}} = -f_{m_{11}} + \frac{m_1}{2} - \frac{m_2}{2}. \tag{6.24}$$

Substituting for $f_{m_{11}}$ from Eq. 6.22 gives the final expression for $f_{m_{12}}$:

$$f_{m_{12}} = -\left(\frac{m_1 + m_2}{4}\right) + \frac{m_1}{2} - \frac{m_2}{2} = \frac{m_1 - 3m_2}{4}. \tag{6.25}$$

We will next find m_{22} by applying unit acceleration to x_2 while holding x_1 stationary. The corresponding forces are shown in Fig. 6.37. The moment summation about the x_1 pivot is:

$$\sum M = f_{m_{22}} \cdot \frac{2l}{3} - m_1 \frac{\ddot{x}_1}{2} \cdot \frac{l}{3} - m_2 \frac{3\ddot{x}_1}{2} \cdot \frac{3l}{3} = f_{m_{22}} \cdot \frac{2l}{3} - m_1 \frac{1}{2} \cdot \frac{l}{3} - m_2 \frac{3}{2} \cdot \frac{3l}{3} = 0. \tag{6.26}$$

Fig. 6.37 Force balance to
determine m_{22}

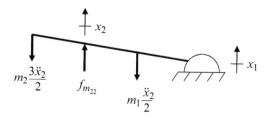

Fig. 6.38 Force balance to
determine m_{21}

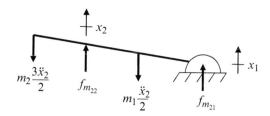

Dividing by $l/3$ and solving for $f_{m_{22}}$ lead to Eq. 6.27:

$$f_{m_{22}} = \frac{m_1 + 9m_2}{4} \qquad (6.27)$$

Let's find m_{21} as a check on our m_{12} result. It is the force required to hold x_1 fixed while applying the unit acceleration to x_2. The forces are displayed in Fig. 6.38 and the force summation gives:

$$\sum f = f_{m_{21}} + f_{m_{22}} - m_1 \frac{\ddot{x}_2}{2} + m_2 \frac{3\ddot{x}_2}{2} = f_{m_{21}} + f_{m_{22}} - m_1 \frac{1}{2} - m_2 \frac{3}{2} = 0. \qquad (6.28)$$

Solving for $f_{m_{21}}$ and substituting for $f_{m_{22}}$ yield the m_{21} value; see Eq. 6.29. As anticipated, it matches the m_{12} result provided in Eq. 6.25. The mass matrix is provided in Eq. 6.30. We note that it is symmetric and, because the off-diagonals are nonzero, the equations of motion are coupled through the mass matrix.

$$f_{m_{21}} = -\left(\frac{m_1 + 9m_2}{4}\right) + \frac{m_1}{2} + \frac{3m_2}{2} = \frac{m_1 - 3m_2}{4} \qquad (6.29)$$

$$m = \begin{bmatrix} m_{11} & m_{12} \\ m_{21} & m_{22} \end{bmatrix} = \begin{bmatrix} \dfrac{m_1 + m_2}{4} & \dfrac{m_1 - 3m_2}{4} \\ \dfrac{m_1 - 3m_2}{4} & \dfrac{m_1 + 9m_2}{4} \end{bmatrix} \qquad (6.30)$$

Now let's populate the stiffness matrix. For the k_{11} on-diagonal term, we need the force required to give a unit displacement to x_1 while holding x_2 stationary. The forces are shown in Fig. 6.39. The moment sum about the motionless x_2 is given by:

$$\sum M = -f_{k_{11}} \cdot \frac{2l}{3} + k_1 x_1 \cdot \frac{2l}{3} = -f_{k_{11}} \cdot \frac{2l}{3} + k_1 \cdot \frac{2l}{3} = 0. \qquad (6.31)$$

Dividing by $2l/3$ and solving for $f_{k_{11}}$ give the stiffness k_{11}.

Fig. 6.39 Force balance to determine k_{11}

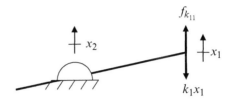

Fig. 6.40 Force balance to determine k_{12}

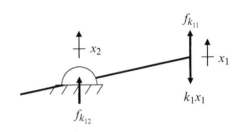

Fig. 6.41 Force balance to determine k_{22}

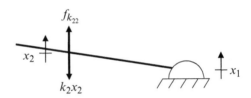

$$f_{k_{11}} = k_1 \tag{6.32}$$

The off-diagonal term, k_{12}, is the force required to hold x_2 motionless. See Fig. 6.40, where the force summation yields:

$$\sum f = f_{k_{11}} + f_{k_{12}} - k_1 x_1 = f_{k_{11}} + f_{k_{12}} - k_1 = 0. \tag{6.33}$$

Substituting k_1 for $f_{k_{11}}$ in Eq. 6.33 enables us to determine the k_{12} stiffness.

$$f_{k_{12}} = 0 \tag{6.34}$$

The other on-diagonal term, k_{22}, is determined from the force required to give a unit displacement to x_2 while holding x_1 fixed. The forces are displayed in Fig. 6.41. Summing the moments about x_1 gives:

$$\sum M = f_{k_{22}} \cdot \frac{2l}{3} - k_2 x_2 \cdot \frac{2l}{3} = f_{k_{22}} \cdot \frac{2l}{3} - k_2 \cdot \frac{2l}{3} = 0. \tag{6.35}$$

Fig. 6.42 Force balance to
determine k_{21}

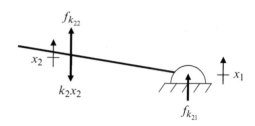

We solve Eq. 6.35 for $f_{k_{22}}$ to determine k_{22}.

$$f_{k_{22}} = k_2 \qquad (6.36)$$

The off-diagonal term, k_{21}, is the force required to hold x_1 stationary. Summing the forces in Fig. 6.42 and substituting for $f_{k_{22}}$ provide the desired result.

$$\sum f = f_{k_{21}} + f_{k_{22}} - k_2 x_2 = f_{k_{21}} + f_{k_{22}} - k_2 = 0$$
$$f_{k_{21}} = -f_{k_{22}} + k_2 = -k_2 + k_2 = 0 \qquad (6.37)$$

The final stiffness matrix is:

$$k = \begin{bmatrix} k_{11} & k_{12} \\ k_{21} & k_{22} \end{bmatrix} = \begin{bmatrix} k_1 & 0 \\ 0 & k_2 \end{bmatrix}. \qquad (6.38)$$

Because the dampers appear at the same locations as the springs, the damping matrix is:

$$c = \begin{bmatrix} c_{11} & c_{12} \\ c_{21} & c_{22} \end{bmatrix} = \begin{bmatrix} c_1 & 0 \\ 0 & c_2 \end{bmatrix}. \qquad (6.39)$$

Chapter Summary

- In practice, it is commonly required that we have an actual dynamic system and would like to build a model that we can use to represent its vibratory behavior.
- The first step in the "backward problem" of starting with a measurement and developing a model is identifying the modal parameters for each of the modes selected for fitting.
- We can use peak picking to determine the modal parameters from the real and imaginary parts of a measured FRF.

- For a measured direct FRF, each mode can be fit independently to find the modal parameters.
- We assume that proportional damping holds when performing the modal fits to the measured FRF.
- We determine the eigenvectors from the ratios of the peak heights between the cross and direct FRFs. The imaginary parts are used to find the peak heights.
- For structures with free-free boundary conditions, rigid body modes exist.
- Nodes are points of zero deflection on the mode shapes.
- The number of direct and cross FRFs must be equal to the number of modeled modes (degrees of freedom) in order to obtain a square modal matrix.
- Alternative lumped-parameter model types were introduced.
- A shortcut method for identifying the mass, damping, and stiffness matrices for lumped-parameter models was described. The on-diagonal and off-diagonal terms were treated separately.

Exercises

1. For a single-degree-of-freedom spring–mass–damper system subject to forced harmonic vibration, the measured FRF is displayed in Figs. P6.1a and P6.1b. Using the peak picking method, determine m (in kg), k (in N/m), and c (in N-s/m).

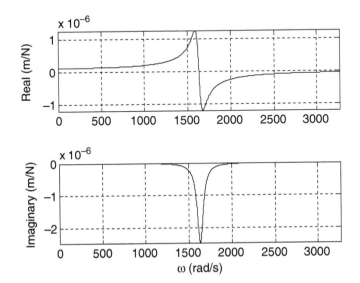

Fig. P6.1a Measured FRF

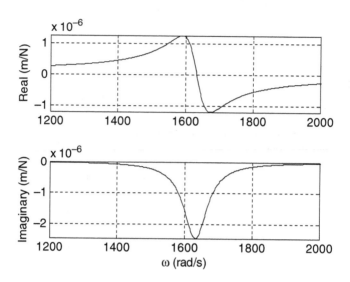

Real (m/N)

x 10^{-6}

1200   1400   1600   1800   2000

Imaginary (m/N)

x 10^{-6}

1200   1400   1600   1800   2000
ω (rad/s)

Fig. P6.1b   Measured FRF (smaller frequency scale)

2. The direct and cross FRFs for the two degree of freedom system shown in
   Fig. P6.2a are provided in Figs. P6.2b and P6.2c.

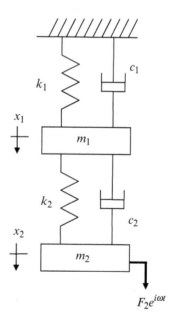

Fig. P6.2a   Two degree of freedom spring-mass-damper system under forced vibration

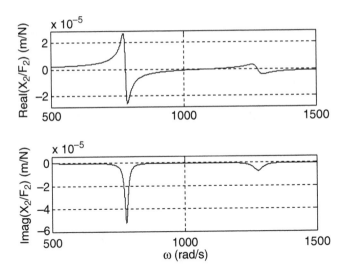

Fig. P6.2b Direct FRF X_2/F_2

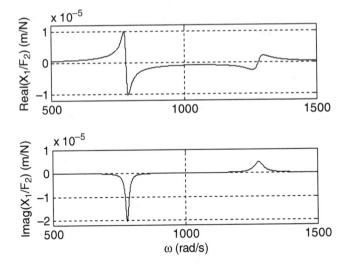

Fig. P6.2c Cross FRF X_1/F_2

(a) If the modal damping ratios are $\zeta_{q1} = 0.01$ and $\zeta_{q2} = 0.016$, determine the modal stiffness values k_{q1} and k_{q2} (N/m) by peak picking.

(b) Determine the mode shapes by peak picking.

3. An FRF measurement was completed to give the two degree of freedom response shown in Fig. P6.3. Use the peak picking approach to identify the modal mass, stiffness, and damping parameters for the two modes. Arrange your results in the 2×2 modal matrices m_q, c_q, and k_q.

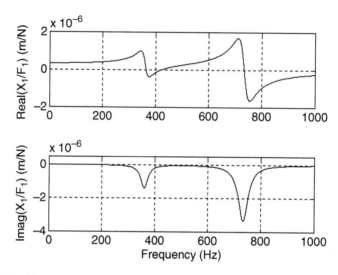

Fig. P6.3 Measured direct FRF for two degree of freedom system

4. An FRF measurement was completed to give the two degree of freedom response shown in Fig. P6.4 (a limited frequency range is displayed to aid in the peak picking activity).

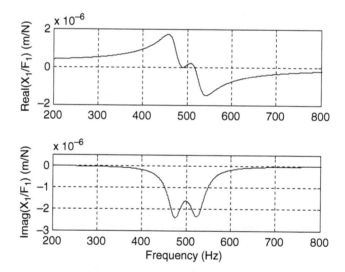

Fig. P6.4 FRF measurement for two degree of freedom system

(a) Use the peak picking approach to identify the modal mass, stiffness, and damping parameters for the two modes. Arrange your results in the 2×2 modal matrices m_q, c_q, and k_q.

(b) The FRF "measurement" in part (a) was defined using the following MATLAB® code.

```
kq1 = 6e6;                    % N/m
kq2 = 6e6;                    % N/m
omega_n1 = 475*2*pi;         % rad/s
omega_n2 = 525*2*pi;
zetaq1 = 0.04;
zetaq2 = 0.04;

omega = 0:1000*2*pi;         % rad/s
r1 = omega/omega_n1;
r2 = omega/omega_n2;

realQ1_R1 = 1/kq1*(1-r1.^2)./((1-r1.^2).^2 + (2*zetaq1*r1).^2);
imagQ1_R1 = 1/kq1*(-2*zetaq1*r1)./((1-r1.^2).^2 + (2*zetaq1*r1).^2);

realQ2_R2 = 1/kq2*(1-r2.^2)./((1-r2.^2).^2 + (2*zetaq2*r2).^2);
imagQ2_R2 = 1/kq2*(-2*zetaq2*r2)./((1-r2.^2).^2 + (2*zetaq2*r2).^2);

realX1_F1 = realQ1_R1 + realQ2_R2;
imagX1_F1 = imagQ1_R1 + imagQ2_R2;

freq = omega/2/pi;

figure(1)
subplot(211)
plot(freq, realX1_F1, 'k')
set(gca,'FontSize', 14)
axis([200 800 -2e-6 2e-6])
ylabel('Real({X_1}/{F_1}) (m/N)')
grid
subplot(212)
plot(freq, imagX1_F1, 'k')
set(gca,'FontSize', 14)
axis([200 800 -3e-6 5e-7])
xlabel('Frequency (Hz)')
ylabel('Imag({X_1}/{F_1}) (m/N)')
grid
```

Plot your modal fit together with the measured FRF and comment on their agreement.

5. Figures P6.5a through P6.5e show direct, X_1/F_1, and cross FRFs, X_2/F_1 through X_5/F_1, measured on a fixed-free beam. They were measured at the beam's free end and in 20 mm increments toward its base; see Fig. 6.36. Determine the mode shape associated with the 200 Hz natural frequency.

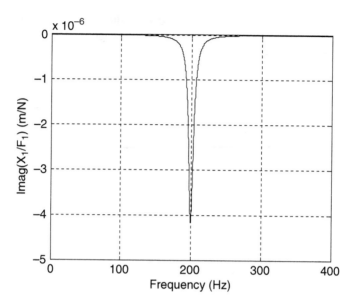

Fig. P6.5a Direct FRF X_1/F_1

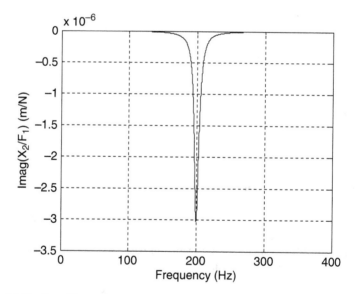

Fig. P6.5b Cross FRF X_2/F_1

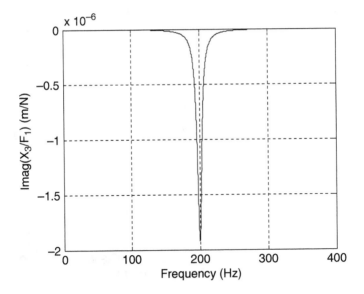

Fig. P6.5c Cross FRF X_3/F_1

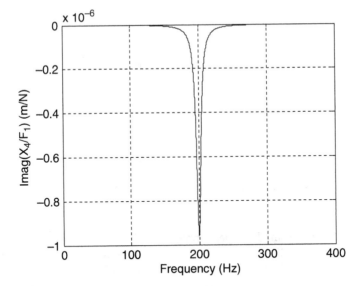

Fig. P6.5d Cross FRF X_4/F_1

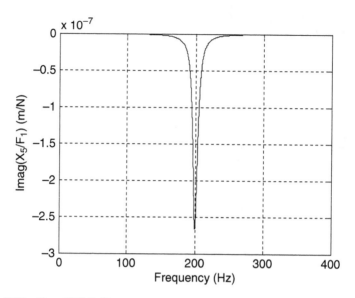

Fig. P6.5e Cross FRF X_5/F_1

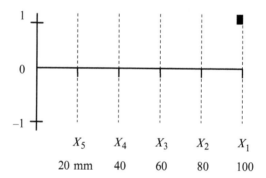

Fig. P6.5f Coordinates for direct and cross FRF measurements

When plotting the mode shape, normalize the free end response at coordinate x_1 to 1 (this is normalizing the mode shape to x_1) and show the relative amplitudes at the other coordinates x_2 through x_5. See Fig. P6.5f.

6. For the same fixed-free beam as Problem 5, the measurement bandwidth was increased so that the first three modes were captured. Again, the direct, X_1/F_1, and cross FRFs, X_2/F_1 through X_5/F_1, were measured. The imaginary part of the direct FRF for the entire bandwidth is shown in Fig. P6.6a. The three natural frequencies are 200, 550, and 1,250 Hz.

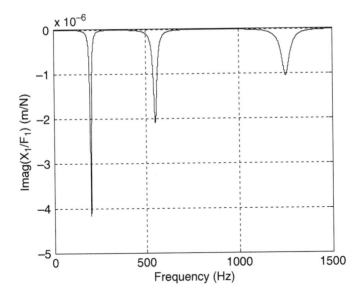

Fig. P6.6a Direct FRF X_1/F_1

Use Figs. P6.6b through P6.6f to identify the mode shape that corresponds to the 1,250 Hz natural frequency. Plot your results using the same approach described in Problem 5.

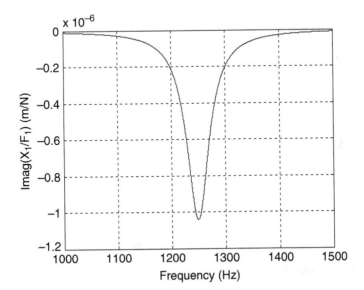

Fig. P6.6b Direct FRF X_1/F_1 (mode 3 only)

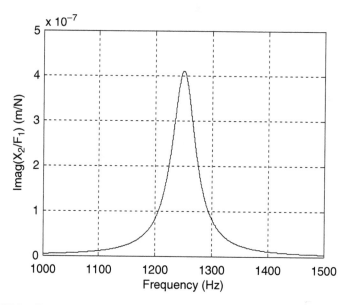

Fig. P6.6c Cross FRF X_2/F_1

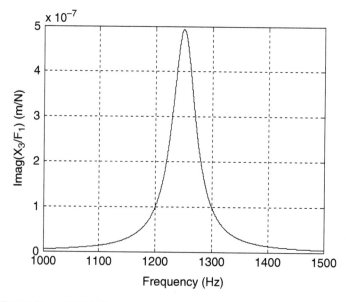

Fig. P6.6d Cross FRF X_3/F_1

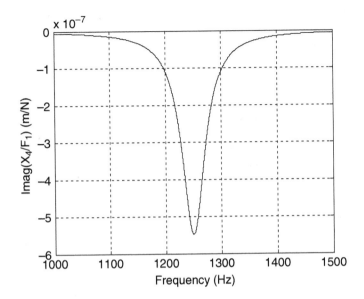

Fig. P6.6e Cross FRF X_4/F_1

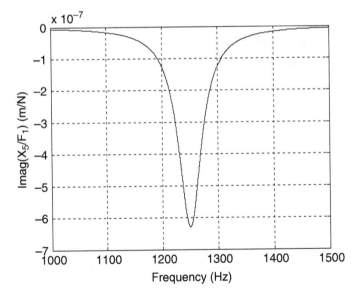

Fig. P6.6f Cross FRF X_5/F_1

7. Find the mass matrix (in local coordinates) for the two degree of freedom system displayed in Fig. P6.7 using the shortcut method described in Sect. 6.6 if $m_1 = 10$ kg and $m_2 = 12$ kg.

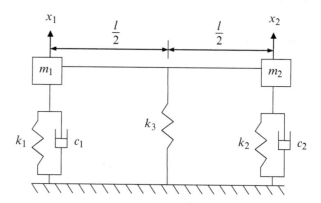

Fig. P6.7 Two degree of freedom system with a rigid, massless bar connecting the two masses, m_1 and m_2

8. Find the stiffness matrix (in local coordinates) for the two degree of freedom system displayed in Fig. P6.7 using the shortcut method described in Sect. 6.6 if $k_1 = 2 \times 10^5$ N/m, $k_2 = 2 \times 10^5$ N/m, and $k_3 = 1 \times 10^5$ N/m.

9. Find the mass matrix (in local coordinates) for the two degree of freedom system displayed in Fig. P6.9 using the shortcut method described in Sect. 6.6 if $m_1 = 10$ kg and $m_2 = 12$ kg.

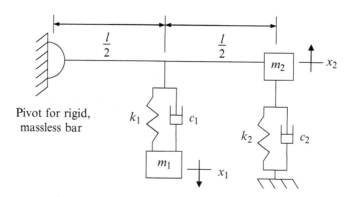

Fig. P6.9 Two degree of freedom system with a rigid, massless bar connecting the mass, m_2, to a fixed pivot

10. Find the stiffness matrix (in local coordinates) for the two degree of freedom system displayed in Fig. P6.9 using the shortcut method described in Sect. 6.6 if $k_1 = 2 \times 10^5$ N/m, $k_2 = 2 \times 10^5$ N/m, and $k_3 = 1 \times 10^5$ N/m.

References

Blevins RD (2001) Formulas for natural frequency and mode shape. Krieger, Malabar (Table 8–1)

Schmitz T, Smith KS (2009) Machining dynamics: Frequency response to improved productivity. Springer, New York

Chapter 7
Measurement Techniques

Man is but a reed, the most feeble thing in nature, but he is a thinking reed.

– Blaise Pascal

7.1 Frequency Response Function Measurement

In Chap. 6, we solved the "backward problem" of starting with frequency response function (FRF) measurements and developing a model. However, we did not describe the measurement procedure. The basic hardware required to measure FRFs is:

- a mechanism for known force input across the desired frequency range (or *bandwidth*)
- a transducer for vibration measurement, again with the required bandwidth
- a *dynamic signal analyzer* to record the time-domain force and vibration inputs and convert these into the desired FRF.

A dynamic signal analyzer includes input channels for the time-domain force and vibration signals and computes the Fourier transform of these signals to convert them to the frequency domain. It then calculates the complex-valued ratio of the frequency-domain vibration signal to the frequency-domain force signal; this ratio is the FRF. A schematic setup of FRF measurement is provided in Fig. 7.1. It includes the time-domain force and vibration (which may take the form of displacement, x, velocity, $\dot{x}$, or acceleration, $\ddot{x}$) inputs and amplifiers for each setup. The amplifiers are used to increase the magnitude of the signals. The force and vibration are continuous in time or *analog*. However, recording these signals with the analyzer requires sampling them at small time intervals, or digitizing them.

T.L. Schmitz and K.S. Smith, *Mechanical Vibrations: Modeling and Measurement*, DOI 10.1007/978-1-4614-0460-6_7, © Springer Science+Business Media, LLC 2012

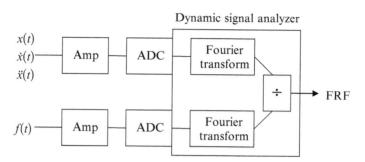

Fig. 7.1 Schematic of FRF measurement setup

This process is completed using an analog-to-digital converter (ADC). These *digital* signals are then used in the FRF calculation by the dynamic signal analyzer. Based on the vibration input type, the FRF may be expressed as:

- receptance or compliance – the ratio of displacement to force
- mobility – the ratio of velocity to force
- accelerance or inertance – the ratio of acceleration to force.

7.2   Force Input

There are three common types of force excitation. These include:

- fixed frequency sine wave – The FRF is determined one frequency at a time. At each frequency within the desired bandwidth, the sinusoidal force is applied, the response to the force input is averaged over a short time interval, and the FRF is calculated. This is referred to as a *sine sweep test*.
- random signal – The frequency content of the random signal may be broad-band (*white noise*) or truncated to a limited range (*pink noise*). Averaging over a fixed period of time is again applied, but all the frequencies within the selected bandwidth are excited in a single test.
- impulse – A short duration impact is used to excite the structure and the corresponding response is measured. This approach enables a broad range of frequencies to be excited in a single, short test. Multiple tests are typically averaged in the frequency domain to improve *coherence* or the correlation between the force and vibration signals.

To generate these different forces, two common types of force input hardware are applied:

- *shaker* (similar to a speaker) – This system includes a harmonically driven armature and a base. The armature may be actuated along its axis by a magnetic coil or hydraulic force. The magnetic coil, or electrodynamics, configurations

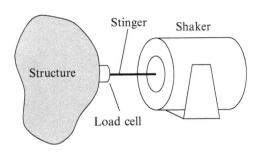

Fig. 7.2 Shaker setup

can provide excitation frequencies of tens of kHz with force levels from tens to thousands of Newton (increased force typically means a lower frequency range). Hydraulic shakers offer high force with the potential for a static preload (i.e., the average, or mean, force is not zero), but relatively lower frequency ranges. In either case, the force is often applied to the structure of interest through a *stinger* or a slender rod that supports axial tension and compression, but not bending or shear. This insures that the force is applied in a single direction only. A load cell is incorporated in the setup to measure the input force. See Fig. 7.2. One consideration is that this load cell adds mass to the system under test, which can alter the FRF for structures with low modal mass values. Finally, the shaker must be isolated from the structure to prevent reaction forces due to the shaker motion from being transmitted through the shaker base to the structure.

- *impact hammer* – An impact hammer incorporates a force transducer located at a metal, plastic, or rubber tip to measure the force input during a hammer strike. When a hammer is used in conjunction with a vibration transducer, the measurement procedure is referred to as *impact testing*. The energy input to the structure is a function of the hammer mass; a larger mass provides more energy (the linear momentum is the product of mass and velocity). Therefore, many sizes are available. Examples are displayed in Fig. 7.3. Also, the excitation bandwidth of the force input depends on the mass and tip stiffness. Stiffer tips tend to excite a wider frequency range, but also spread the input energy over this wider range. Softer tips concentrate the energy over a lower frequency range. This is discussed further in Sect. 7.4. Hard plastic and metal tips provide higher stiffness, while rubber tips give reduced stiffness.

In A NUTSHELL   Impact testing excites the structure with many frequencies, all at the same time. The Fourier transform provides a way to separate them. This type of testing increased in popularity when computers became available to handle the computational load. Prior to that time, shaker measurements dominated.

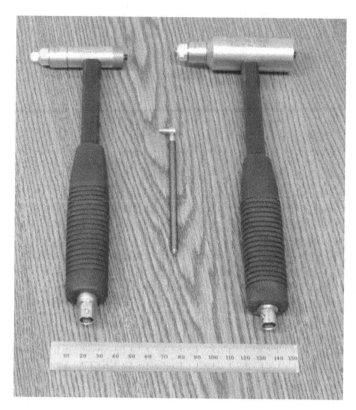

Fig. 7.3 Example impact hammers. A 150 mm steel ruler is included in the photograph to provide scale

7.3   Vibration Measurement

Vibration transducers are available in both noncontact and contact types. While noncontact transducers, such as *capacitance probes* and *laser vibrometers*, are preferred because they do not affect the system dynamics, contacting types, such as *accelerometers*, are often more convenient to implement. As a compromise, low mass accelerometers may be used to minimize the influence on the test structure. They are attached at the location of interest using wax, adhesive, a magnet, or a threaded stud and then removed when the testing is completed. Let's now discuss these different transducers in more detail.

7.3.1   Capacitance Probe

Noncontact capacitive sensors measure changes in capacitance, the ability of a body to hold an electrical charge. When a voltage is applied to two conductors

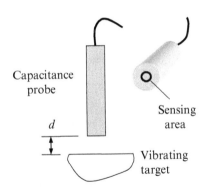

Fig. 7.4 Capacitance probe configuration for vibration measurement

separated by some distance, an electric field is produced between them, and positive and negative charges collect on each conductor. If the polarity of the voltage is reversed, then the charges also reverse.

Capacitive sensors use an alternating voltage which causes the charges to continually reverse their positions. This charge motion generates an alternating electric current which is detected by the sensor. The amount of current flow is determined by the capacitance, which depends on the surface area of the conductors, the distance between them, and the dielectric constant of the material between them (such as air). The capacitance, C, is directly proportional to the surface area, A, and inversely proportional to the distance, d, between them. A larger surface area and a smaller distance produce a larger current. For two parallel plate conductors, the capacitance is given by:

$$C = \varepsilon_r \varepsilon_0 \frac{A}{d}, \tag{7.1}$$

where ε_r is the dielectric constant (or static relative permittivity) and ε_0 is the electric constant (or vacuum permittivity). The value of the dielectric constant is 1 in vacuum and the electric constant is $8.854187817... \times 10^{-12}$ A·s/(V·m).

Typically, the probe is one of the conductors and the measurement target is the other. If the sizes of the sensor and the target and dielectric constant of the material between them are assumed to be constant, then any change in capacitance is due to a change in the distance between the probe and the target (http://www.lionprecision. com/tech-library/technotes/cap-0020-sensor-theory.html). A common capacitance probe configuration is displayed in Fig. 7.4.

IN A NUTSHELL   A capacitance gage requires that the target is a conductor. If the target is not flat and parallel to the tip of the probe, then the distance–capacitance relationship must be calibrated using the target. The capacitance gage makes a relative measurement and there must be a suitable mounting structure for the probe – that is, we do not want to measure the FRF of a flexible probe mount when we are trying to measure the FRF of the structure.

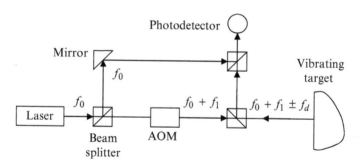

Fig. 7.5 Laser vibrometer schematic

7.3.2 Laser Vibrometer

A vibrometer performs noncontact vibration measurements using the Dopper shift[1] of a laser beam's frequency that occurs due to the motion of the target surface. The vibrometer's output is generally an analog voltage that is directly proportional to the component of the target's velocity in the laser beam direction. A schematic of a vibrometer is shown in Fig. 7.5. In the figure, the laser head emits a single frequency, f_0. A portion of this beam is redirected using a beam splitter and serves as a reference signal. The remainder continues to an acousto-optic modulator (AOM) which upshifts the light frequency to $f_0 + f_1$. When this test beam is reflected from the target surface, it is Doppler shifted by f_d, where the Doppler frequency is directly proportional to the target velocity. The frequency-shifted measurement signal is recombined with the reference signal and the corresponding interference signal is incident on a photodetector. The photodetector current is then used to determine the time-dependent target velocity.

Given the target velocity and excitation force as inputs, the dynamic signal analyzer calculates their frequency-domain ratio, or mobility, $\frac{V}{F}(\omega)$, as the output. To convert from mobility to receptance, $\frac{X}{F}(\omega)$, we simply divide by the product, $i\omega$. To understand this frequency-domain integration, let's write the harmonic displacement as $x(t) = Xe^{i\omega t}$. The corresponding velocity is $\dot{x}(t) = i\omega Xe^{i\omega t} = i\omega \cdot x(t)$. The conversion from mobility to receptance is therefore given by:

$$\frac{X}{F}(\omega) = \frac{X}{V}\frac{V}{F} = \frac{1}{i\omega}\frac{V}{F}. \tag{7.2}$$

---

[1] You may recognize this Doppler frequency shift as the increase in the pitch (frequency) of an approaching automobile's horn and subsequent drop in pitch after the automobile passes you.

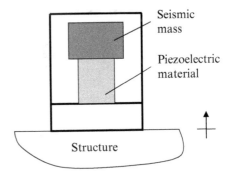

Fig. 7.6 Accelerometer schematic

7.3.3 Accelerometer

Accelerometers for structural vibration measurement typically use the *piezoelectric*[2] effect to generate a voltage signal that is proportional to acceleration. A schematic is provided in Fig. 7.6. An accelerometer includes, at minimum, a seismic mass, piezoelectric material, and package that is connected to the structure under test. The piezoelectric material may be quartz, tourmaline, barium titanate ($BaTiO_3$), or lead zirconate titanate ($Pb[Zr_xTi_{1-x}]O_3$, $0 < x < 1$, or PZT) and produces a charge when strained by the inertial force applied by the mass during motion of the base. The corresponding voltage is equal to this charge divided by the piezoelectric material's capacitance. The output voltage is proportional to the inertial force and, therefore, the acceleration.[3] As shown in Fig. 7.7, the voltage is carried to the dynamic signal analyzer by a cable, which can have a capacitance on the same order as the piezoelectric material. The increased capacitance decreases the voltage for a given charge. In order to eliminate this effect (and reduce the measurement noise), an amplifier is usually located within the accelerometer package. Example accelerometers are pictured in Fig. 7.7.

If we think of the piezoelectric material as having finite stiffness and damping which resist the deformation (strain) imposed by the seismic mass, then we can represent the accelerometer as shown in Fig. 7.8. This emphasizes that the accelerometer is a dynamic system itself with its own natural frequency that could affect the measurement result. For this reason, accelerometers are designed to have high natural frequencies (nearly 100 kHz for low mass versions).

The spring–mass–damper accelerometer representation shown in Fig. 7.9 can be used to determine the equation of motion due to motion of the base structure to which the accelerometer is attached. In the figure, m is the seismic mass, k and c are the piezoelectric material's spring stiffness and viscous damping coefficient, y is the test structure displacement, and x is the seismic mass displacement. The free body

---

[2] The prefix piezo is derived from the Greek word *piezein*, which translates "to squeeze."

[3] This follows from Newton's second law, $F = ma$.

Fig. 7.7 Example accelerometers

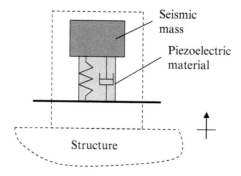

Fig. 7.8 Representation of an accelerometer as a spring–mass–damper system

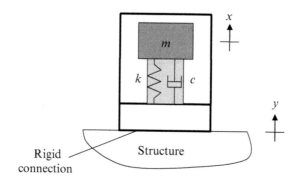

Fig. 7.9 Schematic used to determine the equation of motion for an accelerometer attached to a vibrating structure

Fig. 7.10 Free body diagram
for the accelerometer's
seismic mass with a moving
base structure

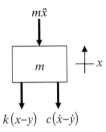

diagram for the mass is provided in Fig. 7.10. The equation of motion from the force
balance in the x direction is:

$$m\ddot{x} + c(\dot{x} - \dot{y}) + k(x - y) = 0. \tag{7.3}$$

This is an example of a *base motion* as we discussed in Sect. 3.6. Let's rewrite
Eq. 7.3 by substituting $z = x - y$ and $\dot{z} = \dot{x} - \dot{y}$. This gives:

$$m\ddot{x} + c\dot{z} + kz = 0. \tag{7.4}$$

If we now let $\ddot{z} = \ddot{x} - \ddot{y}$, then $\ddot{x} = \ddot{z} + \ddot{y}$. Replacing $\ddot{x}$ and rewriting yields:

$$m\ddot{z} + c\dot{z} + kz = -m\ddot{y}. \tag{7.5}$$

If the structure is experiencing harmonic vibration, then we can write $y(t) = Ye^{i\omega t}$. The corresponding acceleration of the structure is $\ddot{y} = -\omega^2 Ye^{i\omega t}$. If we also
let $z(t) = Ze^{i\omega t}$, calculate the velocity and acceleration, and substitute in Eq. 7.5,
then we have:

$$\left(-m\omega^2 + ic\omega + k\right)Ze^{i\omega t} = m\omega^2 Ye^{i\omega t}. \tag{7.6}$$

Equation 7.6 can be written as a ratio of the output (relative vibration) to the
input (base structure motion):

$$\frac{Z}{Y} = \frac{m\omega^2}{-m\omega^2 + ic\omega + k} = \frac{m\omega^2}{(k - m\omega^2) + i(c\omega)} = \frac{\omega^2}{\left(\dfrac{k}{m} - \omega^2\right) + i\left(\dfrac{c}{m}\omega\right)}. \tag{7.7}$$

From Chap. 2, we know that $\frac{k}{m} = \omega_n^2$ and $\frac{c}{m} = 2\zeta\omega_n$. Substituting gives:

$$\frac{Z}{Y} = \frac{\omega^2}{(\omega_n^2 - \omega^2) + i(2\zeta\omega_n\omega)} = \frac{\dfrac{\omega^2}{\omega_n^2}}{\left(1 - \dfrac{\omega^2}{\omega_n^2}\right) + i\left(2\zeta\dfrac{\omega}{\omega_n}\right)}. \tag{7.8}$$

Replacing $\frac{\omega}{\omega_n}$ with the frequency ratio, r, results in:

$$\frac{Z}{Y} = \frac{r^2}{(1 - r^2) + i(2\zeta r)}. \tag{7.9}$$

To eliminate the imaginary part from the denominator, we can multiply both the numerator and denominator by $(1 - r^2) - i(2\zeta r)$.

$$\frac{Z}{Y} = \frac{r^2}{(1 - r^2) + i(2\zeta r)} \cdot \frac{(1 - r^2) - i(2\zeta r)}{(1 - r^2) - i(2\zeta r)} = \frac{r^2((1 - r^2) - i(2\zeta r))}{(1 - r^2)^2 + (2\zeta r)^2} \tag{7.10}$$

The real part of Eq. 7.10 is:

$$\frac{Z}{Y} = \frac{r^2(1 - r^2)}{(1 - r^2)^2 + (2\zeta r)^2} \tag{7.11}$$

and the imaginary part is:

$$\frac{Z}{Y} = \frac{r^2(-2\zeta r)}{(1 - r^2)^2 + (2\zeta r)^2}. \tag{7.12}$$

The magnitude is the square root of the sum of the squares of the real and imaginary parts.

$$\left|\frac{Z}{Y}\right| = \sqrt{Re^2 + Im^2} = \sqrt{\frac{(r^2)^2(1 - r^2)^2 + (r^2)^2(-2\zeta r)^2}{\left((1 - r^2)^2 + (2\zeta r)^2\right)^2}}$$

$$\left|\frac{Z}{Y}\right| = \sqrt{\frac{(r^2)^2\left((1 - r^2)^2 + (2\zeta r)^2\right)}{\left((1 - r^2)^2 + (2\zeta r)^2\right)^2}} = \sqrt{\frac{(r^2)^2}{\left((1 - r^2)^2 + (2\zeta r)^2\right)}}$$

$$= \frac{r^2}{\sqrt{(1 - r^2)^2 + (2\zeta r)^2}} \tag{7.13}$$

Let's replace r^2 with $\frac{\omega^2}{\omega_n^2}$ and rearrange to obtain:

$$\omega_n^2|Z| = \frac{1}{\sqrt{(1 - r^2)^2 + (2\zeta r)^2}}\omega^2|Y|. \tag{7.14}$$

In this equation, $\omega^2|Y|$ represents the base structure's acceleration. The coefficient on this acceleration, $C_A = \frac{1}{\sqrt{(1-r^2)^2+(2\zeta r)^2}}$, defines the bandwidth, or the useful

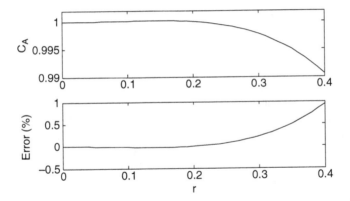

Fig. 7.11 Equation 7.14 coefficient, C_A, plotted versus the frequency ratio, r, for a damping ratio of 0.7. The percent error between 1 and the frequency-dependent C_A value is also shown

frequency range, of the accelerometer. When $r = 0$, the coefficient is 1. Depending on ζ, however, C_A deviates from 1 as r increases. The result of this C_A variation is that the frequency content of the structure's acceleration is not uniformly scaled in the measurement signal; the scaling is frequency dependent. To avoid this effect, the accelerometer bandwidth is limited to a frequency range where C_A is nearly constant. Figure 7.11 displays the variation in C_A with r for $\zeta = 0.7$; the percent error between 1 and the actual value, $(1 - C_A) \cdot 100\%$, is also shown. We observe that for $r < 0.2$, the deviation of C_A from 1 is negligible (Inman 2001). This means that the valid frequency range for an accelerometer with $\zeta = 0.7$ is from zero to approximately one-fifth of its natural frequency. A higher (first) natural frequency for an accelerometer therefore provides a wider measurement bandwidth.

Because the accelerometer produces a voltage that is proportional to accelera-tion, we obtain the accelerance, $\frac{A}{F}(\omega)$, from the dynamic signal analyzer when it calculates the frequency-domain ratio of the accelerometer response to the input force. In order to determine the receptance, $\frac{X}{F}(\omega)$, we follow a similar approach to that described in Sect. 7.3.2 for the vibrometer's mobility data. If the harmonic displacement is written as $x(t) = Xe^{i\omega t}$, then the acceleration is described by $\ddot{x}(t) = (i\omega)^2 Xe^{i\omega t} = -\omega^2 Xe^{i\omega t}$. Therefore, the displacement and acceleration are related by $-\omega^2$. See Eq. 7.15.

$$\frac{X}{F}(\omega) = -\frac{1}{\omega^2}\frac{A}{F} \tag{7.15}$$

This frequency-domain double integration is straightforward to complete, but, like the vibrometer, the zero frequency $(\omega = 0)$ information is lost due to the division by zero. For this reason, accelerometers are not well suited to measuring quasi-static, or slowly changing, signals.

7.4 Impact Testing

As we discussed in Sect. 7.2, impact testing can be used to determine the FRF for a structure. In this approach, an instrumented hammer is used to excite the structure and a transducer is used to record the resulting vibration. We will explore this technique using time-domain simulation (Euler integration) of the displacement due to the impulsive force input. Time-domain simulation provides an alternative to the analytical impulse response function we discussed in Sect. 3.7.

Let's model the case shown in Fig. 7.12 where a single degree of freedom spring–mass–damper system is excited by a hammer impact with a triangular force profile, $f(t)$. The hammer initially contacts the structure at t_1, the force reaches its maximum value at t_2, and the contact is lost at t_3. The differential equation of motion for this case is:

$$m\ddot{x} + c\dot{x} + kx = f(t). \tag{7.16}$$

We can obtain a numerical solution to Eq. 7.16 using Euler integration. As described in Sect. 2.5.2, the displacement is determined at small intervals of time, dt, by numerical (Euler) integration from acceleration to velocity and then velocity to displacement. We begin by solving for acceleration in Eq. 7.17:

$$\ddot{x} = \frac{f(t) - c\dot{x} - kx}{m}. \tag{7.17}$$

The current acceleration value depends on the instantaneous values of the time-dependent force, velocity, and displacement. For the first time step in the simulation, the velocity and displacement are defined by the initial conditions.

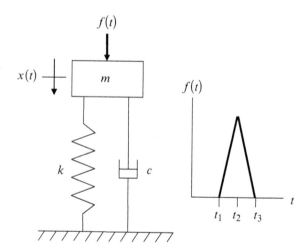

Fig. 7.12 Single degree of freedom spring–mass–damper system excited by an impact force

Fig. 7.13 *By the Numbers 7.2* – Force profile for exciting the single degree of freedom system pictured in Fig. 7.12

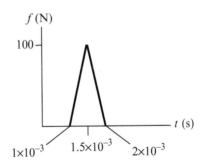

For all subsequent time steps, these are the values from the previous time step. Given the acceleration, the current velocity is determined using:

$$\dot{x} = \dot{x} + \ddot{x}dt. \tag{7.18}$$

In this equation, the velocity on the right hand side is the value from the previous time step (or the initial condition for the first simulation time step) and the acceleration is the value determined using Eq. 7.17. Using this velocity, the current displacement is:

$$x = x + \dot{x}dt \tag{7.19}$$

where the displacement on the right hand side of the equation is again the value from the previous time step. This process is repeated over many time steps to determine the system response. An important consideration for accurate Euler integration is the size of the time step. If the value is too large, inaccurate results are obtained. As a rule of thumb, it is generally acceptable to set *dt* to be at least ten times smaller than the period corresponding to the highest natural frequency for the system in question.

By the Numbers 7.1

Consider a system with a damped natural frequency of $f_d = 1,000$ Hz. The corresponding period of vibration for free vibration is $\tau = \frac{1}{f_d} = \frac{1}{1,000} = 1 \times 10^{-3}$ s. The maximum *dt* value for simulation is $dt = \frac{\tau}{10} = 1 \times 10^{-4}$ s. Smaller values are naturally acceptable (dividing the time constant by 50 or 100, for example), but there is a tradeoff between improved numerical accuracy and execution time. At some point, smaller time steps do not improve the accuracy and only serve to increase the simulation time.

By the Numbers 7.2

Let's use Euler integration to determine the displacement of the single degree of freedom spring–mass–damper system displayed in Fig. 7.12 due to the triangular force profile shown in Fig. 7.13. For the single degree of freedom spring–mass–damper system, we will use $m = 1$ kg, $k = 1 \times 10^6$ N/m, and $c = 80$ N-s/m.

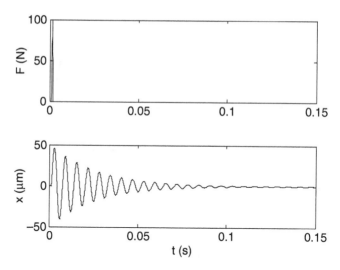

Fig. 7.14 *By the Numbers 7.2* – Time-domain response of the single degree of freedom system to the force profile in Fig. 7.13. The displacement was calculated using Euler integration

Therefore, the undamped natural frequency is $\omega_n = \sqrt{\frac{k}{m}} = \sqrt{\frac{1 \times 10^6}{1}} = 1{,}000$ rad/s, the damping ratio is $\zeta = \frac{c}{2\sqrt{km}} = \frac{80}{2\sqrt{1 \times 10^6(1)}} = 0.04$, the damped natural frequency is $\omega_d = \omega_n\sqrt{1 - \zeta^2} = 1{,}000\sqrt{1 - 0.04^2} = 998.4$ rad/s, and the corresponding period of vibration is $\tau = \frac{1}{f_d} = \frac{2\pi}{\omega_d} = \frac{2\pi}{998.4} = 6.29 \times 10^{-3}$ s. The maximum simulation step size for this vibration period is $dt = \frac{\tau}{10} = 6.3 \times 10^{-4}$. We will choose a smaller, more conservative time step of 5×10^{-5} s and carry out the simulation for 0.15 s (3,000 points).

We can complete the Euler integration described in Eqs. 7.17 through 7.19 using a `for` loop in MATLAB®. For each iteration in the `for` loop, we must: (1) define the time-dependent force; (2) calculate the acceleration; (3) calculate the velocity; and (4) calculate the displacement. The code is provided in MATLAB® MOJO 7.1 and the results are plotted in Fig. 7.14. We see that the 100 N impact force causes an exponentially decaying oscillation with a peak value of 46.1 μm at $t = 0.003$ s. This lags the peak force time of 0.0015 s. The second peak height of 35.9 μm occurs at $t = 0.0093$ s. Using this information, we can verify the damped natural frequency and the damping ratio.

First, the period of free vibration is the difference between the two peak times, $\tau = 0.0093 - 0.003 = 0.0063$. The corresponding oscillating frequency is $\omega_d = \frac{2\pi}{\tau} = \frac{2\pi}{0.0063} = 997$ rad/s. Second, we can estimate the damping ratio using the logarithmic decrement, δ. By Eq. 2.81, $\delta = \ln\frac{x_1}{x_2} = \ln\frac{46.1}{35.9} = 0.25$. According to Eq. 2.86, the damping ratio is then $\zeta = \sqrt{\frac{\delta^2}{4\pi^2 + \delta^2}} = \sqrt{\frac{0.25^2}{4\pi^2 + 0.25^2}} = 0.04$. These values verify the simulation results.

MATLAB® MOJO 7.1

```
% matlab_mojo_7_1.m

clear all
close all
clc

k = 1e6;                % N/m
m = 1;                  % kg
c = 80;                 % N-s/m
fmax = 100;             % N

dt = 5e-5;              % s
total_time = 0.15;      % s
points = round(total_time/dt);

% Initial conditions
dx = 0;
x = 0;

% Predefine vectors used in the 'for' loop
time = zeros(points, 1);
displacement = zeros(points, 1);
force = zeros(points, 1);

for cnt = 1:points
    t = cnt*dt;

    % Define impulse
    if t < 1e-3
        f = 0;
    elseif t >= 1e-3 && t <= 1.5e-3
        f = (t-1e-3)*fmax/0.5e-3;                  % N
    elseif t > 1.5e-3 && t <= 2e-3
        f = fmax - fmax/0.5e-3*(t-1.5e-3);
    else
        f = 0;
    end

    % Perform Euler integration
    ddx = (f - c*dx - k*x)/m;
    dx = dx + ddx*dt;
    x = x + dx*dt;

    % Write results to vectors
    force(cnt) = f;
    displacement(cnt) = x;
    time(cnt) = t;
end

figure(1)
subplot(211)
plot(time, force)
set(gca,'FontSize', 14)
xlim([0 0.15])
ylabel('F (N)')
subplot(212)
plot(time, displacement*1e6)
set(gca,'FontSize', 14)
xlim([0 0.15])
xlabel('t (s)')
ylabel('x (\mum)')
```

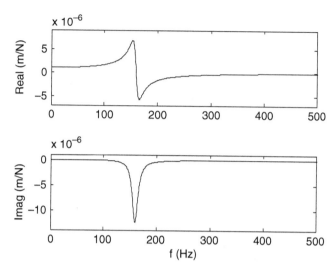

Fig. 7.15 *By the Numbers 7.2* – FRF for the time-domain response displayed in Fig. 7.14

Now that we have the (simulated) time response of a system due to an impact force, we can perform the function of the dynamic signal analyzer and calculate the FRF. To do this, we need to calculate the discrete Fourier transform of the time-domain displacement and force signals.[4] We can use the MATLAB® function fft to do this. The following code is appended to MATLAB® MOJO 7.1 in order to determine the complex-valued force transform, F, and displacement transform, X. The FRF is their ratio; see Fig. 7.15.

```
% Calculate Fourier transform of force
F = fft(force);                                  % m
freq = (0:1/(dt*points):(1-1/(2*points))/dt)';   % Hz
F = F(1:round(points/2+1), :);
freq = freq(1:round(points/2+1), :);

% Calculate Fourier transform of displacement
X = fft(displacement);                           % m
X = X(1:round(points/2+1), :);

FRF = X./F;

figure(1)
subplot(211)
plot(freq, real(FRF))
xlim([0 500])
ylim([-7e-6 9e-6])
ylabel('Real (m/N)')
subplot(212)
plot(freq, imag(FRF))
xlim([0 500])
ylim([-1.4e-5 1e-6])
xlabel('Frequency (Hz)')
ylabel('Imag (m/N)')
```

---

[4] The discrete Fourier transform is applied because our inputs are sampled; they are not continuous in time.

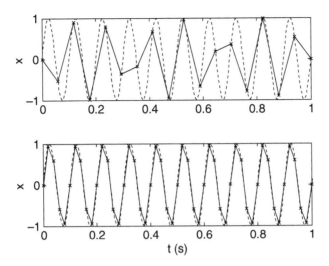

Fig. 7.16 Time-domain example of aliasing. The 10 Hz signal (*dotted line*) is sampled at 17 Hz in the *top panel*. Because the signal frequency is greater than the Nyquist frequency ($\frac{f_s}{2} = \frac{17}{2} = 8.5$ Hz), aliasing occurs. For a 50 Hz sampling frequency with a Nyquist frequency of 25 Hz (*bottom panel*), the signal is captured accurately

In the preceding code, we note that the `fft` function does not return the corresponding frequency vector. This is defined separately as `freq` and depends on both the sampling interval, `dt`, and the number of `points`, points, in the time-domain signals. The frequency resolution for the FRF depends on the ratio `1/(dt*points)`. It is increased (i.e., the frequency step size is decreased) by increasing the number of points in the signals for a fixed *sampling frequency*, $f_s = \frac{1}{dt}$. This is achieved by collecting data over a longer time interval.

We also note that the frequency vector is defined over an interval from zero to half the sampling frequency. This is based on the *Nyquist–Shannon sampling theorem*, which states that only signals with frequencies up to half the sampling frequency, $\frac{f_s}{2}$, can be reconstructed from the sampled data. If a signal with frequencies higher than $\frac{f_s}{2}$ (referred to as the *Nyquist frequency*) is sampled, then the data will be *aliased* and incorrect frequency content will be obtained in the Fourier transform. This is demonstrated in Figs. 7.16 and 7.17. In Fig. 7.16, a sine wave with a unit magnitude and oscillating frequency of 10 Hz is sampled at 17 Hz (top panel) and 50 Hz (bottom panel). It is clear that the 17 Hz sampling frequency is insufficient to capture the behavior of the 10 Hz signal. To see the effect in the frequency domain, the discrete Fourier transform of the sampled signals is provided in Fig. 7.17, which displays the transform magnitudes. We see that when the 10 Hz sine wave is sampled at 17 Hz, the apparent frequency is the difference between the two frequencies, $17 - 10 = 7$ Hz. The signal is aliased because its frequency exceeds the Nyquist frequency of $\frac{f_s}{2} = \frac{17}{2} = 8.5$ Hz. When the same 10 Hz sine wave is sampled at 50 Hz, we observe the expected 10 Hz peak with a unit magnitude. When the maximum frequency content of a signal is unknown (which is

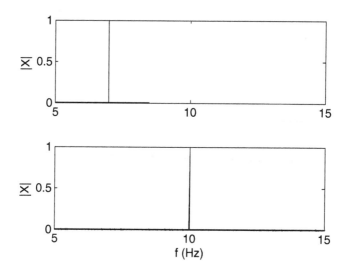

Fig. 7.17 Frequency-domain example of aliasing. In the top panel, the result of sampling a 10 Hz signal with a 17 Hz sampling frequency is shown. The apparent frequency in the aliased signal is their difference, $17 - 10 = 7$ Hz. The Fourier transform magnitude for a sampling frequency of 50 Hz is displayed in the bottom panel. The correct result is obtained

generally the case), an analog *anti-aliasing filter* is applied. This removes content from the continuous signal that exceeds the Nyquist frequency prior to sampling.

 In A NUTSHELL   We have to sample at least twice the frequency of the highest frequency we hope to see. If we want to see a 500 Hz signal, then we need to sample at least 1,000 times per second. In practice, we would like to sample much faster than that. Once the sampling frequency is chosen, then an anti-aliasing filter is used to remove signals beyond our measurement range.

Let's next compare the frequency content of two hammer impacts. We stated in Sect. 7.2 that the impulse excitation bandwidth depends on the hammer mass and tip stiffness. We choose a stiffer tip to excite a wider frequency range and a softer tip to concentrate the energy over a lower frequency range. Let's model the impacts as half-period sine waves and calculate the corresponding frequency content using the discrete Fourier transform. Figure 7.18 shows two example impacts. Both have a maximum value of 100 N, but the total impact durations differ by a factor of 10. A shorter duration of 0.5 ms represents a stiff tip, and a longer duration of 5 ms represents a softer tip. The frequency-domain force magnitudes for the two impacts are provided in Fig. 7.19, where the signals were sampled at 50 kHz for a total of $2^{12} = 4,096$ points. We see that the shorter duration impact excites a wider frequency range, but with a lower force level than the longer impact. Note that it is this broad frequency range excitation that makes the impact test a popular choice

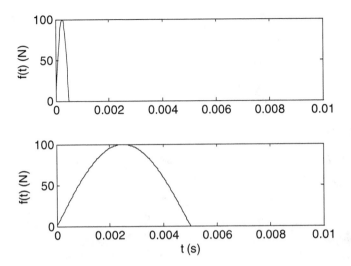

Fig. 7.18 Two example force impacts

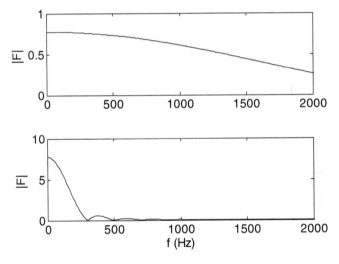

Fig. 7.19 Discrete Fourier transforms of the two force impacts in Fig. 7.18. The *top panel* shows the 0.5 ms duration impact magnitude and the *bottom panel* displays the 5 ms duration impact magnitude

for FRF measurement. With a single hammer impact, the response over a wide range of frequencies is obtained. For the stiff tip spectrum shown in Fig. 7.19, modes with natural frequencies of up to 2,000 Hz would be effectively excited. The softer tip can only excite modes with natural frequencies up to about 200 Hz, but introduces approximately ten times more force into these low frequency modes.

Fig. 7.20 Impact testing setup for the BEP

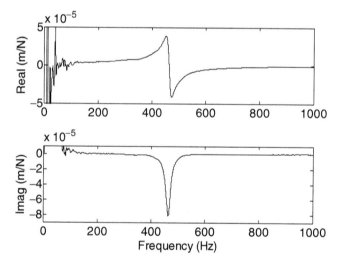

Fig. 7.21 Direct FRF measured at the free end of the BEP's clamped rod

To conclude this section, let's observe the results of an impact test completed on the beam experimental platform (BEP) that was introduced in Sect. 2.6. For these measurements, a low mass accelerometer was attached to the free end of the clamped rod and it was excited using a small impact hammer as shown in Fig. 7.20. The extended length of the steel rod was 130 mm. The direct FRF obtained by exciting and measuring at the rod's free end is provided in Fig. 7.21.

We observe a single mode with a natural frequency of 463.9 Hz within the 1,000 Hz measurement bandwidth. Therefore, if the vibration behavior only up to 1,000 Hz is of interest, then the BEP could be modeled as a single degree of freedom system. (As discussed in the next section, other modes actually exist at higher frequencies outside the measurement bandwidth.) The low frequency behavior observed in Fig. 7.21 is due to the double integration of the accelerance as described in Sect. 7.3.3.

7.5 Modal Truncation

Because FRF measurements always have a finite frequency range and elastic bodies possess an infinite number of degrees of freedom (with increasing natural frequencies), there are necessarily modes that exist outside the measurement range. While all the modes affect the measurement, when we identify a model of the system using peak picking, we only consider the effects of the individual modes within the measurement bandwidth that we selected for fitting. Omitting the higher frequency modes affects the accuracy of the modal fit, particularly the real part of the FRF. This is referred to as *modal truncation*. Equations 7.20 and 7.21, which describe the real and imaginary parts of a single degree of freedom FRF, are included here to demonstrate the effect.

$$\mathrm{Re}\left(\frac{X}{F}\right) = \frac{1}{k}\left(\frac{1 - r^2}{\left(1 - r^2\right)^2 + \left(2\zeta r\right)^2}\right) \tag{7.20}$$

$$\mathrm{Im}\left(\frac{X}{F}\right) = \frac{1}{k}\left(\frac{-2\zeta r}{\left(1 - r^2\right)^2 + \left(2\zeta r\right)^2}\right) \tag{7.21}$$

We see that when the frequency ratio $r = \frac{\omega}{\omega_n}$ is large, or the forcing frequency ω is very high and outside the measurement range, the denominator for the right parenthetical terms in these two equations becomes very large and the response approaches zero. However, when r is very small, the parenthetical term in the real part approaches a value of 1, and the parenthetical term in the imaginary part approaches zero. Therefore, the value of the real part approaches $\frac{1}{k}$ as r approaches zero.[5] Neglecting the modes beyond the measurement bandwidth and, therefore, the associated $\frac{1}{k}$ contribution for each leads to errors in the vertical location of the modal fit's real part. This is demonstrated in *By the Numbers 7.3*.

---

[5] This $\frac{1}{k}$ term can be referred to as the DC compliance.

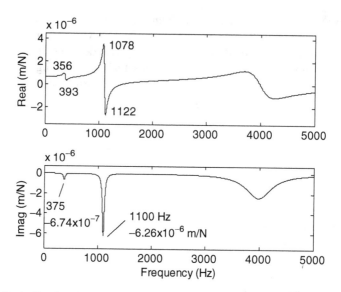

Fig. 7.22 *By the Numbers 7.3* – measured direct FRF. The peak picking values are listed within the 2,000 Hz measurement bandwidth. A 5,000 Hz frequency range is provided to show the truncated 4,000 Hz mode

By the Numbers 7.3

An example FRF is provided in Fig. 7.22. We will presume that the measurement bandwidth was 2,000 Hz, although a 5,000 Hz frequency range is shown for the purposes of this demonstration. Within the 2,000 Hz range, two modes are visible, and peak picking can be applied to determine the associated modal parameters. Using the values from the figure, the modal stiffness, mass, and damping matrix terms may be determined as shown in Sect. 6.2.

$$\zeta_{q1} = \frac{393 - 356}{2 \cdot 375} = 0.049 \quad \zeta_{q2} = \frac{1,122 - 1,078}{2 \cdot 1100} = 0.020$$

$$k_{q1} = \frac{-1}{2 \cdot 0.049 \cdot (-6.74 \times 10^{-7})} = 1.50 \times 10^7 \, \text{N/m}$$

$$k_{q2} = \frac{-1}{2 \cdot 0.020 \cdot (-6.26 \times 10^{-6})} = 3.99 \times 10^6 \, \text{N/m}$$

$$m_{q1} = \frac{1.50 \times 10^7}{(375 \cdot 2\pi)^2} = 2.70 \, \text{kg} \quad m_{q2} = \frac{3.99 \times 10^6}{(1,100 \cdot 2\pi)^2} = 0.084 \, \text{kg}$$

$$c_{q1} = 2 \cdot 0.049 \sqrt{1.50 \times 10^7 \cdot 2.70} = 624 \, \text{N-s/m}$$

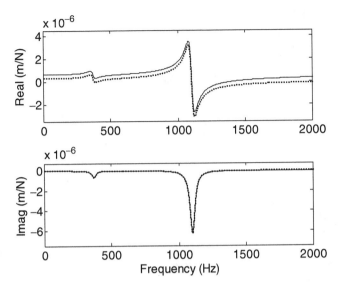

Fig. 7.23 *By the Numbers 7.3* – result of modal fitting. An offset in the real part of the fit (dotted line) is observed because the DC compliance of the 4,000 Hz mode is not included

$$c_{q2} = 2 \cdot 0.020\sqrt{3.99 \times 10^6 \cdot 0.084} = 23.2\,\text{N-s/m}$$

The fit to the measured direct FRF is determined by summing the two contributions in modal coordinates using:

$$\frac{X}{F} = \frac{Q_1}{R_1} + \frac{Q_2}{R_2} = \frac{1}{k_{q1}}\left(\frac{(1 - r_1^2) - i(2\zeta_{q1}r_1)}{(1 - r_1^2)^2 + (2\zeta_{q1}r_1)^2}\right) + \frac{1}{k_{q2}}\left(\frac{(1 - r_2^2) - i(2\zeta_{q2}r_2)}{(1 - r_2^2)^2 + (2\zeta_{q2}r_2)^2}\right),$$

where $r_1 = \frac{f}{375}$ and $r_2 = \frac{f}{1,100}$ and f is given in Hz. It is seen in Fig. 7.23 that, although the shape of the two modes within the 2,000 Hz bandwidth are correctly identified, there is a noticeable offset in the real part of the fit. It appears too stiff (i.e., it is located below the measured FRF) because the DC compliance due to the 4,000 Hz mode has not been considered. Because this mode is outside the measurement frequency range, it is not possible to fit the mode and determine the appropriate modal parameters. However, given the visible offset in Fig. 7.23, the combined contributions of truncated modes can be included by adding an effective DC compliance term to the fit. Specifically, for this example, the fit could be rewritten as:

$$\frac{X}{F} = \frac{1}{k} + \frac{Q_1}{R_1} + \frac{Q_2}{R_2},$$

where the $\frac{Q_j}{R_j}$ terms ($j = 1, 2$) are obtained through peak picking as described previously and the $\frac{1}{k}$ value is selected to move the fit to a vertical overlap with the measured FRF. If a value of $k = 3 \times 10^6$ N/m is applied here, the fit is improved and the result shown in Fig. 7.24 is obtained. Note that this stiffness value is equal to the modal stiffness of the 4,000 Hz mode shown in Fig. 7.22 (for completeness, the modal damping ratio for this mode is 0.07). The code used to produce Figs. 7.22–7.24 is provided in MATLAB® MOJO 7.2.

MATLAB® MOJO 7.2

```
% matlab_mojo_7_2.m

clc
clear all
close all

% Define modal parameters for the "measured" FRF
fn1 = 375;          % Hz
wn1 = fn1*2*pi;     % rad/s
zetaq1 = 0.05;
kq1 = 1.5e7;        % N/m

fn2 = 1100;         % Hz
wn2 = fn2*2*pi;     % rad/s
zetaq2 = 0.02;
kq2 = 4e6;          % N/m

fn3 = 4000;         % Hz
wn3 = fn3*2*pi;     % rad/s
zetaq3 = 0.07;
kq3 = 3e6;          % N/m

% Define the measured FRF
w = (0:0.2:5000)'*2*pi;     % frequency, rad/s
r1 = w/wn1;
r2 = w/wn2;
r3 = w/wn3;
FRF = 1/kq1*((1-r1.^2) - i*(2*zetaq1*r1))./((1-r1.^2).^2 + (2*zetaq1*r1).^2)
+ 1/kq2*((1-r2.^2) - i*(2*zetaq2*r2))./((1-r2.^2).^2 + (2*zetaq2*r2).^2) +
1/kq3*((1-r3.^2) - i*(2*zetaq3*r3))./((1-r3.^2).^2 + (2*zetaq3*r3).^2);

figure(1)
subplot(211)
plot(w/2/pi, real(FRF), 'k')
ylim([-3.5e-6 4.5e-6])
set(gca,'FontSize', 14)
ylabel('Real (m/N)')
subplot(212)
plot(w/2/pi, imag(FRF), 'k')
ylim([-7.5e-6 7.5e-7])
set(gca,'FontSize', 14)
xlabel('Frequency (Hz)')
ylabel('Imag (m/N)')

figure(2)
subplot(211)
plot(w/2/pi, real(FRF), 'k')
axis([0 2000 -3.5e-6 4.5e-6])
set(gca,'FontSize', 14)
ylabel('Real (m/N)')
hold on
subplot(212)
plot(w/2/pi, imag(FRF), 'k')
```

```
axis([0 2000 -7.5e-6 7.5e-7])
set(gca,'FontSize', 14)
xlabel('Frequency (Hz)')
ylabel('Imag (m/N)')
hold on

figure(3)
subplot(211)
plot(w/2/pi, real(FRF), 'k')
axis([0 2000 -3.5e-6 4.5e-6])
set(gca,'FontSize', 14)
ylabel('Real (m/N)')
hold on
subplot(212)
plot(w/2/pi, imag(FRF), 'k')
axis([0 2000 -7.5e-6 7.5e-7])
set(gca,'FontSize', 14)
xlabel('Frequency (Hz)')
ylabel('Imag (m/N)')
hold on

% Perform fit
fn1 = 375;          % Hz
wn1 = fn1*2*pi;     % rad/s
zetaq1 = (393 - 356)*2*pi/(2*wn1);
kq1 = 1/(2*zetaq1*6.74e-7);

fn2 = 1100;         % Hz
wn2 = fn2*2*pi;     % rad/s
zetaq2 = (1122 - 1078)*2*pi/(2*wn2);
kq2 = 1/(2*zetaq2*6.26e-6);

r1 = w/wn1;
r2 = w/wn2;
FRF1 = 1/kq1*((1-r1.^2) - 1i*(2*zetaq1*r1))./((1-r1.^2).^2 +
(2*zetaq1*r1).^2);  % mode 1
FRF2 = 1/kq2*((1-r2.^2) - 1i*(2*zetaq2*r2))./((1-r2.^2).^2 +
(2*zetaq2*r2).^2);  % mode 2
FRF = FRF1 + FRF2;  % modal fit

figure(2)
subplot(211)
plot(w/2/pi, real(FRF), 'k:')
subplot(212)
plot(w/2/pi, imag(FRF), 'k:')

% Add correction for modal truncation
k = 3e6;            % N/m
FRF = 1/k + FRF1 + FRF2;  % new modal fit

figure(3)
subplot(211)
plot(w/2/pi, real(FRF), 'k:')
subplot(212)
plot(w/2/pi, imag(FRF), 'k:')
```

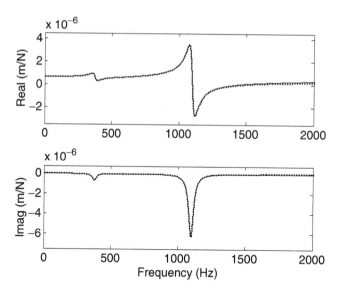

Fig. 7.24 *By the Numbers 7.3* – result of modal fitting with the addition of a DC compliance term to compensate for the truncated mode

Chapter Summary

- FRF measurement requires a mechanism for force input, a transducer to measure the vibration, and a dynamic signal analyzer to compute the vibration to force ratio of the Fourier-transformed signals.
- The FRF may take the form of receptance (displacement), mobility (velocity), or accelerance (acceleration) depending on the vibration measurement technique.
- Force input types include fixed frequency sine waves, random noise, and impulses. These forces can be produced using a shaker or impact hammer.
- Example vibration transducers include capacitance probes, laser vibrometers, and accelerometers. The latter is a contact-type measurement.
- The response of an accelerometer to a structure's motion is an example of base motion.
- Accelerance and mobility can be converted to receptance in the frequency domain using the measurement frequency vector.
- Euler (numerical) integration can be used to solve the differential equation of motion for arbitrary force inputs, including the force impulse in impact testing.
- Care must be exercised in sampling continuous signals to avoid aliasing.
- In modal truncation, modes that exist outside the measurement range affect the accuracy of the modal fit to the measured FRF.

Exercises

1. Complete the following statements.

    (a) Receptance is the frequency-domain ratio of _____ to _____.

    (b) Mobility is the frequency-domain ratio of _____ to _____.

    (c) Accelerance is the frequency-domain ratio of _____ to _____.

2. Find three commercial suppliers of impact hammers for modal testing.
3. Find three commercial suppliers of dynamic signal analyzers for modal testing.
4. Digital data acquisition is to be used to record vibration signals for a particular system. If the highest anticipated frequency in the measurements is 5,000 Hz, select the minimum sampling frequency.
5. An impact test was completed using an instrumented hammer to excite a structure and an accelerometer to measure the vibration response.

    (a) Show how to convert the acceleration-to-force frequency response function (i.e., accelerance) that was obtained to a displacement-to-force frequency response function (i.e., receptance).

    (b) What information is lost in this conversion?

6. As described in Sects. 7.2 and 7.4, FRFs are often measured using impact testing. In this approach, an instrumented hammer is used to excite the structure and a transducer is used to record the resulting vibration.

Fig. P7.6 Impulsive force profile for impact test

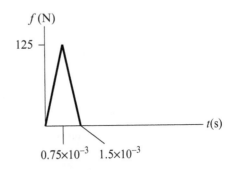

Use Euler integration to determine the displacement due to the triangular impulsive force profile shown in Fig. P7.6. The force excites a single degree of freedom spring–mass–damper system with $m = 2$ kg, $k = 1.1 \times 10^6$ N/m, and $c = 83$ N-s/m. For the Euler integration, use a time step of 1×10^{-5} s and carry out your simulation for 0.2 s (20,000 points).

(a) Plot both the force (N) versus time (s) and displacement (μm) versus time.

(b) Determine the maximum displacement (in μm) and the time at which this displacement occurs.

7. For a particular measurement application, an accelerometer must be selected with a bandwidth or useful frequency range of 5,000 Hz. If the allowable deviation in the scaling coefficient $C_A = \dfrac{1}{\sqrt{(1-r^2)^2+(2\zeta r)^2}}$ is ±1% and the damping ratio is known to be 0.65, determine the minimum required for the natural frequency of the accelerometer.

8. A single degree of freedom spring–mass–damper system which is initially at rest at its equilibrium position is excited by an impulsive force with a magnitude of 250 N over a time interval of 0.5 ms; see Fig. P7.8. If the mass is 3 kg, the stiffness is 3×10^6 N/m, and the viscous damping coefficient is 120 N-s/m, complete the following.

Fig. P7.8 Spring–mass–damper system excited by an impulsive force

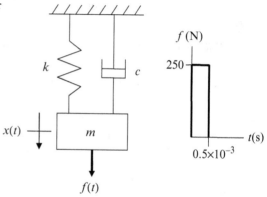

(a) Determine $x(t)$ using Eq. 3.44. Plot the response (in μm) over a time period of 0.3 s with a step size of 1×10^{-4} s in the time vector.

(b) Determine $x(t)$ using Euler integration. Use a time step of 1×10^{-4} s and carry out your simulation for 0.3 s (30,000 points). Plot $x(t)$ (in μm) versus time.

9. Determine the FRF for the system described in Problem 8 using Euler integration to calculate the time-domain displacement due to the impulsive input force. To increase the FRF frequency resolution, use a total simulation time of 1 s. Given the time-domain displacement and force vectors, use the MATLAB® function fft to calculate the complex-valued force transform, F, and displacement transform, X. Plot the real and imaginary parts (in m/N) of their ratio, X/F, versus frequency (in Hz). Use axis limits of axis([0 500 -5e-6 5e-6]) for the real plot and axis limits of axis([0 500 -1e-5 1e-6]) for the imaginary plot.

10. The existence of modes with frequencies higher than the measurement bandwidth leads to an effect referred to as _____ when performing a modal fit to the measured FRF.

References

http://www.lionprecision.com/tech-library/technotes/cap-0020-sensor-theory.html
Inman D (2001) Engineering vibration, 2nd edn. Prentice Hall, Upper Saddle River

Chapter 8
Continuous Beam Modeling

Continuity in everything is unpleasant.

– Blaise Pascal

8.1 Beam Bending

In Chaps. 1 through 5 we discussed the solution of discrete, lumped-parameter models. For multiple degree of freedom systems, we employed modal analysis to enable us to transform the coupled equations of motion in local (model) coordinates into modal coordinates. In this coordinate frame, the equations of motion were uncoupled and we could apply single degree of freedom solution techniques. In Chap. 6 we shifted our attention to the "backwards problem," which is representative of a common task for vibration engineers. In this problem, we begin with measurements of an existing structure and use this information to develop a model. We again used discrete models to describe the system behavior.

An alternative to lumped-parameter models is continuous models. In this case, the mass is distributed throughout the structure, rather than localized at the model coordinates. This modeling approach can be very effective and can offer good accuracy. As an example, the modal truncation effects we discussed in Sect. 7.5 are non-existent when a continuous model that includes the effects of all modes is implemented.

T.L. Schmitz and K.S. Smith, *Mechanical Vibrations: Modeling and Measurement*, DOI 10.1007/978-1-4614-0460-6_8, © Springer Science+Business Media, LLC 2012

Fig. 8.1 Coordinate
definitions for continuous
beam transverse deflection.
The deflection, y, depends on
the location along the beam
axis, x

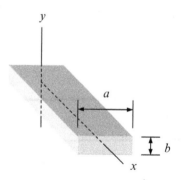

IN A NUTSHELL   Continuous solutions can be derived for
many simple geometries. These solutions offer insight into the way
that vibrating systems behave and describe the distributed mass,
stiffness, and damping elements that are often encountered.
Continuous models become complicated for any but the simplest
systems, but it is useful to understand the parallels between the continuous and the
discrete parameter models. In addition, the continuous models form the basis for
deriving finite element representations, a useful technique for discretizing
continuous systems.

To begin this discussion, let's consider the bending of beams using *Euler–Bernoulli
beam theory* and see how we can expand the analysis to derive the vibration response.
The fundamental relationship between the transverse deflection, y, and a load per
unit length, q, applied to a *beam*[1] with constant cross section and material properties
is provided in Eq. 8.1. In this equation E is the elastic modulus, I is the second
moment of area, and x is the (continuous) position along the beam. Figure 8.1
displays the x and y coordinate axes for a rectangular beam. In this case, the second
moment of area is $I = ab^3/12$.

$$\frac{q}{EI} = \frac{d^4y}{dx^4} \tag{8.1}$$

To determine the static (non-vibratory) deflection of a continuous beam under a
selected loading condition, we integrate Eq. 8.1 four successive times. Let's apply
the loading condition shown in Fig. 8.2, where a simply supported[2] beam with
length, l, is loaded by a force per unit length, w. Due to symmetry, the reaction
force, $wl/2$, is the same at both ends. Equation 8.1 can be rewritten as shown in
Eq. 8.2 for this case.

---

[1] A beam can be described as a structure where one dimension is much larger than the other two
dimensions.

[2] A simply supported beam is pinned at one end and has a rolling support at the other.

$$\frac{q}{EI} = \frac{d^4y}{dx^4} = \frac{-w}{EI}. \tag{8.2}$$

Integrating Eq. 8.2 gives Eq. 8.3, where C_1 is the integration constant and V is the position-dependent shear force acting on the beam.

$$\frac{V}{EI} = \frac{d^3y}{dx^3} = \frac{-w}{EI}x + C_1 \tag{8.3}$$

Integrating a second time yields the moment, M, equation shown in Eq. 8.4. Now there are two integration constants.

$$\frac{M}{EI} = \frac{d^2y}{dx^2} = \frac{-w^2}{2EI}x + C_1x + C_2 \tag{8.4}$$

The third integration gives the beam slope, θ. See Eq. 8.5.

$$\theta = \frac{dy}{dx} = \frac{-w^3}{6EI}x + \frac{C_1x^2}{2} + C_2x + C_3 \tag{8.5}$$

Finally, the deflection equation is obtained by a fourth integration as shown in Eq. 8.6.

$$y = \frac{-w^4}{24EI}x + \frac{C_1x^3}{6} + \frac{C_2x^2}{2} + C_3x + C_4 \tag{8.6}$$

In order to obtain the deflection profile from Eq. 8.6, we must determine the four integration constants. We identify these using the beam's *boundary conditions*. For the simply supported beam in Fig. 8.2, the following boundary conditions apply.

1. The shear force at the left end ($x = 0$) is equal to the reaction force. We can express this relationship as:

$$\left.\frac{V}{EI}\right|_{x=0} = \frac{wl}{2EI}. \tag{8.7}$$

2. The ends points of the beam are free to rotate, so the moment at the left end is zero.

$$\left.\frac{M}{EI}\right|_{x=0} = 0 \tag{8.8}$$

Fig. 8.2 Simply supported
beam loaded by the force per
unit length, *w*

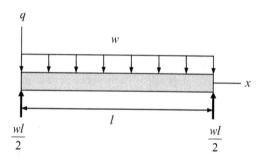

3. The slope at the beam's midpoint is zero due to symmetry.

$$\theta|_{x=\frac{l}{2}} = 0 \tag{8.9}$$

4. The deflection at the left end is zero.

$$y|_{x=0} = 0 \tag{8.10}$$

Substitution of Eqs. 8.7 through 8.10 into Eqs. 8.3 through 8.6 and simultaneous solution yields the following four relationships for the beam in Fig. 8.2.

$$\frac{V}{EI} = \frac{w}{EI}\left(\frac{l}{2} - x\right) \tag{8.11}$$

$$\frac{M}{EI} = \frac{w}{2EI}\left(lx - x^2\right) \tag{8.12}$$

$$\theta = \frac{w}{24EI}\left(6lx^2 - 4x^3 - l^3\right) \tag{8.13}$$

$$y = \frac{w}{24EI}\left(2lx^3 - l^3x - x^4\right) \tag{8.14}$$

When implementing Euler–Bernoulli beam theory, there are two underlying assumptions: (1) shear deformations are negligible; and (2) planar cross sections remain planar and normal to the beam axis during deformation. These assumptions limit the accuracy of the model when the beam is not long and slender (approximately ten times longer than the largest cross-sectional dimension).

By the Numbers 8.1

Let's determine the shear force, moment, slope, and deflection diagrams for a steel, 10 mm (0.01 m) square cross section, simply supported beam with a length of 200 mm (0.2 m). The loading conditions follow Fig. 8.2 with $w = 100$ N/m. We will assume a steel modulus of 200 GPa. The area moment of inertia is determined for the square cross section using:

$$I = \frac{(0.01)^4}{12} = 8.3 \times 10^{-10} \, \text{m}^4.$$

The results are shown in Fig. 8.3. We see that the maximum deflection of 12.5 μm occurs at the beam midpoint. The code used to produce Fig. 8.3 is provided in MATLAB® MOJO .8.1

MATLAB® MOJO 8.1

```
% matlab_mojo_8_1.m

clc
clear all
close all

% Define beam parameters
s = 10e-3;          % m
E = 200e9;          % N/m^2
I = s^4/12;         % m^4
w = 100;            % N/m
l = 200e-3;         % m
x = 0:1e-3:l;       % m

% Shear force
V = w*(l/2 - x);
% Moment
M = w*(l*x - x.^2);
% Slope
theta = w/(24*E*I)*(6*l*x.^2 - 4*x.^3 - l^3);
% Deflection
y = w/(24*E*I)*(2*l*x.^3 - l^3*x - x.^4);

figure(1)
subplot(411)
plot(x*1e3, V, 'k')
set(gca,'FontSize', 14)
ylabel('V (N)')
subplot(412)
plot(x*1e3, M, 'k')
set(gca,'FontSize', 14)
ylim([0 1.2])
ylabel('M (N/m)')
subplot(413)
plot(x*1e3, theta, 'k')
set(gca,'FontSize', 14)
ylim([-2.5e-4 2.5e-4])
ylabel('\theta (rad)')
subplot(414)
plot(x*1e3, y*1e6, 'k')
set(gca,'FontSize', 14)
ylim([-13 0])
xlabel('x (mm)')
ylabel('y (\mum)')
```

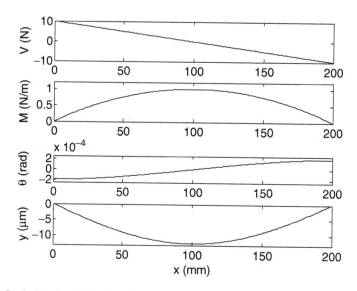

Fig. 8.3 *By the Numbers 8.1* – shear force, moment, slope, and deflection diagrams for the simply supported steel beam subjected to self-loading

8.2 Transverse Vibration Equation of Motion

While the analysis in Sect. 8.1 is useful, it only gives the static solution. It does not describe the dynamic, or vibratory, behavior of a beam. For the beam vibration, we need to determine the time-dependent deflection. We will now build on the previous analysis to determine the required differential equation of motion for the beam. Let us begin with Fig. 8.4, which shows the forces and moments acting on a section of a vibrating Euler–Bernoulli beam with an infinitesimal length, ∂x.

According to Newton's second law, the sum of the forces in the y direction is equal to the product of the section mass and acceleration, $\sum F_y = m\frac{\partial^2 y}{\partial t^2}$. See Eq. 8.15, where the mass is rewritten as the product of the density, ρ, cross-sectional area, A, and section length, ∂x.

$$q\partial x + V(x) - V(x + \partial x) = \rho A \partial x \frac{\partial^2 y}{\partial t^2} \tag{8.15}$$

We can rewrite Eq. 8.15 by substituting $\partial V = V(x + \partial x) - V(x)$ and dividing each term by ∂x. The result is provided in Eq. 8.16.

$$q - \frac{\partial V}{\partial x} = \rho A \frac{\partial^2 y}{\partial t^2} \tag{8.16}$$

Fig. 8.4 Forces and moments acting on a small section of a vibrating Euler–Bernoulli beam

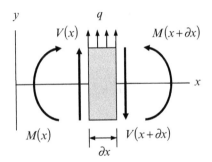

From Eq. 8.3, the shear force is $V = EI\frac{\partial^3 y}{\partial x^3}$. Calculating the partial derivative with respect to x gives:

$$\frac{\partial V}{\partial x} = EI\frac{\partial^4 y}{\partial x^4}. \tag{8.17}$$

Substituting Eq. 8.17 into 8.16 gives the differential equation of motion for the transverse vibration of a uniform cross section Euler–Bernoulli beam.

$$\rho A\frac{\partial^2 y}{\partial t^2} + EI\frac{\partial^4 y}{\partial x^4} = q \tag{8.18}$$

For free vibration, the external transverse load, q, is zero, so we can write:

$$\rho A\frac{\partial^2 y}{\partial t^2} + EI\frac{\partial^4 y}{\partial x^4} = 0. \tag{8.19}$$

8.3   Frequency Response Function for Transverse Vibration

Our next task is to determine the beam's frequency response function using Eq. 8.19. A general solution to this equation is:

$$y(x, t) = Y(x)\sin(\omega t), \tag{8.20}$$

where $Y(x)$ is a function that describes the position-dependent vibration behavior and ω is frequency (Bishop and Johnson 1960). Let's calculate the required partial derivatives of Eq. 8.20 that appear in Eq. 8.19.

$$\frac{\partial^2 y}{\partial t^2} = Y(x)\left(-\omega^2\right)\sin(\omega t) \tag{8.21}$$

$$\frac{\partial^4 y}{\partial x^4} = \frac{\partial^4 Y}{\partial x^4}\sin(\omega t) \tag{8.22}$$

Substituting Eqs. 8.21 and 8.22 into Eq. 8.19 gives:

$$\left(\rho A \left(-\omega^2 Y\right) + EI \frac{\partial^4 Y}{\partial x^4} \right) \sin(\omega t) = 0. \tag{8.23}$$

Rewriting Eq. 8.23 yields:

$$\frac{\partial^4 Y}{\partial x^4} - \omega^2 \frac{\rho A}{EI} Y = 0. \tag{8.24}$$

Letting $\lambda^4 = \omega^2 \frac{\rho A}{EI}$, we now have:

$$\frac{\partial^4 Y}{\partial x^4} - \lambda^4 Y = 0. \tag{8.25}$$

The time-dependence of Eq. 8.19 has been eliminated in Eq. 8.25. In addition, frequency has been introduced through the λ^4 term. A general solution to this new equation is:

$$Y(x) = A\cos(\lambda x) + B\sin(\lambda x) + C\cosh(\lambda x) + D\sinh(\lambda x). \tag{8.26}$$

To determine the coefficients A, B, C, and D, we must apply the beam's boundary conditions. Given these coefficients, the continuous beam's direct and cross FRFs can be determined. In the next two sections, we will derive the FRFs for beams with fixed-free (cantilever) and free-free boundary conditions.

IN A NUTSHELL   The sinh and cosh functions are the hyperbolic sine and hyperbolic cosine, respectively. These functions are not often encountered by most engineers, but they are analogs of the more familiar sine and cosine functions and they frequently appear in continuous structure solutions. They are defined by $\sinh(x) = \frac{1}{2} \times (e^x - e^{-x})$ and $\cosh(x) = \frac{1}{2}(e^x + e^{-x})$.

8.3.1   Fixed-Free Beam

Figure 8.5 shows a fixed-free beam. In order to determine the FRF at the free end, coordinate 1, we need to apply a force, $F_1 = F\sin(\omega t)$, at this location. Note that the force and vibration, $y(x, t) = Y(x)\sin(\omega t)$, expressions both have the same sinusoidal form. The cantilever base coordinate is labeled as 2.

We need to identify four boundary conditions for this beam in order to determine the coefficients A though D in Eq. 8.26. For the free end, where $x = l$, no moment is supported (it is free to rotate), so we can modify Eq. 8.4 to be:

Fig. 8.5 Fixed-free beam model with a harmonic force applied at the free end

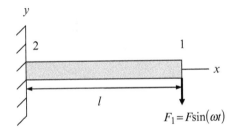

$$\left.\frac{M}{EI}\right|_{x=l} = \left.\frac{d^2y}{dx^2}\right|_{x=l} = 0. \qquad (8.27)$$

We know that the shear force at the free end is $F_1 = F\sin(\omega t)$ so we can define the corresponding boundary condition (see Eq. 8.3).

$$\left.\frac{V}{EI}\right|_{x=l} = \left.\frac{d^3y}{dx^3}\right|_{x=l} = -\frac{F}{EI}\sin(\omega t) \qquad (8.28)$$

At the fixed end, $x = 0$, both the deflection and the slope are zero. You can visualize this by considering the shape of a swimming pool's diving board when you stand at the free end.

$$\left.\frac{dy}{dx}\right|_{x=0} = 0 \qquad (8.29)$$

$$y|_{x=0} = 0 \qquad (8.30)$$

Let'snow use Eqs. 8.27 through 8.30 to determine the coefficients in Eq. 8.26. At $x = 0$, we obtain:

$$Y(0) = A\cos(0) + B\sin(0) + C\cosh(0) + D\sinh(0) = A + C = 0 \qquad (8.31)$$

and

$$\frac{\partial Y}{\partial x}(0) = \lambda(-A\sin(0) + B\cos(0) + C\sinh(0) + D\cosh(0)) = \lambda(B + D) = 0. \qquad (8.32)$$

These two equations specify that $A = -C$ and $B = -D$. At $x = l$, applying the boundary conditions gives:

$$\frac{\partial^2 Y}{\partial x^2}(l) = \lambda^2(-A\cos(\lambda l) - B\sin(\lambda l) + C\cosh(\lambda l) + D\sinh(\lambda l)) = 0 \qquad (8.33)$$

and

$$\frac{\partial^3 Y}{\partial x^3}(l) = \lambda^3(A \sin(\lambda l) - B \cos(\lambda l) + C \sinh(\lambda l) + D \cosh(\lambda l)) = -\frac{F}{EI} \sin(\omega t).$$

$$(8.34)$$

Using the relationships determined from Eqs. 8.31 and 8.32, we can substitute for A and B in Eqs. 8.33 and 8.34 to obtain a system of two equations with two unknowns.

$$C \cos(\lambda l) + D \sin(\lambda l) + C \cosh(\lambda l) + D \sinh(\lambda l) = 0 \qquad (8.35)$$

$$-C \sin(\lambda l) + D \cos(\lambda l) + C \sinh(\lambda l) + D \cosh(\lambda l) = -\frac{F}{\lambda^3 EI} \sin(\omega t) \quad (8.36)$$

Combining terms in these two equations yields:

$$C(\cos(\lambda l) + \cosh(\lambda l)) + D(\sin(\lambda l) + \sinh(\lambda l)) = 0 \qquad (8.37)$$

and

$$C(-\sin(\lambda l) + \sinh(\lambda l)) + D(\cos(\lambda l) + \cosh(\lambda l)) = -\frac{F}{\lambda^3 EI} \sin(\omega t). \qquad (8.38)$$

Let's arrange Eqs. 8.37 and 8.38 into matrix form.

$$\begin{bmatrix} \cos(\lambda l) + \cosh(\lambda l) & \sin(\lambda l) + \sinh(\lambda l) \\ -\sin(\lambda l) + \sinh(\lambda l) & \cos(\lambda l) + \cosh(\lambda l) \end{bmatrix} \begin{Bmatrix} C \\ D \end{Bmatrix} = \begin{Bmatrix} 0 \\ -\frac{F}{\lambda^3 EI} \sin(\omega t) \end{Bmatrix} \qquad (8.39)$$

There are a number of solution methods available for Eq. 8.39. We could apply matrix inversion, for example. However, let's use *Cramer's rule* to determine C and D (Chapra and Canale 1985). To describe this technique, let's rewrite Eq. 8.39 in generic form:

$$\begin{bmatrix} a_{11} & a_{12} \\ a_{21} & a_{22} \end{bmatrix} \begin{Bmatrix} x_1 \\ x_2 \end{Bmatrix} = \begin{Bmatrix} b_1 \\ b_2 \end{Bmatrix}. \qquad (8.40)$$

According to Cramer's rule, we determine x_1 and x_2 using ratios of determinants as shown in Eqs. 8.41 and 8.42.

$$x_1 = \frac{\begin{vmatrix} b_1 & a_{12} \\ b_2 & a_{22} \end{vmatrix}}{\begin{vmatrix} a_{11} & a_{12} \\ a_{21} & a_{22} \end{vmatrix}} = \frac{b_1 a_{22} - b_2 a_{12}}{a_{11} a_{22} - a_{21} a_{12}} \qquad (8.41)$$

$$x_2 = \frac{\begin{vmatrix} a_{11} & b_1 \\ a_{21} & b_2 \end{vmatrix}}{\begin{vmatrix} a_{11} & a_{12} \\ a_{21} & a_{22} \end{vmatrix}} = \frac{a_{11}b_2 - a_{21}b_1}{a_{11}a_{22} - a_{21}a_{12}} \tag{8.42}$$

Using Eqs. 8.39 and 8.41, we can determine C.

$$C = \frac{\begin{vmatrix} 0 & \sin(\lambda l) + \sinh(\lambda l) \\ -\dfrac{F}{\lambda^3 EI}\sin(\omega t) & \cos(\lambda l) + \cosh(\lambda l) \end{vmatrix}}{\begin{vmatrix} \cos(\lambda l) + \cosh(\lambda l) & \sin(\lambda l) + \sinh(\lambda l) \\ -\sin(\lambda l) + \sinh(\lambda l) & \cos(\lambda l) + \cosh(\lambda l) \end{vmatrix}} \tag{8.43}$$

$$C = \frac{\dfrac{F}{\lambda^3 EI}(\sin(\lambda l) + \sinh(\lambda l))}{(\cos(\lambda l) + \cosh(\lambda l))^2 - (-\sin(\lambda l) + \sinh(\lambda l))(\sin(\lambda l) + \sinh(\lambda l))}\sin(\omega t)$$
$$\tag{8.44}$$

$$C = \frac{F(\sin(\lambda l) + \sinh(\lambda l))}{2\lambda^3 EI(1 + \cos(\lambda l)\cosh(\lambda l))}\sin(\omega t) \tag{8.45}$$

Using Eqs. 8.39 and 8.42, we can find D.

$$D = \frac{\begin{vmatrix} \cos(\lambda l) + \cosh(\lambda l) & 0 \\ -\sin(\lambda l) + \sinh(\lambda l) & -\dfrac{F}{\lambda^3 EI}\sin(\omega t) \end{vmatrix}}{\begin{vmatrix} \cos(\lambda l) + \cosh(\lambda l) & \sin(\lambda l) + \sinh(\lambda l) \\ -\sin(\lambda l) + \sinh(\lambda l) & \cos(\lambda l) + \cosh(\lambda l) \end{vmatrix}} \tag{8.46}$$

$$D = \frac{-\dfrac{F}{\lambda^3 EI}(\cos(\lambda l) + \cosh(\lambda l))}{(\cos(\lambda l) + \cosh(\lambda l))^2 - (-\sin(\lambda l) + \sinh(\lambda l))(\sin(\lambda l) + \sinh(\lambda l))}\sin(\omega t) \tag{8.47}$$

$$D = -\frac{F(\cos(\lambda l) + \cosh(\lambda l))}{2\lambda^3 EI(1 + \cos(\lambda l)\cosh(\lambda l))}\sin(\omega t) \tag{8.48}$$

To find Y_1 due to the harmonic force F_1 (see Fig. 8.5), we substitute for A, B, C, and D and let $x = l$ in Eq. 8.26.

$$Y_1 = A\cos(\lambda l) + B\sin(\lambda l) + C\cosh(\lambda l) + D\sinh(\lambda l) \tag{8.49}$$

$$Y_1 = -C\cos(\lambda l) - D\sin(\lambda l) + C\cosh(\lambda l) + D\sinh(\lambda l) \tag{8.50}$$

$$Y_1 = C(-\cos(\lambda l) + \cosh(\lambda l)) + D(-\sin(\lambda l) + \sinh(\lambda l)) \tag{8.51}$$

$$Y_1 = \frac{F(\sin(\lambda l) + \sinh(\lambda l))}{2\lambda^3 EI(1 + \cos(\lambda l)\cosh(\lambda l))} \sin(\omega t)(-\cos(\lambda l) + \cosh(\lambda l))$$
$$-\frac{F(\cos(\lambda l) + \cosh(\lambda l))}{2\lambda^3 EI(1 + \cos(\lambda l)\cosh(\lambda l))} \sin(\omega t)(-\sin(\lambda l) + \sinh(\lambda l)) \tag{8.52}$$

Expanding and simplifying Eq. 8.52 gives:

$$Y_1 = \frac{\sin(\lambda l)\cosh(\lambda l) - \cos(\lambda l)\sinh(\lambda l)}{\lambda^3 EI(1 + \cos(\lambda l)\cosh(\lambda l))} F \sin(\omega t). \tag{8.53}$$

Because $F_1 = F\sin(\omega t)$, the direct FRF at the beam's free end is:

$$\frac{Y_1}{F_1} = \frac{\sin(\lambda l)\cosh(\lambda l) - \cos(\lambda l)\sinh(\lambda l)}{\lambda^3 EI(1 + \cos(\lambda l)\cosh(\lambda l))}, \tag{8.54}$$

where $\lambda^4 = \omega^2 \rho A/EI$. This gives the frequency dependence.

What about the cross FRF, Y_2/F_1, for the fixed-free beam? This is the response at the base due to the harmonic force at the free end. Our intuition should tell us that the cross FRF is zero since, by definition, the base does not move regardless of the force input. Let's verify this using Eq. 8.26. At $x = 0$ we have:

$$Y_2 = A\cos(0) + B\sin(0) + C\cosh(0) + D\sinh(0) = A + C. \tag{8.55}$$

However, based on the boundary conditions (Eq. 8.31), we already know that $A = -C$. Substitution in Eq. 8.55 gives $Y_2 = 0$. Therefore, $Y_2/F_1 = 0$ as expected.

8.3.2   Free-Free Beam

A free-free beam is displayed in Fig. 8.6. To determine the FRF at the right end, coordinate 1 ($x = l$), we apply a force, $F_1 = F\sin(\omega t)$, at this location. The left end is coordinate 2 ($x = 0$). The boundary conditions at the right end are:

$$\left.\frac{M}{EI}\right|_{x=l} = \left.\frac{d^2 y}{dx^2}\right|_{x=l} = 0 \tag{8.56}$$

Fig. 8.6 Free-free beam
model with a harmonic force
applied at coordinate 1

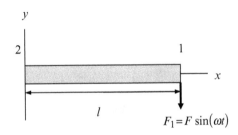

and

$$\frac{V}{EI}\bigg|_{x=l} = \frac{d^3y}{dx^3}\bigg|_{x=l} = -\frac{F}{EI}\sin(\omega t). \qquad (8.57)$$

At the left end, $x = 0$, the beam is again free to rotate, but there is no shear force, so we have:

$$\frac{M}{EI}\bigg|_{x=0} = \frac{d^2y}{dx^2}\bigg|_{x=0} = 0 \qquad (8.58)$$

and

$$\frac{V}{EI}\bigg|_{x=0} = \frac{d^3y}{dx^3}\bigg|_{x=0} = 0. \qquad (8.59)$$

Let's next use the four boundary conditions described by Eqs. 8.56 through 8.59 to determine the coefficients in Eq. 8.26. At $x = 0$, we obtain:

$$\frac{\partial^2 Y}{\partial x^2}(0) = \lambda^2(-A\cos(0) - B\sin(0) + C\cosh(0) + D\sinh(0)) = \lambda^2(-A + C) = 0 \qquad (8.60)$$

and

$$\frac{\partial^3 Y}{\partial x^3}(0) = \lambda^3(A\sin(0) - B\cos(0) + C\sinh(0) + D\cosh(0)) = \lambda^3(-B + D) = 0. \qquad (8.61)$$

From these two equations, we see that $A = C$ and $B = D$. At $x = l$, applying the boundary conditions gives:

$$\frac{\partial^2 Y}{\partial x^2}(l) = \lambda^2(-A\cos(\lambda l) - B\sin(\lambda l) + C\cosh(\lambda l) + D\sinh(\lambda l)) = 0 \qquad (8.62)$$

and

$$\frac{\partial^3 Y}{\partial x^3}(l) = \lambda^3(A\sin(\lambda l) - B\cos(\lambda l) + C\sinh(\lambda l) + D\cosh(\lambda l)) = -\frac{F}{EI}\sin(\omega t).$$

$$(8.63)$$

Substituting for A and B in Eqs. 8.62 and 8.63 yields a system of two equations with two unknowns.

$$-C\cos(\lambda l) - D\sin(\lambda l) + C\cosh(\lambda l) + D\sinh(\lambda l) = 0 \qquad (8.64)$$

$$C\sin(\lambda l) - D\cos(\lambda l) + C\sinh(\lambda l) + D\cosh(\lambda l) = -\frac{F}{\lambda^3 EI}\sin(\omega t) \qquad (8.65)$$

Combining terms gives:

$$C(-\cos(\lambda l) + \cosh(\lambda l)) + D(-\sin(\lambda l) + \sinh(\lambda l)) = 0 \qquad (8.66)$$

and

$$C(\sin(\lambda l) + \sinh(\lambda l)) + D(-\cos(\lambda l) + \cosh(\lambda l)) = -\frac{F}{\lambda^3 EI}\sin(\omega t). \qquad (8.67)$$

Arranging Eqs. 8.66 and 8.67 into matrix form results in Eq. 8.68.

$$\begin{bmatrix} -\cos(\lambda l) + \cosh(\lambda l) & -\sin(\lambda l) + \sinh(\lambda l) \\ \sin(\lambda l) + \sinh(\lambda l) & -\cos(\lambda l) + \cosh(\lambda l) \end{bmatrix} \begin{Bmatrix} C \\ D \end{Bmatrix} = \begin{Bmatrix} 0 \\ -\dfrac{F}{\lambda^3 EI}\sin(\omega t) \end{Bmatrix}$$

$$(8.68)$$

Using Cramer's rule, we determine C.

$$C = \frac{\begin{vmatrix} 0 & -\sin(\lambda l) + \sinh(\lambda l) \\ -\dfrac{F}{\lambda^3 EI}\sin(\omega t) & -\cos(\lambda l) + \cosh(\lambda l) \end{vmatrix}}{\begin{vmatrix} -\cos(\lambda l) + \cosh(\lambda l) & -\sin(\lambda l) + \sinh(\lambda l) \\ \sin(\lambda l) + \sinh(\lambda l) & -\cos(\lambda l) + \cosh(\lambda l) \end{vmatrix}} \qquad (8.69)$$

$$C = \frac{\dfrac{F}{\lambda^3 EI}(-\sin(\lambda l) + \sinh(\lambda l))}{(-\cos(\lambda l) + \cosh(\lambda l))^2 - (\sin(\lambda l) + \sinh(\lambda l))(-\sin(\lambda l) + \sinh(\lambda l))}\sin(\omega t)$$

$$(8.70)$$

$$C = \frac{F(-\sin(\lambda l) + \sinh(\lambda l))}{2\lambda^3 EI(1 - \cos(\lambda l)\cosh(\lambda l))}\sin(\omega t) \qquad (8.71)$$

Let's now calculate D.

$$D = \frac{\begin{vmatrix} -\cos(\lambda l) + \cosh(\lambda l) & 0 \\ \sin(\lambda l) + \sinh(\lambda l) & -\dfrac{F}{\lambda^3 EI}\sin(\omega t) \end{vmatrix}}{\begin{vmatrix} -\cos(\lambda l) + \cosh(\lambda l) & -\sin(\lambda l) + \sinh(\lambda l) \\ \sin(\lambda l) + \sinh(\lambda l) & -\cos(\lambda l) + \cosh(\lambda l) \end{vmatrix}} \tag{8.72}$$

$$D = \frac{-\dfrac{F}{\lambda^3 EI}(-\cos(\lambda l) + \cosh(\lambda l))}{(-\cos(\lambda l) + \cosh(\lambda l))^2 - (\sin(\lambda l) + \sinh(\lambda l))(-\sin(\lambda l) + \sinh(\lambda l))} \sin(\omega t) \tag{8.73}$$

$$D = \frac{F(\cos(\lambda l) - \cosh(\lambda l))}{2\lambda^3 EI(1 - \cos(\lambda l)\cosh(\lambda l))}\sin(\omega t) \tag{8.74}$$

To find Y_1 due to the harmonic force F_1 (see Fig. 8.6), we substitute for $A, B, C,$ and D and let $x = l$ in Eq. 8.26.

$$Y_1 = A\cos(\lambda l) + B\sin(\lambda l) + C\cosh(\lambda l) + D\sinh(\lambda l) \tag{8.75}$$

$$Y_1 = C\cos(\lambda l) + D\sin(\lambda l) + C\cosh(\lambda l) + D\sinh(\lambda l) \tag{8.76}$$

$$Y_1 = C(\cos(\lambda l) + \cosh(\lambda l)) + D(\sin(\lambda l) + \sinh(\lambda l)) \tag{8.77}$$

$$\begin{aligned} Y_1 = &\frac{F(-\sin(\lambda l) + \sinh(\lambda l))}{2\lambda^3 EI(1 - \cos(\lambda l)\cosh(\lambda l))}\sin(\omega t)(\cos(\lambda l) + \cosh(\lambda l)) \\ + &\frac{F(\cos(\lambda l) - \cosh(\lambda l))}{2\lambda^3 EI(1 - \cos(\lambda l)\cosh(\lambda l))}\sin(\omega t)(\sin(\lambda l) + \sinh(\lambda l)) \end{aligned} \tag{8.78}$$

Expanding and simplifying Eq. 8.78 gives:

$$Y_1 = \frac{\cos(\lambda l)\sinh(\lambda l) - \sin(\lambda l)\cosh(\lambda l)}{\lambda^3 EI(1 - \cos(\lambda l)\cosh(\lambda l))}F\sin(\omega t). \tag{8.79}$$

The direct FRF at the beam's free end is:

$$\frac{Y_1}{F_1} = \frac{\cos(\lambda l)\sinh(\lambda l) - \sin(\lambda l)\cosh(\lambda l)}{\lambda^3 EI(1 - \cos(\lambda l)\cosh(\lambda l))}. \tag{8.80}$$

For the cross FRF, Y_2/F_1, we require the expression for Y_2. This is the response at coordinate 2 due to the harmonic force at coordinate 1. At $x = 0$ we have:

$$Y_2 = A\cos(0) + B\sin(0) + C\cosh(0) + D\sinh(0) = A + C. \tag{8.81}$$

Based on the free boundary condition (Eq. 8.60), we know that $A = C$. Therefore, $Y_2 = 2C$. Using Eq. 8.71, we obtain:

$$Y_2 = \frac{(-\sin(\lambda l) + \sinh(\lambda l))}{\lambda^3 EI(1 - \cos(\lambda l)\cosh(\lambda l))} F\sin(\omega t). \tag{8.82}$$

The cross FRF is finally:

$$\frac{Y_2}{F_1} = \frac{(-\sin(\lambda l) + \sinh(\lambda l))}{\lambda^3 EI(1 - \cos(\lambda l)\cosh(\lambda l))}. \tag{8.83}$$

8.4   Solid Damping in Beam Models

At this point we have not yet included damping in our continuous beam transverse vibration FRFs. As we discussed in Sect. 2.4, various damping models are available. These include: viscous damping, which relates the damping force to velocity; Coulomb damping, which represents the energy dissipation due to dry sliding between two surfaces; and *solid damping*, which occurs due to internal energy dissipation within the material of the vibrating body. Viscous damping is convenient mathematically, but solid damping makes the most intuitive sense for modeling the vibration of continuous beams.

Solid damping is included in the differential equation of motion as a complex stiffness term. For the differential equation of forced harmonic motion, we include the unitless solid damping factor, η, as:

$$m\ddot{x} + k(1 + i\eta)x = Fe^{i\omega t}. \tag{8.84}$$

In our continuous beam model, this material-dependent damping is conveniently incorporated in the elastic modulus which, together with the second moment of area and beam length, defines the beam's stiffness. The new *complex modulus*, E_s, is:

$$E_s = E(1 + i\eta). \tag{8.85}$$

Damping is therefore incorporated in the continuous beam FRFs by replacing E with E_s. Note that this substitution also holds for $\lambda^4 = \omega^2 \frac{\rho A}{E_s I}$.

Let's now explore the relationship between η and the viscous damping ratio, ζ. For the forced harmonic motion described by Eq. 8.84, we can assume a solution of the form $x(t) = Xe^{i\omega t}$. Substitution yields:

$$\left(-m\omega^2 + k(1 + i\eta)\right)Xe^{i\omega t} = Fe^{i\omega t}. \tag{8.86}$$

Rearranging to identify the FRF gives:

$$\frac{X}{F}(\omega) = \frac{1}{k - m\omega^2 + ik\eta}. \tag{8.87}$$

The value at resonance,[3] where $\omega = \omega_n$, is:

$$\frac{X}{F}(\omega_n) = \frac{1}{k - m\omega_n^2 + ik\eta} = \frac{1}{k - m\frac{k}{m} + ik\eta} = \frac{1}{ik\eta}. \tag{8.88}$$

The corresponding FRF magnitude is $1/k\eta$. For viscous damping, we can complete an equivalent analysis. The differential equation of motion is now:

$$m\ddot{x} + c\dot{x} + kx = Fe^{i\omega t}. \tag{8.89}$$

Again, assuming a solution of the form $x(t) = Xe^{i\omega t}$ gives:

$$\left(-m\omega^2 + i\omega c + k\right)Xe^{i\omega t} = Fe^{i\omega t}. \tag{8.90}$$

The FRF is:

$$\frac{X}{F}(\omega) = \frac{1}{-m\omega^2 + i\omega c + k}. \tag{8.91}$$

As we saw in Sect. 3.2, we can rewrite this equation as:

$$\frac{X}{F}(\omega) = \frac{1}{k}\left(\frac{1}{\left(1 - \left(\frac{\omega}{\omega_n}\right)^2\right) + i2\zeta\left(\frac{\omega}{\omega_n}\right)}\right). \tag{8.92}$$

---

[3] We consider the resonant case because this is where damping has the most significant effect. Its influence is less at frequencies far from resonance.

Fig. 8.7 *By the Numbers*
8.2 – Dimensions for a steel
machinist's scale with fixed-
free boundary conditions

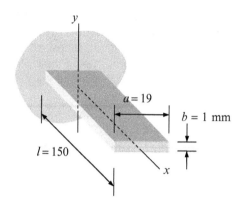

At resonance, the FRF value is:

$$\frac{X}{F}(\omega_n) = \frac{1}{k}\left(\frac{1}{\left(1 - \left(\dfrac{\omega_n}{\omega_n}\right)^2\right) + i2\zeta\left(\dfrac{\omega_n}{\omega_n}\right)}\right) = \frac{1}{i2k\zeta}. \qquad (8.93)$$

The FRF magnitude at resonance for viscous damping is therefore $1/2k\zeta$. Equating this result with the resonant magnitude for solid damping gives the relationship between η and ζ at resonance.

$$\zeta = \frac{\eta}{2} \qquad (8.94)$$

Solid damping factors are quite low. For steel alloys, typical values are between 0.001 and 0.002. According to Eq. 8.93, the equivalent damping ratio at resonance is 0.0005–0.001 (0.05–0.1%). These small values emphasize that most damping in structures is introduced at the connections between individual beam members rather than within the beams themselves.

By the Numbers 8.2

Let's consider an example fixed-free beam and calculate the direct FRF at its free end. We will use the dimensions and material properties for a typical steel ruler/ machinist's scale; see Fig. 8.7. For the steel beam, we will use an elastic modulus of 200 GPa and a density of 7,800 kg/m³. The second moment of area is:

$$I = \frac{ab^3}{12} = \frac{0.019(0.001)^3}{12} = 1.583 \times 10^{-12}\,\mathrm{m}^4$$

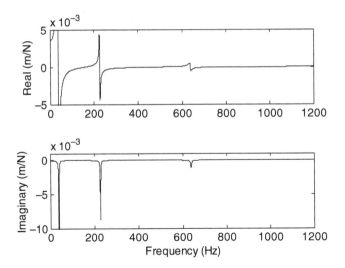

Fig. 8.8 *By the Numbers 8.2* – Direct FRF for the free end of the beam depicted in Fig. 8.7

and the cross-sectional area, A, is:

$$A = ab = 0.019(0.001) = 1.9 \times 10^{-5} \, \text{m}^2.$$

To plot the free end direct FRF, we use Eq. 8.54 and replace E with the complex modulus $E_s = E(1 + i\eta)$ from Eq. 8.85.

$$\frac{Y_1}{F_1} = \frac{\sin(\lambda l)\cosh(\lambda l) - \cos(\lambda l)\sinh(\lambda l)}{\lambda^3 E(1 + i\eta)I(1 + \cos(\lambda l)\cosh(\lambda l))} \tag{8.95}$$

Let's choose a solid damping factor of 0.01 for display purposes. Even though this value is approximately an order of magnitude too large, it reduces the "sharpness" of the real and imaginary parts of the FRF and enables us to view them more clearly. Figure 8.8 shows the first three bending modes for a bandwidth of 1,200 Hz; note that the scale was selected to be able to observe the third mode. If we increased the frequency range, we would see additional modes.[4] These first three modes, which appear at 36.4, 227.8, and 637.9 Hz, correspond to the first three bending mode shapes shown in Fig. 6.19.

For Euler–Bernoulli beams, a closed-form expression for the natural frequencies has been developed for various boundary conditions (Blevins 2001). See Eq. 8.96, where $i = 1, 2, 3...$ indicates the natural frequency numbers in ascending order.

---

[4] For a continuous beam there are an infinite number of modes for an infinite bandwidth.

Table 8.1 β_i values for
fixed-free Euler–Bernoulli
beam natural frequency
calculations (Blevins 2001)

i	β_i
1	1.87510107
2	4.69409113
3	7.85475744
4	10.99554073
5	14.13716839
>5	$\pi/2(2i-1)$

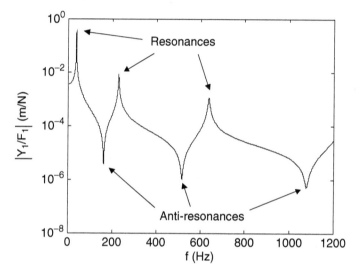

Fig. 8.9 *By the Numbers 8.2* – Semi-logarithmic plot of the direct FRF for the free end of the beam depicted in Fig. 8.7

$$f_{n,i} = \frac{\beta_i^2}{2\pi l^2} \sqrt{\frac{EI}{\rho A}} \ (\text{Hz}) \tag{8.96}$$

For the fixed-free beam in this example, the β_i values are provided in Table 8.1 (Blevins 2001). Substitution of the beam geometry and material properties gives the same first three natural frequencies we determined using Figs. 8.8 and 8.9.

The direct FRF is presented in a semi-logarithmic format in Fig. 8.9. We again observe the resonant peaks at 36.4, 227.8, and 637.9 Hz, but we also see local minima at 159.4, 516.6, and 1,077.8 Hz. These are referred to as *anti-resonant frequencies* and represent frequencies where the beam response is small even for a large force input. While this definition sounds similar to the node description provided in Sect. 6.5, in this case we are describing frequencies, not spatial locations (nodes), where the response is small. The code used to produce Figs. 8.8 and 8.9 is provided in MATLAB® MOJO 8.2.

MATLAB® MOJO 8.2

```
% matlab_mojo_8_2.m

clc
clear all
close all

f = 1:0.1:1200;          % Hz
omega = f*2*pi;          % rad/s

%  Define beam
a = 19e-3;               % m
b = 1e-3;                % m
l = 150e-3;              % m
E = 2e11;                % N/m^2
density = 7800;          % kg/m^3
eta = 0.01;
I = a*b^3/12;            % m^4
EI = E*I;                % N-m^2
EI = EI*(1+1i*eta);      % N-m^2
A = a*b;                 % m^2

lambda = (omega.^2*density*A/EI).^0.25;

% Direct FRF for the free end of the fixed-free beam
Y1_F1 = (sin(lambda*l).*cosh(lambda*l)-
cos(lambda*l).*sinh(lambda*l))./(lambda.^3*EI.*(1+cos(lambda*l).*cosh(lambda*
l)));

figure(1)
subplot(211)
plot(f, real(Y1_F1), 'k')
set(gca,'FontSize', 14)
ylim([-0.005 0.005])
ylabel('Real (m/N)')
subplot(212)
plot(f, imag(Y1_F1), 'k')
set(gca,'FontSize', 14)
ylim([-0.01 0.001])
xlabel('Frequency (Hz)')
ylabel('Imaginary (m/N)')

figure(2)
semilogy(f, abs(Y1_F1), 'k')
set(gca,'FontSize', 14)
xlabel('f (Hz)')
ylabel('|Y_1/F_1| (m/N)')
```

What would happen if we increased the width of the beam in Fig. 8.7? Let's consider a 20% increase so that $a = 0.019(1.2) = 0.0228$ m. The beam stiffness depends on the modulus (unchanged), length (unchanged), and second moment of area. The new second moment of area is:

$$I = \frac{ab^3}{12} = \frac{0.0228(0.001)^3}{12} = 1.9 \times 10^{-12} \text{m}^4.$$

Because the stiffness increases with I, we might anticipate that the natural frequencies will increase with a. The new direct FRF, together with the previous result from Fig. 8.8 (dotted line), is provided in Fig. 8.10. Surprisingly, the natural

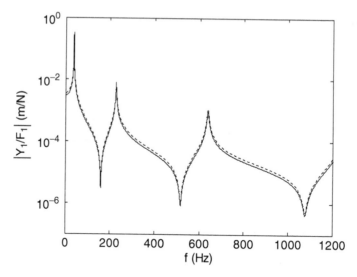

Fig. 8.10 *By the Numbers 8.2* – Direct FRF for the beam with a 20% width increase (the dotted line shows the FRF for the original width)

frequencies did not change! While the wider beam is indeed stiffer, it is also heavier. These effects serve to offset each other so that the natural frequencies are not affected. We can observe this in Eq. 8.96. Substituting for I and A (based on the rectangular beam in Fig. 8.7), we see that a cancels.

$$f_{n,i} = \frac{\beta_i^2}{2\pi l^2}\sqrt{\frac{E\frac{ab^3}{12}}{\rho ab}} = \frac{\beta_i^2}{2\pi l^2}\sqrt{\frac{E\frac{b^2}{12}}{\rho}}(\text{Hz}) \qquad (8.97)$$

The same result is obtained from Eq. 8.95. The roots of the denominator give the natural frequencies and these roots depend on the product:

$$\lambda l = \left(\omega^2 \frac{\rho A}{E_s I}\right)^{\frac{1}{4}} l = \left(\omega^2 \frac{\rho ab}{E_s \frac{ab^3}{12}}\right)^{\frac{1}{4}} l = \left(\omega^2 \frac{\rho}{E_s \frac{b^2}{12}}\right)^{\frac{1}{4}} l.$$

Again, the natural frequencies of the rectangular beam do not depend on a after substituting and simplifying. While the resonant peaks appear at the same frequencies for both FRFs in Fig. 8.10, the new FRF with the increased width is stiffer (its FRF appears below the original FRF). The source of the change in magnitude is also observed using Eq. 8.95. We see that I appears in the denominator. It does not change the roots, but does serve to scale the FRF. As I (and the denominator) increases with a, the magnitude decreases.

8.5   Rotation Frequency Response Functions

In Sects. 8.3.1 and 8.3.2 we derived the transverse vibration FRFs, Y/F, for beams with fixed-free and free-free boundary conditions. We are not limited to these boundary conditions of course. We can complete the same analysis using the boundary conditions summarized in Table 8.2. In this table we see one unfamiliar entry – the boundary condition due to a harmonic bending couple is included. Let's now extend our analysis to consider not only transverse deflection, $y(x, t)$, but also rotation of the beam in the bending plane, $\theta(x, t)$; see Fig. 8.11.

To determine the rotation FRF, Θ_1/F_1, for a fixed-free beam, we return to Eq. 8.26 and substitute $A = -C$ and $B = -D$ (from Eqs. 8.31 and 8.32).

$$Y = C(-\cos(\lambda x) + \cosh(\lambda x)) + D(-\sin(\lambda x) + \sinh(\lambda x)) \qquad (8.98)$$

We obtain rotation by differentiating Y with respect to x, $\Theta = dY/dx$. We then evaluate this expression at $x = l$ (coordinate 1) and substitute for C and D from Eqs. 8.45 and 8.48. Finally, we divide this expression by F_1. The result is provided in Eq. 8.99.

$$\frac{\Theta_1}{F_1} = \frac{\sin(\lambda L)\sinh(\lambda L)}{\lambda^2 EI(1 + \cos(\lambda L)\cosh(\lambda L))} \qquad (8.99)$$

In addition to the deflection and rotation responses due to a harmonic force, we can also determine the responses due to a harmonic bending couple, Y_1/M_1 and Θ_1/M_1. We find these terms by applying the harmonic bending couple $M_1 = M\sin(\omega t)$ at coordinate 1 as shown in Fig. 8.12. The boundary conditions at coordinate 2 $(x = 0)$ are $y = 0$ and $\partial y/\partial x = 0$. The boundary conditions at

Table 8.2 Boundary conditions for beam FRF calculations	End description	Boundary conditions
	Fixed	$y = 0$, $\partial y/\partial x = 0$
	Free	$\partial^2 y/\partial x^2 = 0$, $\partial^3 y/\partial x^3 = 0$
	Pinned	$y = 0$, $\partial^2 y/\partial x^2 = 0$
	Sliding	$\partial y/\partial x = 0$, $\partial^3 y/\partial x^3 = 0$
	Harmonic force $F\sin(\omega t)$	$\partial^3 y/\partial x^3 = -(F/EI)\sin(\omega t)$
	Harmonic bending couple $M\sin(\omega t)$	$\partial^2 y/\partial x^2 = (M/EI)\sin(\omega t)$

Fig. 8.11 Coordinates for both transverse deflection and rotation vibration in the bending plane

Fig. 8.12 Fixed-free beam with a harmonic bending couple applied at the free end

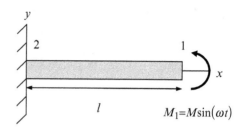

coordinate 1 ($x = l$) are $\partial^2 y/\partial x^2 = M/EI \sin(\omega t)$ and $\partial^3 y/\partial x^3 = 0$. We find the coefficients A, B, C, and D from Eq. 8.26 in the same manner as described previously.

At $x = 0$, the situation is identical to the force application case shown in Fig. 8.5, so we obtain $A = -C$ and $B = -D$. At $x = l$, we first use $\partial^2 y/\partial x^2 = M/EI \sin(\omega t)$ as demonstrated in Eq. 8.100.

$$\left.\frac{\partial^2 y}{\partial x^2}\right|_{x=l} = \lambda^2 \left(\begin{array}{c} -A\cos(\lambda l) - B\sin(\lambda l)+ \\ C\cosh(\lambda l) + D\sinh(\lambda l) \end{array} \right) = \frac{M}{EI} \sin(\omega t) \qquad (8.100)$$

Substitution for A and B in Eq. 8.100 gives:

$$C(\cos(\lambda L) + \cosh(\lambda L)) + D(\sin(\lambda L) + \sinh(\lambda L)) = \frac{M}{\lambda^2 EI} \sin(\omega t). \qquad (8.101)$$

We next apply $\partial^3 y/\partial x^3 = 0$ (at $x = l$) and substitute for A and B to get:

$$C(-\sin(\lambda L) + \sinh(\lambda L)) + D(\cos(\lambda L) + \cosh(\lambda L)) = 0. \qquad (8.102)$$

Expressing Eqs. 8.101 and 8.102 in matrix form yields:

$$\begin{bmatrix} \cos(\lambda L) + \cosh(\lambda L) & \sin(\lambda L) + \sinh(\lambda L) \\ -\sin(\lambda L) + \sinh(\lambda L) & \cos(\lambda L) + \cosh(\lambda L) \end{bmatrix} \begin{Bmatrix} C \\ D \end{Bmatrix} = \begin{Bmatrix} \dfrac{M}{\lambda^2 EI} \\ 0 \end{Bmatrix} \sin(\omega t).$$

$$(8.103)$$

Applying Cramer's rule, we solve for C and D.

$$C = \frac{M(\cos(\lambda L) + \cosh(\lambda L))}{2\lambda^2 EI(1 + \cos(\lambda L)\cosh(\lambda L))} \sin(\omega t) \qquad (8.104)$$

$$D = -\frac{M(-\sin(\lambda L) + \sinh(\lambda L))}{2\lambda^2 EI(1 + \cos(\lambda L)\cosh(\lambda L))} \sin(\omega t) \qquad (8.105)$$

We find Y_1 by substituting Eqs. 8.104 and 8.105, together with the relationships $A = -C$ and $B = -D$ in Eq. 8.26. We also set $x = l$. The result is provided in Eq. 8.106.

$$Y_1 = -\left(\frac{\dfrac{(\cos(\lambda L) + \cosh(\lambda L))(\cos(\lambda L) - \cosh(\lambda L))}{2\lambda^2 EI(1 + \cos(\lambda L)\cosh(\lambda L))} +}{\dfrac{(\sin(\lambda L) - \sinh(\lambda L))(\sin(\lambda L) - \sinh(\lambda L))}{2\lambda^2 EI(1 + \cos(\lambda L)\cosh(\lambda L))}} \right) M\sin(\omega t) \quad (8.106)$$

We obtain the Y_1/M_1 FRF at the free end of the fixed-free beam by dividing Eq. 8.106 by M_1 and simplifying; see Eq. 8.107. A comparison of Eqs. 8.107 and 8.99 shows us that the Y_1/M_1 and Θ_1/F_1 FRFs are identical.

$$\frac{Y_1}{M_1} = \frac{-\left(\dfrac{\dfrac{(\cos(\lambda L) + \cosh(\lambda L))(\cos(\lambda L) - \cosh(\lambda L))}{2\lambda^2 EI(1 + \cos(\lambda L)\cosh(\lambda L))} +}{\dfrac{(\sin(\lambda L) - \sinh(\lambda L))(\sin(\lambda L) - \sinh(\lambda L))}{2\lambda^2 EI(1 + \cos(\lambda L)\cosh(\lambda L))}} \right) M\sin(\omega t)}{M\sin(\omega t)}$$

$$\frac{Y_1}{M_1} = -\left(\frac{\dfrac{(\cos(\lambda L) + \cosh(\lambda L))(\cos(\lambda L) - \cosh(\lambda L))}{2\lambda^2 EI(1 + \cos(\lambda L)\cosh(\lambda L))} +}{\dfrac{(\sin(\lambda L) - \sinh(\lambda L))(\sin(\lambda L) - \sinh(\lambda L))}{2\lambda^2 EI(1 + \cos(\lambda L)\cosh(\lambda L))}} \right)$$

$$\frac{Y_1}{M_1} = \frac{\sin(\lambda L)\sinh(\lambda L)}{\lambda^2 EI(1 + \cos(\lambda L)\cosh(\lambda L))} \quad (8.107)$$

To determine the Θ_1/M_1 FRF, we return to Eq. 8.26 and substitute $A = -C$ and $B = -D$ (from Eqs. 8.31 and 8.32).

$$Y = (C(-\cos(\lambda x) + \cosh(\lambda x)) + D(-\sin(\lambda x) + \sinh(\lambda x))) \quad (8.108)$$

We then find Θ_1/M_1 by: (1) differentiating Y with respect to x to obtain rotation $\Theta = dY/dx$; (2) evaluating this expression at $x = l$; (3) substituting for C and D from Eqs. 8.104 and 8.105; and (4) dividing this result by M_1. See Eq. 8.109.

$$\frac{\Theta_1}{M_1} = \frac{\sin(\lambda L)\cosh(\lambda L) + \cos(\lambda L)\sinh(\lambda L)}{\lambda EI(1 + \cos(\lambda L)\cosh(\lambda L))} \quad (8.109)$$

This process can be repeated for any of the boundary conditions shown in Table 8.2. The FRFs for fixed-free and free-free boundary conditions are summarized in Table 8.3, where both direct and cross FRFs are included for the free-free beam. As discussed previously, no cross FRFs are shown for the fixed-free beam because the response at the free end is zero for any excitation at the fixed end and the response is always zero at the fixed end.

Table 8.3 Euler–Bernoulli beam FRFs for fixed-free and free-free boundary conditions (Bishop and Johnson 1960). Coordinate 2 is the fixed end for the fixed-free beam

$\frac{Y_2}{F_2}$	$\frac{\Theta_2}{F_2},\frac{Y_2}{M_2}$	$\frac{Y_1}{F_2},\frac{Y_2}{F_1}$	$\frac{\Theta_1}{F_2},\frac{Y_2}{M_1}$	$\frac{\Theta_2}{M_2}$	$\frac{\Theta_2}{F_1},\frac{Y_1}{M_2}$	$\frac{\Theta_1}{M_2},\frac{\Theta_2}{M_1}$	$\frac{Y_1}{F_1}$	$\frac{\Theta_1}{F_1},\frac{Y_1}{M_1}$	$\frac{\Theta_1}{M_1}$
				Free-free					
$\frac{-c_1}{\lambda^3 c_7}$	$\frac{-c_2}{\lambda^2 c_7}$	$\frac{c_3}{\lambda^3 c_7}$	$\frac{c_4}{\lambda^2 c_7}$	$\frac{c_5}{\lambda c_7}$	$\frac{-c_4}{\lambda^2 c_7}$	$\frac{c_6}{\lambda c_7}$	$\frac{-c_1}{\lambda^3 c_7}$	$\frac{c_2}{\lambda^2 c_7}$	$\frac{c_5}{\lambda c_7}$
				Fixed-free					
							$\frac{-c_1}{\lambda^3 c_8}$	$\frac{c_2}{\lambda^2 c_8}$	$\frac{c_5}{\lambda c_8}$

Terms c_1 through c_8

$c_1 = \cos(\lambda l)\sinh(\lambda l) - \sin(\lambda l)\cosh(\lambda l)$	$c_5 = \cos(\lambda l)\sinh(\lambda l) + \sin(\lambda l)\cosh(\lambda l)$
$c_2 = \sin(\lambda l)\sinh(\lambda l)$	$c_6 = \sin(\lambda l) + \sinh(\lambda l)$
$c_3 = \sin(\lambda l) - \sinh(\lambda l)$	$c_7 = E_s I(\cos(\lambda l)\cosh(\lambda l) - 1)$
$c_4 = \cos(\lambda l) - \cosh(\lambda l)$	$c_8 = E_s I(\cos(\lambda l)\cosh(\lambda l) + 1)$

8.6   Transverse Vibration FRF Measurement Comparisons

8.6.1   Fixed-Free Beam

To apply the FRFs defined in Table 8.3, let's compare these models with measurements completed using the BEP. In Sect. 7.4, we performed an impact test on the BEP with the cantilevered steel rod extended 130 mm beyond the base; the measurement setup is displayed in Fig. 7.20. Let's now model the rod as a fixed-free beam with a length of 130 mm and compare the analytical prediction with the measured result (shown previously in Fig. 7.21).

The relevant equation from Table 8.2 is:

$$\frac{Y_1}{F_1} = \frac{\sin(\lambda l)\cosh(\lambda l) - \cos(\lambda l)\sinh(\lambda l)}{\lambda^3 E(1+i\eta)I(\cos(\lambda l)\cosh(\lambda l) + 1)},$$

where $\lambda^4 = \omega^2 \frac{\rho A}{E(1+i\eta)I}$. The beam in this case is a 12.7-mm-diameter cylinder, so the second moment of area is:

$$I = \frac{\pi d^4}{64} = \frac{\pi(0.0127)^4}{64} = 1.277 \times 10^{-9}\,\text{m}^4,$$

and the cross-sectional area, A, is:

$$A = \frac{\pi d^2}{4} = \frac{\pi(0.0127)^2}{4} = 1.267 \times 10^{-4}\,\text{m}^2.$$

For the steel rod, we can use $\rho = 7,800\,\text{kg/m}^3$ and $E = 200\,\text{GPa}$. The measured (dotted line) and predicted (solid line) FRFs are shown in Fig. 8.13. The solid

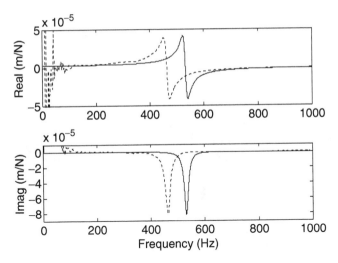

Fig. 8.13 Comparison between measured (*dotted line*) and predicted (*solid line*) fixed-free FRFs on the BEP

damping factor was selected to be 0.034 to match the measured and predicted FRF magnitudes. Given this very large solid damping factor (a reasonable value of 0.001–0.002 was suggested for steel in Sect. 8.4) and the natural frequency mismatch (the predicted natural frequency is too high), we can make an observation about the BEP setup. A reasonable explanation for the experimental behavior is that the split-clamp used to secure the rod in the BEP holder does not provide an ideal fixed boundary condition. Small relative motion between the rod and base at the split-clamp connection would explain the additional damping. Also, the effective beam length may be slightly more than the measured value of 130 mm due to the radius/chamfer at the edge of the hole used to clamp the rod. This result is not unique; in practice, it is quite difficult to realize a fixed boundary condition.

8.6.2   Free-Free Beam

Let's now remove the steel rod from the BEP and measure its response alone (the rod's length is 152.5 mm). We can approximate free-free boundary conditions by supporting the rod on a soft foam base as shown in Fig. 8.14. The appropriate FRF equation from Table 8.3 is now:

$$\frac{Y_1}{F_1} = \frac{\sin(\lambda l)\cosh(\lambda l) - \cos(\lambda l)\sinh(\lambda l)}{\lambda^3 E(1 + i\eta)I(\cos(\lambda l)\cosh(\lambda l) + 1)}.$$

A comparison between the measured (dotted line) and predicted (solid line) FRFs is provided in Fig. 8.15 (5,000 Hz measurement bandwidth). The solid

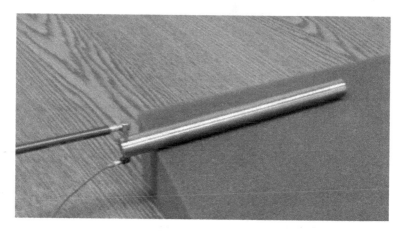

Fig. 8.14 Experimental impact testing setup for the free-free beam measurement

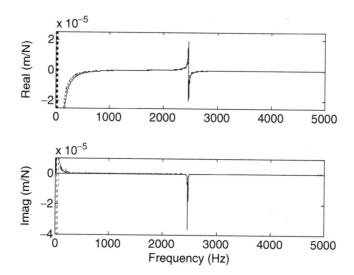

Fig. 8.15 Comparison between measured (*dotted line*) and predicted (*solid line*) free-free FRFs

damping factor for the prediction is 0.003. This significantly reduced value supports our theory that the clamping conditions for the rod served as a primary source of energy dissipation (damping) for the fixed-free setup. We do note, however, that $\eta = 0.003$ is still slightly larger than the anticipated value. In this case, it is the foam base that contributes the additional damping.

If we examine the FRFs in Fig. 8.15 carefully, we observe a behavior unique to free-free FRFs. The absolute value of the real parts gets very large as the frequency approaches zero. This is due to *rigid body modes* of the free-free beam. As we discussed in Sect. 6.5, if we imagine the rod floating in space, then applying a force at the center of mass will cause it to translate as a rigid body in the direction

	i	β_i
Table 8.4 β_i values for free-free Euler–Bernoulli beam natural frequency calculations (Blevins 2001)	1	4.73004074
	2	7.85320462
	3	10.9956078
	4	14.1371655
	5	17.2787597
	>5	$\pi/2(2i-1)$

of the force. Alternately, applying a force at any other location will also cause it to rotate, again as a rigid body. Because there is no oscillation associated with these modes, they occur at zero frequency and are real-valued. Note that the free-free FRF expressions in Table 8.3 include these rigid body modes (as seen in the Fig. 8.15 prediction).

8.6.3   Natural Frequency Uncertainty

As a final consideration regarding the prediction of a beam's dynamic behavior, we need to recognize that, because the inputs for any model are uncertain (the true value of any quantity is never known), the output inherently includes uncertainty. As we discussed in Sect. 2.4.7, we can perform a first-order Taylor series expansion of the model (provided an analytical expression is available) to determine the output uncertainty as a function of the input uncertainties. In this way, we propagate the input uncertainties through the model to find the output uncertainty. Let's consider the sensitivity of the natural frequency equation shown in Eq. 8.96 to the beam's length. To do so, let's rewrite Eq. 8.96 to isolate the beam length, l:

$$f_{n,i} = \frac{1}{l^2}\frac{\beta_i^2}{2\pi}\sqrt{\frac{EI}{\rho A}}(\text{Hz}).$$

For the free-free beam discussed in Sect. 8.6.2, the β_i values are provided in Table 8.4 (Blevins 2001). While there is uncertainty associated with E, I, ρ, A, if we consider only l, then the natural frequency uncertainty is determined using:

$$u^2\left(f_{n,i}\right) = \left(\frac{\partial f_{n,i}}{\partial l}\right)^2 u^2(l) = \left(\frac{-2}{l^3}\frac{\beta_i^2}{2\pi}\sqrt{\frac{EI}{\rho A}}\right)^2 u^2(l),$$

where $u(l)$ is the beam length uncertainty (the square of the uncertainty, $u^2(l)$, is referred to as the *variance*) and the average (or mean) values of the input variables are applied to evaluate the natural frequency uncertainty. For the free-free 12.7-mm-diameter steel rod with a nominal length of 152.5 mm and an associated uncertainty of 0.1 mm, the uncertainty in the first natural frequency is $u\left(f_{n,1}\right) = 3.2$ Hz. This value represents one standard deviation and means that we would

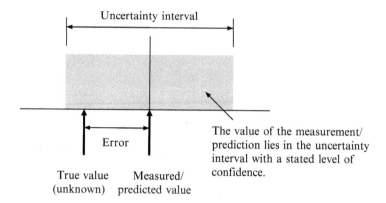

Fig. 8.16 Relationships between error, uncertainty, the measured/predicted value, and the unknown true value

expect our predicted natural frequency to be within ±3.2 Hz of the true value 68% of the time for a normal or *Gaussian distribution*. The relationships between error, uncertainty, the measured/predicted value, and the (unknown) true value are presented graphically in Fig. 8.16.[5] A complete measurement/prediction description must include not only the mean value but also the associated uncertainty.

8.7 Torsion Vibration

The equation of motion for a uniform beam with a circular cross section under torsion vibration is provided in Eq. 8.110, where G is the beam's shear modulus and $\phi(x, t)$ is the rotation about the beam's axis. The following assumptions apply: (1) radial lines extending from the beam center to its outer diameter remain straight after an external torque is applied; and (2) shear is the only significant stress (Bishop and Johnson 1960).

$$G\frac{\partial^2 \phi}{\partial x^2} = \rho\frac{\partial^2 \phi}{\partial t^2} \tag{8.110}$$

For harmonic torsion vibration due to an external torque $T\sin(\omega t)$, a general solution to Eq. 8.110 is given by $\phi(x, t) = \Phi(x)\sin(\omega t)$, where $\Phi(x)$ is a function that describes the position-dependent vibration behavior and ω is the forcing frequency (rad/s). Calculating the second-order partial derivatives of this general solution with respect to x and t and substituting in Eq. 8.110 yields Eq. 8.111, where the $\sin(\omega t)$ term appears on both sides of the equality and is not shown.

---

[5] The authors credit Dr. W.T. Estler (retired, National Institute of Standards and Technology) with this figure.

Fig. 8.17 Circular cross section free-free beam with a torque applied at coordinate ϕ_1

$$G\frac{\partial^2 \Phi}{\partial x^2} = \left(-\omega^2 \rho\right)\Phi \qquad (8.111)$$

A general solution to Eq. 8.111 is given in Eq. 8.112, where $\lambda = \omega\sqrt{\rho/G}$. Using this equation, we can determine the rotation FRFs for a free-free beam due to an external torque. Two boundary conditions are required. First, $\partial\Phi/\partial x = 0$ at a free boundary. Second, for an external torque application at the beam's end, $\partial\Phi/\partial x = \frac{T}{GJ}\sin(\omega t)$, where T is the harmonic torque magnitude and J is the second polar moment of area for the beam's cross section. This boundary condition follows from the relationship between the shear stress, τ, at a radius r and the shear strain, γ, $\tau = G\gamma$, or $Tr/J = rG\frac{d\phi}{dx}$.

$$\Phi(x) = A\cos(\lambda x) + B\sin(\lambda x) \qquad (8.112)$$

In Fig. 8.17, an external torque $T_1 = T\sin(\omega t)$ is applied at the right end of the free-free beam, labeled as coordinate ϕ_1, where $x = l$ and l is the beam length ($x = 0$ at the left end of the beam). The corresponding boundary conditions are provided in Eq. 8.113.

$$\left.\frac{\partial\Phi}{\partial x}\right|_{x=0} = 0 \quad \left.\frac{\partial\Phi}{\partial x}\right|_{x=l} = \frac{T}{GJ}\sin(\omega t) \qquad (8.113)$$

We determine the coefficients A and B in Eq. 8.112 by calculating $\partial\Phi/\partial x$ and applying the two boundary conditions from Eq. 8.113. This gives $A = \frac{-T}{GJ\lambda\sin(\lambda l)} \times \sin(\omega t)$ and $B = 0$. Substitution of these coefficient values in Eq. 8.112 gives Eq. 8.114.

$$\Phi(x) = \frac{-T}{GJ\lambda\sin(\lambda l)}\cos(\lambda x)\sin(\omega t) \qquad (8.114)$$

Finally, we write the direct torsion FRF at coordinate 1, Φ_1/T_1, as shown in Eq. 8.115 by substituting $T_1 = T\sin(\omega t)$. Similarly, we find the cross FRF, Φ_2/T_1, by substituting $x = 0$ in Eq. 8.114; see Eq. 8.116.

$$\frac{\Phi_1}{T_1} = \frac{-\cos(\lambda l)}{GJ\lambda\sin(\lambda l)} = \frac{-\cot(\lambda l)}{GJ\lambda} \qquad (8.115)$$

$$\frac{\Phi_2}{T_1} = \frac{-\cos(\lambda \cdot 0)}{GJ\lambda\sin(\lambda l)} = \frac{-1}{GJ\lambda\sin(\lambda l)} = \frac{-\csc(\lambda l)}{GJ\lambda} \qquad (8.116)$$

Fig. 8.18 Free-free beam
with an axial force P_1 applied
at coordinate v_1

To determine the direct and cross FRFs due to a torque applied at the other end
of the beam, Φ_2/T_2 and Φ_1/T_2, we repeat the process. These results are given in
Eqs. 8.117 and 8.118. In order to introduce damping in the component responses,
we again incorporate solid damping, but this time using a complex shear modulus
$G_s = G(1 + i\eta)$. Note that λ is a function of G as well.

$$\frac{\Phi_2}{T_2} = \frac{-\cot(\lambda l)}{GJ\lambda} \tag{8.117}$$

$$\frac{\Phi_1}{T_2} = \frac{-\csc(\lambda l)}{GJ\lambda} \tag{8.118}$$

8.8 Axial Vibration

The equation of motion for a uniform cross section beam[6] under axial, or longitudinal,
vibration is provided in Eq. 8.119, where γ is the deflection along the beam axis and
Poisson effects are neglected[7] (Bishop and Johnson 1960).

$$E\frac{\partial^2 \gamma}{\partial x^2} = \rho\frac{\partial^2 \gamma}{\partial t^2} \tag{8.119}$$

Because the equation of motion has the same form as Eq. 8.110 for torsion,
the free-free receptance development is similar to that provided in Sect. 8.7. For the
free-free beam of length l shown in Fig. 8.18, the application of a harmonic axial
force $P_1 = P\sin(\omega t)$ enables us to determine the FRFs Γ_1/P_1 and Γ_2/P_1.
See Eqs. 8.120 and 8.121, where $\lambda = \omega\sqrt{\rho/E}$. As before, determining the FRFs
requires that two boundary conditions be applied. In this case, these are $\partial\Gamma/\partial x = 0$
at a free end and $\partial\Gamma/\partial x = \frac{P}{EA}\sin(\omega t)$ for an external axial force application at the
beam's end, where A is the beam's cross-sectional area. The latter boundary
conditions follows from the relationship between the axial stress, σ, and axial
strain, ε, $\sigma = E\varepsilon$, or $\frac{P}{A} = E(d\gamma/dx)$. The other two FRFs for the beam in Fig. 8.14
are determined by applying P_2 to γ_2. See Eqs. 8.122 and 8.123. Again, in order to

---

[6] The beam's cross section is not required to be circular as in the torsion vibration analysis in Sect. 8.7.

[7] This means that the beam's expansion and contraction in the directions normal to the oscillating
axial deflection are ignored.

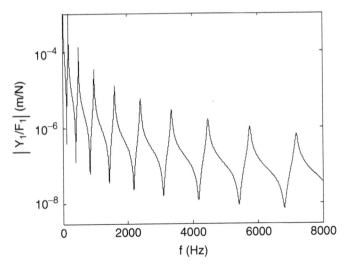

Fig. 8.19 *By the Numbers 8.3* – Semi-logarithmic plot of the transverse deflection FRF for the 10 mm diameter, 500 mm long free-free steel beam

introduce damping in the component responses, we apply solid damping by replacing the elastic modulus with the complex elastic modulus $E_s = E(1 + i\eta)$. Note that λ is a function of E as well.

$$\frac{\Gamma_1}{P_1} = \frac{-\cot(\lambda l)}{EA\lambda} \tag{8.120}$$

$$\frac{\Gamma_2}{P_1} = \frac{-\csc(\lambda l)}{EA\lambda} \tag{8.121}$$

$$\frac{\Gamma_2}{P_2} = \frac{-\cot(\lambda l)}{EA\lambda} \tag{8.122}$$

$$\frac{\Gamma_1}{P_2} = \frac{-\csc(\lambda l)}{EA\lambda} \tag{8.123}$$

By the Numbers 8.3

Let's now compare the transverse deflection, torsion, and axial FRFs for a cylindrical beam. We will consider a 10 mm diameter steel beam that is 500 mm long with free-free boundary conditions. Steel's material properties are $\rho = 7,800\,\text{kg/m}^3$, $E = 200\,\text{GPa}$, Poisson's ratio is $\nu = 0.29$, and $G = E/2(1 + \nu)$. Also, for the cylindrical beam, $A = \pi d^2/4$, $I = \pi d^4/64$, and $J = \pi d^2/32$. For plotting purposes, let's select $\eta = 0.01$. The relevant FRF equations are Eq. 8.80 for transverse deflection, Eq. 8.115 for torsion, and Eq. 8.120 for axial deflection. The FRFs are plotted using a semi-logarithmic scale in Figs. 8.19–8.21. We see that the axial

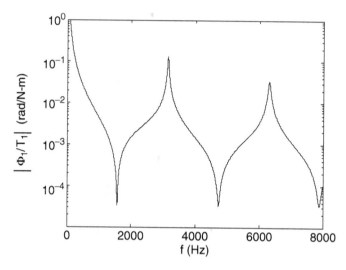

Fig. 8.20 *By the Numbers 8.3* – Semi-logarithmic plot of the torsion FRF for the 10 mm diameter, 500 mm long free-free steel beam

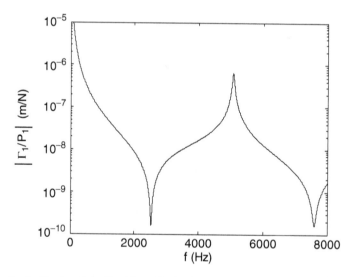

Fig. 8.21 *By the Numbers 8.3* – Semi-logarithmic plot of the axial deflection FRF for the 10 mm diameter, 500 mm long free-free steel beam

deflection natural frequencies are higher than the torsion natural frequencies which are, in turn, higher than the transverse deflection natural frequencies. We can also note that the increasing magnitude as the responses approach zero frequency is due to the rigid body modes for the free-free beam. The code used to produce Figs. 8.19–8.21 is provided in MATLAB® Mojo 8.3.

MATLAB® MOJO 8.3

```
% matlab_mojo_8_3.m

clc
clear all
close all

f = 1:0.1:8000;          % Hz
omega = f*2*pi;          % rad/s

% Define beam
d = 10e-3;               % m
l = 500e-3;              % m
E = 2e11;                % N/m^2
nu = 0.29;
G = E/(2*(1+nu));        % N/m^2
density = 7800;          % kg/m^3
eta = 0.01;
A = pi*d^2/4;            % m^2
I = pi*d^4/64;           % m^4
J = pi*d^4/32;           % m^4
E = E*(1+1i*eta);        % N-m^2
G = G*(1+1i*eta);        % N-m^2

% Free-free beam FRFs
% Transverse deflection FRF
lambda = (omega.^2*density*A/(E*I)).^0.25;
Y1_F1 = (cos(lambda*l).*sinh(lambda*l)-
sin(lambda*l).*cosh(lambda*l))./(lambda.^3*E*I.*(1-
cos(lambda*l).*cosh(lambda*l)));

figure(1)
semilogy(f, abs(Y1_F1), 'k')
set(gca,'FontSize', 14)
ylim([3e-9 1e-3])
xlabel('f (Hz)')
ylabel('|Y_1/F_1|  (m/N)')

% Torsion FRF
lambda = omega*(density/G)^0.5;
Phi1_T1 = -cot(lambda*l)./(G*J*lambda);

figure(2)
semilogy(f, abs(Phi1_T1), 'k')
ylim([1e-5 1])
set(gca,'FontSize', 14)
xlabel('f (Hz)')
ylabel('|\Phi_1/T_1|  (rad/N-m)')

% Axial deflection FRF
lambda = omega*(density/E)^0.5;
Tau1_P1 = -cot(lambda*l)./(E*A*lambda);

figure(3)
semilogy(f, abs(Tau1_P1), 'k')
ylim([1e-10 1e-5])
set(gca,'FontSize', 14)
xlabel('f (Hz)')
ylabel('|\Gamma_1/P_1|  (m/N)')
```

8.9   Timoshenko Beam Model

While the closed-form Euler–Bernoulli beam transverse vibration FRFs provided in Table 8.3 are convenient to apply, accurate solutions are obtained only for beams which exhibit small cross sectional area-to-length ratios (i.e., long slender beams). An alternative for beams that do not meet this criterion is the *Timoshenko beam model* (Weaver et al. 1990). The corresponding differential equation is given by:

$$\left(\frac{\partial^2 y}{\partial t^2} + \frac{EI}{\rho A}\frac{\partial^4 y}{\partial x^4}\right) + \left(\frac{\rho I}{\hat{k}AG}\frac{\partial^4 y}{\partial t^4} + \frac{EI}{\hat{k}AG}\frac{\partial^4 y}{\partial x^2 \partial t^2}\right) - \left(\frac{I}{A}\frac{\partial^4 y}{\partial x^2 \partial t^2}\right) = 0, \quad (8.124)$$

where $\hat{k}$ is a shape factor that depends on the beam cross section (Hutchinson 2001). Equation 8.124 is grouped into three sections (i.e., three parenthetical expressions). We see that the first section matches the Euler–Bernoulli beam equation provided in Eq. 8.19. The second and third sections account for shear deformations and rotary inertia, respectively. While these additional terms improve the model accuracy (particularly at higher frequencies), the tradeoff is that a closed-form solution to Eq. 8.124 is not available. Finite element calculations may be applied, but at the expense of increased computation time.

Chapter Summary

- In continuous models, the mass is distributed throughout the structure, rather than localized at the coordinates as in discrete models.
- Euler–Bernoulli beam theory can be used to derive transverse deflection FRFs for continuous cross section beams.
- The beam's boundary conditions, such as fixed, free, or pinned, determine the FRF behavior.
- Solid damping can be incorporated in continuous beam FRFs using a complex modulus.
- The solid damping factor is equal to two times the viscous damping ratio at resonance.
- Anti-resonant frequencies represent frequencies where the beam response is small even for a large force input.
- Rotation FRFs are related to the transverse beam FRFs and describe the harmonic rotation of a beam within the bending plane due to a harmonic force or moment.
- Torsion FRFs describe the ratio of the frequency-domain rotation about the beam's axis to a harmonic torque applied to the beam.
- Axial FRFs, which provide the axial deflection response due to a harmonic axial force, have a similar form to torsion FRFs.
- The Timoshenko beam model may be implemented when the accuracy of the Euler–Bernoulli beam model is not adequate.

Exercises

1. Consider a uniform cross section fixed-free (i.e., clamped-free or cantilever) beam.

   (a) Sketch the first bending mode shape (lowest natural frequency).
   (b) Sketch the second mode shape (next lowest natural frequency).
   (c) On your sketches in parts (a) and (b), identify any node location(s).

2. In describing beam vibrations using Euler–Bernoulli beam theory, we derived the equation of motion $(\partial^4 Y / \partial x^4) - \lambda^4 Y = 0$.

   (a) In the equation of motion, what does x represent physically?
   (b) In the equation of motion, what does Y represent physically?
   (c) Write the equation for λ (it replaces several other variables) and describe what each variable represents (include the SI units).

3. Consider a fixed-free beam. The general solution to the equation of motion can be written as $Y(x) = A\cos(\lambda x) + B\sin(\lambda x) + C\cosh(\lambda x) + D\sinh(\lambda x)$. To determine the four coefficients, A through D, four boundary conditions are required. Write the four boundary conditions (in the table) as a function of x and y for the beam shown in Fig. P8.3.

Fig. P8.3 Fixed-free beam model

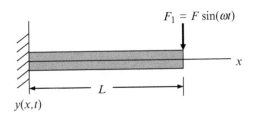

$F_1 = F\sin(\omega t)$

L

$y(x,t)$

At $x = 0$	At $x = L$
1.	3.
2.	4.

4. Consider the free-sliding beam shown in Fig. P8.4a. Direct and cross FRFs were measured at six locations and the imaginary parts are provided for the frequency interval near its second bending natural frequency of 350 Hz in Figs. P8.4b–P8.4g. Given the FRF data, sketch the mode shape corresponding to the second natural

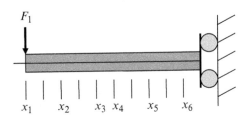

F_1

Fig. P8.4a Free-sliding beam model

$x_1 \quad x_2 \quad x_3 \ x_4 \quad x_5 \quad x_6$

Fig. P8.4b Direct FRF $\frac{X_1}{F_1}$ for the free-sliding beam

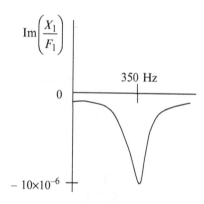

frequency. Normalize the mode shape to a value of 1 at the free end.

Fig. P8.4c Cross FRF $\frac{X_2}{F_1}$ for the free-sliding beam

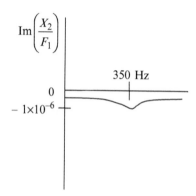

Fig. P8.4d Cross FRF $\frac{X_3}{F_1}$ for the free-sliding beam

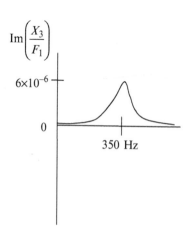

Fig. P8.4e Cross FRF $\frac{X_4}{F_1}$ for the free-sliding beam

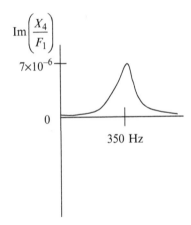

Fig. P8.4f Cross FRF $\frac{X_5}{F_1}$ for the free-sliding beam

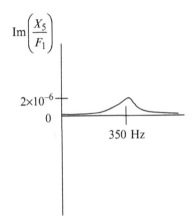

Fig. P8.4g Cross FRF $\frac{X_6}{F_1}$ for the free-sliding beam

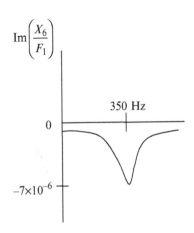

5. Complete the following for the transverse deflection of a free-free cylindrical beam. The beam's diameter is 15 mm and it is 480 mm long. The beam material is 6061-T6 aluminum with $\rho = 2{,}700 \, \text{kg/m}^3$, $E = 70 \, \text{GPa}$, $v = 0.35$, $G = E/2(1 + v)$, and $\eta = 0.002$.

   (a) Plot the transverse deflection FRF over a frequency range of 10,000 Hz. Use a semi-logarithmic scale.
   (b) How many modes are captured in this bandwidth (excluding the rigid body modes)?
   (c) What is the natural frequency of the first (non-rigid) bending mode?

6. Complete the following for the torsion vibration of a free-free cylindrical beam. The beam's diameter is 15 mm and it is 480 mm long. The beam material is 6061-T6 aluminum with $\rho = 2{,}700 \, \text{kg/m}^3$, $E = 70 \, \text{GPa}$, $v = 0.35$, $G = E/2(1 + v)$, and $\eta = 0.002$.

   (a) Plot the torsion FRF over a frequency range of 10,000 Hz. Use a semi-logarithmic scale.
   (b) How many modes are captured in this bandwidth (excluding the rigid body mode)?
   (c) What is the natural frequency of the first (non-rigid) torsion mode?

7. Complete the following for the axial vibration of a free-free cylindrical beam. The beam's diameter is 15 mm and it is 480 mm long. The beam material is 6061-T6 aluminum with $\rho = 2{,}700 \, \text{kg/m}^3$, $E = 70 \, \text{GPa}$, $v = 0.35$, $G = E/2(1 + v)$, and $\eta = 0.002$.

   (a) Plot the axial FRF over a frequency range of 10,000 Hz. Use a semi-logarithmic scale.
   (b) How many modes are captured in this bandwidth (excluding the rigid body mode)?
   (c) What is the natural frequency of the first (non-rigid) torsion mode?

8. Consider the transverse vibration of a free-free cylindrical beam. If the diameter of a solid beam is d, determine the outer diameter, d_o, of a hollow beam with the same length and material properties to give the same natural frequencies as the solid beam if the inner diameter, d_i, is one-half of the outer diameter, $d_i = 0.5d_o$.

9. For a 25 mm diameter 6061-T6 aluminum rod ($\rho = 2{,}700 \, \text{kg/m}^3$ and $E = 70 \, \text{GPa}$) with a nominal length of 190 mm and an associated uncertainty of 0.2 mm, determine the uncertainty in the second bending natural frequency, $u(f_{n,2})$ (in Hz), if free-free boundary conditions are imposed. You may neglect the uncertainty in E, ρ, and d.

10. The Timoshenko beam model is more accurate than the Euler–Bernoulli beam model because it includes the effects of _____ and _____.

References

Bishop R, Johnson D (1960) The mechanics of vibration. Cambridge University Press, Cambridge
Blevins RD (2001) Formulas for natural frequency and mode shape. Krieger, Malabar (Table 8–1)
Chapra S, Canale R (1985) Numerical methods for engineers with personal computer applications. McGraw-Hill, New York (Section 7.1)
Hutchinson J (2001) Shear coefficients for Timoshenko beam theory. J Appl Mech 68:87–92
Weaver W Jr, Timoshenko S, Young D (1990) Vibration problems in engineering, 5th edn. Wiley, New York (Section 5.12)

Chapter 9
Receptance Coupling

*I have made this letter longer than usual, only because I have
not had the time to make it shorter.*

— Blaise Pascal

9.1 Introduction

In Chap. 1 through 8 we discussed both discrete and continuous beam models that
can be used to describe the behavior of vibrating systems. We also detailed
experimental techniques that we can use to identify these models. In this chapter,
we will introduce an approach to combine models or measurements of individual
components in order to predict the assembly's frequency response function (FRF).
This method is referred to as receptance coupling (Bishop and Johnson 1960); recall
from Sect. 7.1 that a receptance is a type of FRF.

IN A NUTSHELL   Receptance coupling enables the connection of
measurements to measurements, models to models, or models to
measurements. Sometimes it is easy to make the measurements and
sometimes it is difficult or impossible, such as when the component to
be added only exists as a model.

9.2 Two-Component Rigid Coupling

Let's begin with the rigid coupling of two *components, or substructures*. Our goal is
to predict the direct and cross FRFs for the *assembly* based on the direct and cross
FRFs for the two components. As a practical example, we could consider

T.L. Schmitz and K.S. Smith, *Mechanical Vibrations: Modeling and Measurement*,
DOI 10.1007/978-1-4614-0460-6_9, © Springer Science+Business Media, LLC 2012

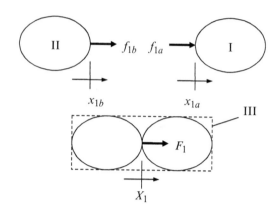

Fig. 9.1 Rigid coupling of components I and II to form assembly III. The force F_1 is applied to the assembly in order to determine H_{11}

performing experiments to identify the free-free FRFs for an airplane's wing and fuselage separately and then coupling the wing to the fuselage mathematically to predict the response for the entire structure. This concept is demonstrated in Fig. 9.1; where the two components, I and II, are rigidly coupled to form assembly III. Note that the direct and cross FRFs for the two components I and II could be derived from: (1) measurements; (2) discrete models; or (3) continuous beam models. The coupling coordinates are x_{1a} and x_{1b} for the two substructures I and II, respectively. The corresponding assembly coordinate, X_1, is located at the same physical location as x_{1a} and x_{1b} after they are joined.[1] An attractive aspect of receptance coupling is that the component FRFs are only required at the coupling locations and any point where the assembly response is to be predicted. Therefore, the assembly's direct receptance at X_1 due to a harmonic force applied at that location, $H_{11} = \frac{X_1}{F_1}$, can be fully described using the direct component receptances $h_{1a1a} = \frac{x_{1a}}{f_{1a}}$ and $h_{1b1b} = \frac{x_{1b}}{f_{1b}}$ obtained from harmonic forces applied to the components at x_{1a} and x_{1b}, respectively.

IN A NUTSHELL   Recall that the FRFs (H_{11}, h_{1a1a}, and h_{1b1b}) are complex functions of frequency. At each frequency, the FRF has the form $a + i \cdot b$. The function may be continuous (e.g., based on a beam model) or it may be discrete (i.e., a measurement known only at a number of frequencies over the measurement frequency range).

To determine the assembly response, we must first describe the *compatibility condition*, $x_{1b} - x_{1a} = 0$, which represents the *rigid coupling* between component coordinates x_{1a} and x_{1b}. We can therefore write $x_{1b} = x_{1a} = X_1$ due to our decision to locate assembly coordinate X_1 at the (rigid) coupling point. We must also define the *equilibrium condition*, $f_{1a} + f_{1b} = F_1$, which equates the internal (component)

---

[1] We will follow this lower case/upper case notation to differentiate between component and assembly coordinates throughout the chapter.

and external (assembly) forces. Let's substitute for the displacements in the compatibility equation.

$$x_{1b} - x_{1a} = 0$$
$$h_{1b1b}f_{1b} - h_{1a1a}f_{1a} = 0 \qquad (9.1)$$

We next use the equilibrium condition, rewritten as $f_{1a} = F_1 - f_{1b}$, to eliminate f_{1a} in Eq. 9.1. Rearranging enables us to solve for f_{1b}.

$$h_{1b1b}f_{1b} - h_{1a1a}F_1 + h_{1a1a}f_{1b} = 0$$
$$(h_{1a1a} + h_{1b1b})f_{1b} = h_{1a1a}F_1$$
$$f_{1b} = (h_{1a1a} + h_{1b1b})^{-1}h_{1a1a}F_1 \qquad (9.2)$$

Now that we have f_{1b}, we can again use the equilibrium condition to determine f_{1a}.

$$f_{1a} = F_1 - f_{1b}$$
$$f_{1a} = \left(1 - (h_{1a1a} + h_{1b1b})^{-1}h_{1a1a}\right)F_1 \qquad (9.3)$$

We solve for H_{11}, as shown in Eq. 9.4. This equation gives the direct assembly response at the coupling coordinate, X_1, as a function of the component receptances. These frequency-dependent, complex-valued receptances may have any number of modes. There are no restrictions on the relationship between the number of modes and coordinates as with modal analysis (i.e., we saw in Sect. 6.5 that the number of modeled modes and coordinates must be equal to obtain square matrices when using modal analysis).

$$H_{11} = \frac{X_1}{F_1} = \frac{x_{1a}}{F_1} = \frac{h_{1a1a}f_{1a}}{F_1} = h_{1a1a} - h_{1a1a}(h_{1a1a} + h_{1b1b})^{-1}h_{1a1a} \qquad (9.4)$$

IN A NUTSHELL   Coupling FRFs requires that the mathematical operations (such as addition or multiplication) be completed in a frequency-by-frequency manner. For example, if h_{1a1a} is given by $a + i \cdot b$ and h_{1b1b} is $c + i \cdot d$ at frequency ω_1, then $(h_{1a1a} + h_{1b1b}) = (a + c) + i \cdot (b + d)$ at ω_1. It is therefore required that the component FRFs are defined at the same frequency values. Inverting an FRF means that we compute $\frac{1}{a+i \cdot b}$ at each particular frequency. This inversion is completed by rationalizing the complex number (i.e., multiplying the numerator and denominator by its complex conjugate). The result at the selected frequency is $\frac{a-i \cdot b}{a^2+b^2}$, which has a real part $\frac{a}{a^2+b^2}$ and an imaginary part $\frac{-b}{a^2+b^2}$. This computation is repeated at every frequency. Multiplying two FRFs means that we multiply the two complex numbers at each frequency, $(a + i \cdot b)(c + i \cdot d) = (ac - bd) + i \cdot (bc + ad)$.

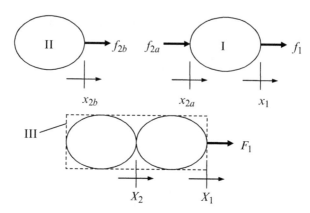

Fig. 9.2 Example showing rigid coupling of components I and II to form assembly III. The force F_1 is applied to the assembly at coordinate X_1 in order to determine H_{11} and H_{21}

Similarly, we can predict the assembly response at another coordinate, not coincident with the coupling point, by defining the component receptance at the desired location. Consider Fig. 9.2, where the direct assembly response at X_1 is again desired, but this location is now at another point on component I. We again assume x_1 and X_1 are collocated before and after coupling. The new coupling coordinates at the rigid coupling point are x_{2a} and x_{2b}. The component direct and cross receptances corresponding to Fig. 9.2 are $h_{11} = \frac{x_1}{f_1}$, $h_{2a2a} = \frac{x_{2a}}{f_{2a}}$, $h_{12a} = \frac{x_1}{f_{2a}}$, and $h_{2a1} = \frac{x_{2a}}{f_1}$ for I and $h_{2b2b} = \frac{x_{2b}}{f_{2b}}$ for II. The compatibility condition for the rigid coupling is $x_{2b} - x_{2a} = 0$ and we can therefore write $x_{2a} = x_{2b} = X_2$. Also, $x_1 = X_1$. The equilibrium conditions are $f_{2a} + f_{2b} = 0$ (because there is no external force at the coupling point in this case) and $f_1 = F_1$.

To determine $H_{11} = \frac{X_1}{F_1}$, we will first write the component displacements. For I, we now have two forces acting on the body, so the frequency-domain displacements are:

$$x_1 = h_{11}f_1 + h_{12a}f_{2a} \text{ and } x_{2a} = h_{2a1}f_1 + h_{2a2a}f_{2a}. \tag{9.5}$$

For II, we have $x_{2b} = h_{2b2b}f_{2b}$. Substitution into the compatibility condition gives:

$$x_{2b} - x_{2a} = h_{2b2b}f_{2b} - h_{2a1}f_1 - h_{2a2a}f_{2a} = 0. \tag{9.6}$$

We apply the equilibrium conditions to replace f_1 with F_1 and eliminate f_{2a} $(f_{2a} = -f_{2b})$.

$$h_{2b2b}f_{2b} - h_{2a1}F_1 + h_{2a2a}f_{2b} = 0. \tag{9.7}$$

This enables us to group terms and solve for f_{2b}. Specifically, we have that $f_{2b} = (h_{2a2a} + h_{2b2b})^{-1}h_{2a1}F_1$. Therefore, we can also write $f_{2a} = -(h_{2a2a} + h_{2b2b})^{-1}h_{2a1}F_1$. Substitution of this force value into the H_{11} expression

Fig. 9.3 Example showing rigid coupling of components I and II to form assembly III. The force F_2 is applied to the assembly at coordinate X_2 in order to determine H_{22} and H_{12}

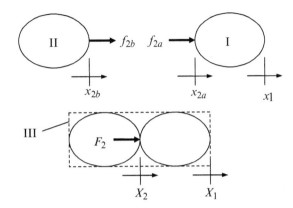

gives us the desired result; see Eq. 9.8. Again, the assembly response is written as a function of the direct (h_{11}, h_{2a2a}, and h_{2b2b}) and cross (h_{12a} and h_{2a1}) receptances for the two components.

$$H_{11} = \frac{X_1}{F_1} = \frac{x_1}{F_1} = \frac{h_{11}f_1 + h_{12a}f_{2a}}{F_1} = \frac{h_{11}f_1 - h_{12a}(h_{2a2a} + h_{2b2b})^{-1}h_{2a1}F_1}{F_1}$$

$$H_{11} = \frac{h_{11}F_1 - h_{12a}(h_{2a2a} + h_{2b2b})^{-1}h_{2a1}F_1}{F_1} = h_{11} - h_{12a}(h_{2a2a} + h_{2b2b})^{-1}h_{2a1}$$

$$(9.8)$$

We can also use f_{2a} to determine the cross receptance H_{21}. See Eq. 9.9.

$$H_{21} = \frac{X_2}{F_1} = \frac{x_{2a}}{F_1} = \frac{h_{2a1}f_1 + h_{2a2a}f_{2a}}{F_1} = \frac{h_{2a1}f_1 - h_{2a2a}(h_{2a2a} + h_{2b2b})^{-1}h_{2a1}F_1}{F_1}$$

$$H_{21} = \frac{h_{2a1}F_1 - h_{2a2a}(h_{2a2a} + h_{2b2b})^{-1}h_{2a1}F_1}{F_1} = h_{2a1} - h_{2a2a}(h_{2a2a} + h_{2b2b})^{-1}h_{2a1}$$

$$(9.9)$$

We determine the direct and cross receptances, H_{22} and H_{12}, respectively, by applying a force to the assembly coordinate X_2. See Fig. 9.3. The component receptances are again $h_{11} = \frac{x_1}{f_1}$, $h_{2a2a} = \frac{x_{2a}}{f_{2a}}$, $h_{12a} = \frac{x_1}{f_{2a}}$, and $h_{2a1} = \frac{x_{2a}}{f_1}$ for I and $h_{2b2b} = \frac{x_{2b}}{f_{2b}}$ for II. The compatibility condition for the rigid coupling remains as $x_{2b} - x_{2a} = 0$. However, the equilibrium condition is $f_{2a} + f_{2b} = F_2$ because the force is applied to the coupling coordinate.

To determine $H_{22} = \frac{X_2}{F_2}$, we begin by writing the component displacements. For I, the displacements are:

$$x_1 = h_{12a}f_{2a} \text{ and } x_{2a} = h_{2a2a}f_{2a}.$$

$$(9.10)$$

For II, we have $x_{2b} = h_{2b2b}f_{2b}$. Substitution in the compatibility condition gives:

$$h_{2b2b}f_{2b} - h_{2a2a}f_{2a} = 0. \tag{9.11}$$

We apply the equilibrium condition, $f_{2a} = F_2 - f_{2b}$, to eliminate f_{2a} in Eq. 9.11.

$$h_{2b2b}f_{2b} - h_{2a2a}F_2 + h_{2a2a}f_{2b} = 0 \tag{9.12}$$

This enables us to group terms and solve for f_{2b}. We find that $f_{2b} = (h_{2a2a} + h_{2b2b})^{-1}h_{2a2a}F_2$. Again using the equilibrium condition, we can write $f_{2a} = \left(1 - (h_{2a2a} + h_{2b2b})^{-1}h_{2a2a}\right)F_2$. Equation 9.13 gives the desired H_{22} expression.

$$H_{22} = \frac{X_2}{F_2} = \frac{x_{2a}}{F_2} = \frac{h_{2a2a}f_{2a}}{F_2} = \frac{h_{2a2a}\left(1 - (h_{2a2a} + h_{2b2b})^{-1}h_{2a2a}\right)F_2}{F_2}$$

$$H_{22} = \frac{h_{2a2a}F_2 - h_{2a2a}(h_{2a2a} + h_{2b2b})^{-1}h_{2a2a}F_2}{F_2} = h_{2a2a} - h_{2a2a}(h_{2a2a} + h_{2b2b})^{-1}h_{2a2a}$$

$$\tag{9.13}$$

We use f_{2a} to find the cross receptance H_{12} as well. See Eq. 9.14.

$$H_{12} = \frac{X_1}{F_2} = \frac{x_1}{F_2} = \frac{h_{12a}f_{2a}}{F_2} = \frac{h_{12a}\left(1 - (h_{2a2a} + h_{2b2b})^{-1}h_{2a2a}\right)F_2}{F_2} \tag{9.14}$$

$$H_{12} = h_{12a} - h_{12a}(h_{2a2a} + h_{2b2b})^{-1}h_{2a2a}$$

9.3   Two-Component Flexible Coupling

Let's continue with the system shown in Fig. 9.1, but now couple the two components through a linear spring, described by the constant k. This case is depicted in Fig. 9.4 and represents the situation where the connection between two components is not rigid. As an example, perhaps the joint is a bolted connection where the elastic deformation of the bolts due to an external harmonic force acts as a spring that couples the two components; see Fig. 9.5. In Fig. 9.4 the component receptances are $h_{1a1a} = \frac{x_{1a}}{f_{1a}}$ and $h_{1b1b} = \frac{x_{1b}}{f_{1b}}$ and the equilibrium condition is $f_{1a} + f_{1b} = F_{1a}$. These are analogous to the rigid coupling case. However, the compatibility condition now becomes:

$$k(x_{1b} - x_{1a}) = -f_{1b}. \tag{9.15}$$

for the *flexible coupling* case. This relationship follows Hooke's Law, $F = kx$.

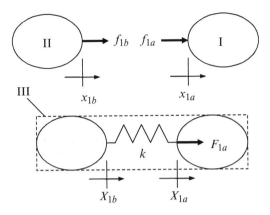

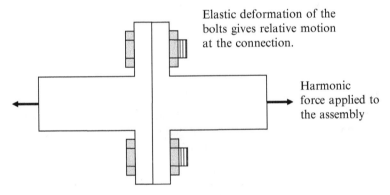

Fig. 9.4 Flexible coupling of components I and II to form assembly III. The force F_{1a} is applied to the assembly at coordinate X_{1a} in order to determine H_{1a1a} and H_{1b1a}

Fig. 9.5 A bolted connection where the elastic strain in the bolts due to the applied harmonic force effectively serves as a spring that connects the two components

> IN A NUTSHELL   We recognize that components cannot, in practice, be rigidly connected. Sometimes rigid connection is a good approximation, but when it is not, connection through a spring is often more accurate.

Because the component and assembly coordinates are coincident, we have that $x_{2a} = X_{2a}$, and $x_{2b} = X_{2b}$. To determine $H_{1a1a} = \frac{X_{1a}}{F_{1a}}$, we first substitute the component displacements in the compatibility condition. See Eq. 9.16.

$$k(h_{1b1b}f_{1b} - h_{1a1a}f_{1a}) = -f_{1b} \qquad (9.16)$$

Using the equilibrium condition, $f_{1a} = F_{1a} - f_{1b}$, we can eliminate f_{1a} to obtain the equation for f_{1b}.

Fig. 9.6 Flexible coupling of components I and II to form assembly III. The force F_{1b} is applied to the assembly at coordinate X_{1b} in order to determine H_{1b1b} and H_{1a1b}

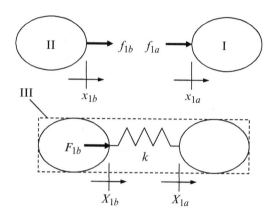

$$k(h_{1b1b}f_{1b} - h_{1a1a}F_{1a} + h_{1a1a}f_{1b}) = -f_{1b}$$

$$kh_{1b1b}f_{1b} - kh_{1a1a}F_{1a} + kh_{1a1a}f_{1b} = -f_{1b}$$

$$\left(h_{1a1a} + h_{1b1b} + \frac{1}{k}\right)f_{1b} = h_{1a1a}F_{1a}$$

$$f_{1b} = \left(h_{1a1a} + h_{1b1b} + \frac{1}{k}\right)^{-1}h_{1a1a}F_{1a} \qquad (9.17)$$

Using f_{1b} and the equilibrium condition, we find that $f_{1a} = (1 - (h_{1a1a} + h_{1b1b} + \frac{1}{k})^{-1}h_{1a1a})F_{1a}$. Substitution then yields the direct assembly receptance $H_{1a1a,}$ as shown in Eq. 9.18. We can see that this equation simplifies to Eq. 9.4 as k approaches infinity (rigid connection).

$$H_{1a1a} = \frac{X_{1a}}{F_{1a}} = \frac{x_{1a}}{F_{1a}} = \frac{h_{1a1a}f_{1a}}{F_{1a}} = \frac{h_{1a1a}\left(1 - \left(h_{1a1a} + h_{1b1b} + \frac{1}{k}\right)^{-1}h_{1a1a}\right)F_{1a}}{F_{1a}}$$

$$H_{1a1a} = h_{1a1a} - h_{1a1a}\left(h_{1a1a} + h_{1b1b} + \frac{1}{k}\right)^{-1}h_{1a1a} \qquad (9.18)$$

The cross receptance due to the force F_{1a} is provided in Eq. 9.19.

$$H_{1b1a} = \frac{X_{1b}}{F_{1a}} = \frac{x_{1b}}{F_{1a}} = \frac{h_{1b1b}f_{1b}}{F_{1a}} = \frac{h_{1b1b}\left(h_{1a1a} + h_{1b1b} + \frac{1}{k}\right)^{-1}h_{1a1a}F_{1a}}{F_{1a}}$$

$$H_{1b1a} = h_{1b1b}\left(h_{1a1a} + h_{1b1b} + \frac{1}{k}\right)^{-1}h_{1a1a} \qquad (9.19)$$

As shown in Fig. 9.6, we can alternately apply the assembly force to coordinate X_{1b}. The component receptances and displacements are unchanged, but the equilibrium condition is $f_{1a} + f_{1b} = F_{1b}$. Similarly, we modify the compatibility condition to be:

$$k(x_{1a} - x_{1b}) = -f_{1a}. \tag{9.20}$$

Substitution for the component displacements and f_{1b} (from the equilibrium condition) yields the expression for f_{1a}.

$$k(h_{1a1a}f_{1a} - h_{1b1b}F_{1b} + h_{1b1b}f_{1a}) = -f_{1a}$$
$$kh_{1a1a}f_{1a} - kh_{1b1b}F_{1b} + kh_{1b1b}f_{1a} = -f_{1a}$$
$$\left(h_{1a1a} + h_{1b1b} + \frac{1}{k}\right)f_{1a} = h_{1b1b}F_{1b}$$
$$f_{1a} = \left(h_{1a1a} + h_{1b1b} + \frac{1}{k}\right)^{-1} h_{1b1b}F_{1b} \tag{9.21}$$

Again applying the equilibrium condition, $f_{1b} = F_{1b} - f_{1a}$, we obtain $f_{1b} = \left(1 - \left(h_{1a1a} + h_{1b1b} + \frac{1}{k}\right)^{-1} h_{1b1b}\right)F_{1b}$. Substitution then gives the assembly direct and cross receptances due to F_{1b}.

$$H_{1b1b} = \frac{X_{1b}}{F_{1b}} = \frac{x_{1b}}{F_{1b}} = \frac{h_{1b1b}f_{1b}}{F_{1b}} = \frac{h_{1b1b}\left(1 - \left(h_{1a1a} + h_{1b1b} + \frac{1}{k}\right)^{-1} h_{1b1b}\right)F_{1b}}{F_{1b}}$$
$$H_{1b1b} = h_{1b1b} - h_{1b1b}\left(h_{1a1a} + h_{1b1b} + \frac{1}{k}\right)^{-1} h_{1b1b} \tag{9.22}$$

$$H_{1a1b} = \frac{X_{1a}}{F_{1b}} = \frac{x_{1a}}{F_{1b}} = \frac{h_{1a1a}f_{1a}}{F_{1b}} = \frac{h_{1a1a}\left(h_{1a1a} + h_{1b1b} + \frac{1}{k}\right)^{-1} h_{1b1b}F_{1b}}{F_{1b}}$$
$$H_{1a1b} = h_{1a1a}\left(h_{1a1a} + h_{1b1b} + \frac{1}{k}\right)^{-1} h_{1b1b} \tag{9.23}$$

Similar to the rigid connection example depicted in Fig. 9.2, we can again add another coordinate, not located at the coupling location, and apply the external force at that point. See Fig. 9.7. The component displacements are again $x_1 = h_{11}f_1 + h_{12a}f_{2a}$ and $x_{2a} = h_{2a1}f_1 + h_{2a2a}f_{2a}$ for substructure I and $x_{2b} = h_{2b2b}f_{2b}$ for substructure II. The equilibrium conditions are $f_{2a} + f_{2b} = 0$ and $f_1 = F_1$. The compatibility condition is:

$$k(x_{2b} - x_{2a}) = -f_{2b}. \tag{9.24}$$

Fig. 9.7 Flexible coupling of components I and II to form assembly III. The force F_1 is applied to the assembly at coordinate X_1 in order to determine H_{11}, H_{2a1}, and H_{2b1}

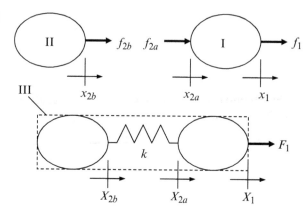

As before, the component and assembly coordinates are coincident, so we have that $x_1 = X_1$, $x_{2a} = X_{2a}$, and $x_{2b} = X_{2b}$. To determine $H_{11} = \frac{X_1}{F_1}$, we first substitute the component displacements in the compatibility condition. See Eq. 9.25.

$$k(h_{2b2b}f_{2b} - h_{2a1}f_1 - h_{2a2a}f_{2a}) = -f_{2b} \tag{9.25}$$

Using the equilibrium conditions, we can eliminate f_{2a} and replace f_1 with F_1 to obtain the equation for f_{2b}.

$$k(h_{2b2b}f_{2a} - h_{2a1}F_1 + h_{2a2a}f_{2b}) = -f_{2b}$$
$$kh_{2b2b}f_{2b} - kh_{2a1}F_1 + kh_{2a2a}f_{2b} = -f_{2b}$$
$$\left(h_{2a2a} + h_{2b2b} + \frac{1}{k}\right)f_{2b} = h_{2a1}F_1$$
$$f_{2b} = \left(h_{2a2a} + h_{2b2b} + \frac{1}{k}\right)^{-1}h_{2a1}F_1 \tag{9.26}$$

Applying the equilibrium condition $f_{2a} = -f_{2b}$, we obtain:

$$f_{2a} = -\left(h_{2a2a} + h_{2b2b} + \frac{1}{k}\right)^{-1}h_{2a1}F_1. \tag{9.27}$$

This enables us to write the direct and cross receptances as shown in Eqs. 9.28 and 9.29, respectively. We note that these equations simplify to the rigid coupling results provided in Eqs. 9.8 and 9.9 as k approaches infinity. The assembly cross receptance at coordinate X_{2b} is given by Eq. 9.30.

$$H_{11} = \frac{X_1}{F_1} = \frac{x_1}{F_1} = \frac{h_{11}f_1 + h_{12a}f_{2a}}{F_1} = \frac{h_{11}f_1 - h_{12a}\left(h_{2a2a} + h_{2b2b} + \frac{1}{k}\right)^{-1}h_{2a1}F_1}{F_1}$$
$$H_{11} = \frac{h_{11}F_1 - h_{12a}\left(h_{2a2a} + h_{2b2b} + \frac{1}{k}\right)^{-1}h_{2a1}F_1}{F_1} = h_{11} - h_{12a}\left(h_{2a2a} + h_{2b2b} + \frac{1}{k}\right)^{-1}h_{2a1}$$

$$\tag{9.28}$$

Fig. 9.8 Flexible coupling of components I and II to form assembly III. The force F_{2a} is applied to the assembly at coordinate X_{2a} in order to determine H_{2a2a}, H_{2b2a}, and H_{12a}

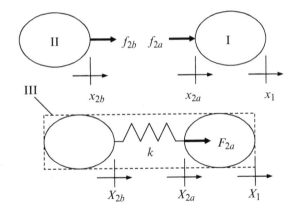

$$H_{2a1} = \frac{X_{2a}}{F_1} = \frac{x_{2a}}{F_1} = \frac{h_{2a1}f_1 + h_{2a2a}f_{2a}}{F_1} = \frac{h_{2a1}F_1 - h_{2a2a}\left(h_{2a2a} + h_{2b2b} + \frac{1}{k}\right)^{-1}h_{2a1}F_1}{F_1}$$

$$H_{2a1} = h_{2a1} - h_{2a2a}\left(h_{2a2a} + h_{2b2b} + \frac{1}{k}\right)^{-1}h_{2a1}$$

$$(9.29)$$

$$H_{2b1} = \frac{X_{2b}}{F_1} = \frac{x_{2b}}{F_1} = \frac{h_{2b2b}f_{2b}}{F_1} = \frac{h_{2b2b}\left(h_{2a2a} + h_{2b2b} + \frac{1}{k}\right)^{-1}h_{2a1}F_1}{F_1}$$

$$H_{2b1} = h_{2b2b}\left(h_{2a2a} + h_{2b2b} + \frac{1}{k}\right)^{-1}h_{2a1}$$

$$(9.30)$$

Let's now apply the external force, F_{2a}, to coordinate X_{2a} as shown in Fig. 9.8 in order to determine the assembly receptances H_{2a2a}, H_{2b2a}, and H_{12a}. The component displacements are $x_1 = h_{12a}f_{2a}$ and $x_{2a} = h_{2a2a}f_{2a}$ for substructure I and $x_{2b} = h_{2b2b}f_{2b}$ for substructure II. The equilibrium condition is $f_{2a} + f_{2b} = F_{2a}$ and the compatibility condition is:

$$k(x_{2b} - x_{2a}) = -f_{2b}.$$

$$(9.31)$$

We first determine the force f_{2b} by substituting the component displacements in Eq. 9.31 and replacing f_{2a} with $F_{2a} - f_{2b}$.

$$k(h_{2b2b}f_{2b} - h_{2a2a}F_{2a} + h_{2a2a}f_{2b}) = -f_{2b}$$

$$kh_{2b2b}f_{2b} - kh_{2a2a}F_{2a} + kh_{2a2a}f_{2b} = -f_{2b}$$

$$\left(h_{2a2a} + h_{2b2b} + \frac{1}{k}\right)f_{2b} = h_{2a2a}F_{2a}$$

$$f_{2b} = \left(h_{2a2a} + h_{2b2b} + \frac{1}{k}\right)^{-1}h_{2a2a}F_{2a}$$

$$(9.32)$$

Again using the equilibrium condition we find the equation for f_{2a}.

$$f_{2a} = F_{2a} - f_{2b} = \left(1 - \left(h_{2a2a} + h_{2b2b} + \frac{1}{k}\right)^{-1} h_{2a2a}\right) F_{2a} \qquad (9.33)$$

The direct and cross receptances for this situation (depicted in Fig. 9.8) are provided in Eqs. 9.34 through 9.36.

$$H_{2a2a} = \frac{X_{2a}}{F_{2a}} = \frac{x_{2a}}{F_{2a}} = \frac{h_{2a2a} f_{2a}}{F_{2a}} = \frac{h_{2a2a}\left(1 - \left(h_{2a2a} + h_{2b2b} + \frac{1}{k}\right)^{-1} h_{2a2a}\right) F_{2a}}{F_{2a}}$$

$$H_{2a2a} = h_{2a2a} - h_{2a2a}\left(h_{2a2a} + h_{2b2b} + \frac{1}{k}\right)^{-1} h_{2a2a} \qquad (9.34)$$

$$H_{2b2a} = \frac{X_{2b}}{F_{2a}} = \frac{x_{2b}}{F_{2a}} = \frac{h_{2b2b} f_{2b}}{F_{2b}} = \frac{h_{2b2b}\left(h_{2a2a} + h_{2b2b} + \frac{1}{k}\right)^{-1} h_{2a2a} F_{2a}}{F_{2a}}$$

$$H_{2b2a} = h_{2b2b}\left(h_{2a2a} + h_{2b2b} + \frac{1}{k}\right)^{-1} h_{2a2a} \qquad (9.35)$$

$$H_{12a} = \frac{X_1}{F_{2a}} = \frac{x_1}{F_{2a}} = \frac{h_{12a} f_{2a}}{F_{2a}} = \frac{h_{12a}\left(1 - \left(h_{2a2a} + h_{2b2b} + \frac{1}{k}\right)^{-1} h_{2a2a}\right) F_{2a}}{F_{2a}}$$

$$H_{12a} = h_{12a} - h_{12a}\left(h_{2a2a} + h_{2b2b} + \frac{1}{k}\right)^{-1} h_{2a2a} \qquad (9.36)$$

IN A NUTSHELL   While these computations are straightforward, they are more complicated than the rigid connection case. Increased accuracy often comes at the expense of increased computational complexity. The essence of engineering is to determine the required accuracy of the measurement/model/desired outcome and proceeding accordingly.

Our final scenario for the two-component flexible coupling is shown in Fig. 9.9. Here, we apply the external force F_{2b} to coordinate X_{2b} to obtain the direct and cross-assembly receptances H_{2b2b}, H_{2a2b}, and H_{12b}. The component displacements are the same as the previous case: $x_1 = h_{12a} f_{2a}$ and $x_{2a} = h_{2a2a} f_{2a}$ for substructure I and $x_{2b} = h_{2b2b} f_{2b}$ for substructure II. However, the equilibrium condition is modified to be $f_{2a} + f_{2b} = F_{2b}$ and the compatibility condition is rewritten as:

$$k(x_{2a} - x_{2b}) = -f_{2a}. \qquad (9.37)$$

Fig. 9.9 Flexible coupling of components I and II to form assembly III. The force F_{2b} is applied to the assembly at coordinate X_{2b} in order to determine H_{2b2b}, H_{2a2b}, and H_{12b}

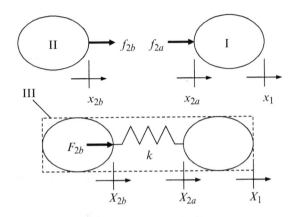

We find f_{2a} by substituting the component displacements in Eq. 9.37 and replacing f_{2b} with $F_{2a} - f_{2a}$.

$$k(h_{2a2a}f_{2a} - h_{2b2b}F_{2b} + h_{2b2b}f_{2a}) = -f_{2a}$$

$$kh_{2a2a}f_{2a} - kh_{2b2b}F_{2b} + kh_{2b2b}f_{2a} = -f_{2a}$$

$$\left(h_{2a2a} + h_{2b2b} + \frac{1}{k}\right)f_{2a} = h_{2b2b}F_{2b}$$

$$f_{2a} = \left(h_{2a2a} + h_{2b2b} + \frac{1}{k}\right)^{-1}h_{2b2b}F_{2b} \qquad (9.38)$$

Again using the equilibrium condition we find the equation for f_{2b}.

$$f_{2b} = F_{2b} - f_{2a} = \left(1 - \left(h_{2a2a} + h_{2b2b} + \frac{1}{k}\right)^{-1}h_{2b2b}\right)F_{2b} \qquad (9.39)$$

The direct and cross receptances for the case shown in Fig. 9.9 are given in Eqs. 9.40 through 9.42.

$$H_{2b2b} = \frac{X_{2b}}{F_{2b}} = \frac{x_{2b}}{F_{2b}} = \frac{h_{2b2b}f_{2b}}{F_{2b}} = \frac{h_{2b2b}\left(1 - \left(h_{2a2a} + h_{2b2b} + \frac{1}{k}\right)^{-1}h_{2b2b}\right)F_{2b}}{F_{2b}}$$

$$H_{2b2b} = h_{2b2b} - h_{2b2b}\left(h_{2a2a} + h_{2b2b} + \frac{1}{k}\right)^{-1}h_{2b2b} \qquad (9.40)$$

$$H_{2a2b} = \frac{X_{2a}}{F_{2b}} = \frac{x_{2a}}{F_{2b}} = \frac{h_{2a2a}f_{2a}}{F_{2b}} = \frac{h_{2a2a}\left(h_{2a2a} + h_{2b2b} + \frac{1}{k}\right)^{-1}h_{2b2b}F_{2b}}{F_{2b}}$$

$$H_{2a2b} = h_{2a2a}\left(h_{2a2a} + h_{2b2b} + \frac{1}{k}\right)^{-1}h_{2b2b} \qquad (9.41)$$

$$H_{12b} = \frac{X_1}{F_{2b}} = \frac{x_1}{F_{2b}} = \frac{h_{12a}f_{2a}}{F_{2a}} = \frac{h_{12a}\left(h_{2a2a} + h_{2b2b} + \dfrac{1}{k}\right)^{-1} h_{2b2b}F_{2b}}{F_{2b}}$$

$$H_{12b} = h_{12a}\left(h_{2a2a} + h_{2b2b} + \frac{1}{k}\right)^{-1} h_{2b2b} \tag{9.42}$$

9.4   Two-Component Flexible-Damped Coupling

As we discussed in Sect. 1.3, damping is always present in mechanical systems. Therefore, as a final step in our receptance coupling of components I and II to predict the assembly III response, we can expand the model in Fig. 9.7 to include viscous damping at the coupling interface. See Fig. 9.10. If we again consider the bolted connection in Fig. 9.5, the elastic deflection of the bolts could cause rubbing between the bolts and their mating surfaces. This would lead to energy dissipation that we could model as viscous damping.

The expressions for the component displacements and equilibrium conditions remain unchanged relative to the flexible coupling derivation when we add damping. However, the compatibility condition is now:

$$k(x_{2b} - x_{2a}) + i\omega c(x_{2b} - x_{2a}) = -f_{2b}, \tag{9.43}$$

where we have assumed harmonic motion so that the velocity-dependent damping forces can be expressed in the form $i\omega cx$. Equation 9.43 can be rewritten as:

$$(k + i\omega c)(x_{2b} - x_{2a}) = -f_{2b}. \tag{9.44}$$

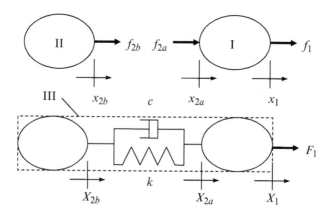

Fig. 9.10 Viscously damped, flexible coupling of components I and II to form assembly III. As with the flexible coupling case, the force F_1 is applied to the assembly at coordinate X_1 in order to determine H_{11}, H_{2a1}, and H_{2b1}

Table 9.1 Direct and cross receptances for two component coupling. The connection type is either rigid, R, or flexible, F. The receptance type is either direct, D, or cross, C. The corresponding figure and equation numbers are also included

R/F	Component coordinates		Receptances		
	I	II	D/C	Figures	Equations
R	x_{1a}	x_{1b}	D	$H_{11} = h_{1a1a} - h_{1a1a}(h_{1a1a} + h_{1b1b})^{-1} h_{1a1a}$	9.1	9.4
R	x_1, x_{2a}	x_{2b}	D	$H_{11} = h_{11} - h_{12a}(h_{2a2a} + h_{2b2b})^{-1} h_{2a1}$	9.2	9.8
			C	$H_{21} = h_{2a1} - h_{2a2a}(h_{2a2a} + h_{2b2b})^{-1} h_{2a1}$		9.9
			D	$H_{22} = h_{2a2a} - h_{2a2a}(h_{2a2a} + h_{2b2b})^{-1} h_{2a2a}$	9.3	9.13
			C	$H_{12} = h_{12a} - h_{12a}(h_{2a2a} + h_{2b2b})^{-1} h_{2a2a}$		9.14
F	x_{1a}	x_{1b}	D	$H_{1a1a} = h_{1a1a} - h_{1a1a}\left(h_{1a1a} + h_{1b1b} + \frac{1}{k}\right)^{-1} h_{1a1a}$	9.4	9.18
			C	$H_{1b1a} = h_{1b1b}\left(h_{1a1a} + h_{1b1b} + \frac{1}{k}\right)^{-1} h_{1a1a}$		9.19
			D	$H_{1b1b} = h_{1b1b} - h_{1b1b}\left(h_{1a1a} + h_{1b1b} + \frac{1}{k}\right)^{-1} h_{1b1b}$	9.6	9.22
			C	$H_{1a1b} = h_{1a1a}\left(h_{1a1a} + h_{1b1b} + \frac{1}{k}\right)^{-1} h_{1b1b}$		9.23
F	x_1, x_{2a}	x_{2b}	D	$H_{11} = h_{11} - h_{12a}\left(h_{2a2a} + h_{2b2b} + \frac{1}{k}\right)^{-1} h_{2a1}$	9.7	9.28
			C	$H_{2a1} = h_{2a1} - h_{2a2a}\left(h_{2a2a} + h_{2b2b} + \frac{1}{k}\right)^{-1} h_{2a1}$		9.29
			C	$H_{2b1} = h_{2b2b}\left(h_{2a2a} + h_{2b2b} + \frac{1}{k}\right)^{-1} h_{2a1}$		9.30
			D	$H_{2a2a} = h_{2a2a} - h_{2a2a}\left(h_{2a2a} + h_{2b2b} + \frac{1}{k}\right)^{-1} h_{2a2a}$	9.8	9.34
			C	$H_{2b2a} = h_{2b2b}\left(h_{2a2a} + h_{2b2b} + \frac{1}{k}\right)^{-1} h_{2a2a}$		9.35
			C	$H_{12a} = h_{12a} - h_{12a}\left(h_{2a2a} + h_{2b2b} + \frac{1}{k}\right)^{-1} h_{2a2a}$		9.36
			D	$H_{2b2b} = h_{2b2b} - h_{2b2b}\left(h_{2a2a} + h_{2b2b} + \frac{1}{k}\right)^{-1} h_{2b2b}$	9.9	9.40
			C	$H_{2a2b} = h_{2a2a}\left(h_{2a2a} + h_{2b2b} + \frac{1}{k}\right)^{-1} h_{2b2b}$		9.41
			C	$H_{12b} = h_{12a}\left(h_{2a2a} + h_{2b2b} + \frac{1}{k}\right)^{-1} h_{2b2b}$		9.42

If we substitute the complex, frequency-dependent variable k' for $(k + i\omega c)$, then we see that the compatibility equation takes the same form as shown in Eq. 9.24. Therefore, we can simply replace k in Eq. 9.28 with k' to obtain Eq. 9.45. This defines the direct FRF at coordinate X_1 on assembly III in Fig. 9.10. The same substitution can be made in the other assembly receptances derived for the two-component flexible coupling in order to obtain the two component flexible-damped coupling results.

$$H_{11} = h_{11} - h_{12a}\left(h_{2a2a} + h_{2b2b} + \frac{1}{k'}\right)^{-1} h_{2a1} \tag{9.45}$$

Before proceeding with a numerical example in the next section, we present Table 9.1 which summarizes the receptance coupling equations developed in the previous sections.

9.5   Comparison of Assembly Modeling Techniques

Let's now complete an example where we compare receptance coupling to the modal analysis and complex matrix inversion methods we discussed in Chap. 5. As shown in Fig. 9.11, two single degree of freedom spring–mass–damper systems, I and II, are to be connected using the spring, k_c, to form the new two degree of freedom assembly, III. The assembled system's equations of motion are determined using the appropriate free body diagrams. The matrix representation of these equations, after substituting the assumed harmonic form of the solution, is provided in Eq. 9.46. This equation takes the form:

$$\left(s^2[m] + s[c] + [k]\right)\{\vec{X}\}e^{st} = \{\vec{F}\}e^{st},$$

where we have used the Laplace variable s to represent the product $i\omega$ and $[m]$, $[c]$, and $[k]$ are the assembly's lumped parameter mass, damping, and stiffness matrices in local coordinates, respectively.

$$\left(s^2\begin{bmatrix} m_1 & 0 \\ 0 & m_2 \end{bmatrix} + s\begin{bmatrix} c_1 & 0 \\ 0 & c_2 \end{bmatrix} + \begin{bmatrix} k_1 + k_c & -k_c \\ -k_c & k_2 + k_c \end{bmatrix}\right)\begin{Bmatrix} X_{1a} \\ X_{1b} \end{Bmatrix} = \begin{Bmatrix} 0 \\ F_{1b} \end{Bmatrix}$$

$$\begin{bmatrix} m_1 s^2 + c_1 s + (k_1 + k_c) & -k_c \\ -k_c & m_2 s^2 + c_2 s + (k_2 + k_c) \end{bmatrix}\begin{Bmatrix} X_{1a} \\ X_{1b} \end{Bmatrix} = \begin{Bmatrix} 0 \\ F_{1b} \end{Bmatrix} \qquad (9.46)$$

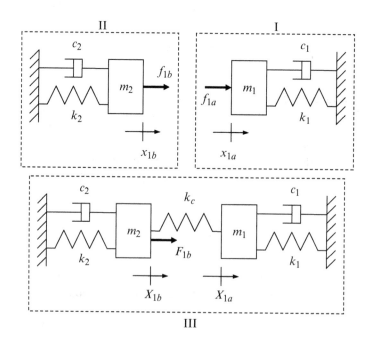

Fig. 9.11 Flexible coupling of spring–mass–damper systems I and II to form the two degree of freedom assembly III

9.5.1 Modal Analysis

We can use the equations of motion shown in Eq. 9.46 to find the modal solution for the assembled system. If we assume that proportional damping exists (i.e., $[c] = \alpha[m] + \beta[k]$, where α and β are real numbers), damping can be neglected in the modal solution. Note that this solution is also independent of the external force, F_{1b}. We write the characteristic equation for this system as shown in Eq. 9.47. The quadratic roots of this fourth order equation, s_1^2 and s_2^2, give the two eigenvalues ($s_1^2 = -\omega_{n1}^2$ and $s_2^2 = -\omega_{n2}^2$, where $\omega_{n1} < \omega_{n2}$) for the two degree of freedom system.

$$\left(m_1 s^2 + (k_1 + k_c)\right)\left(m_2 s^2 + (k_2 + k_c)\right) - k_c^2 = 0$$

$$m_1 m_2 s^4 + (m_1(k_2 + k_c) + m_2(k_1 + k_c))s^2 + (k_1 + k_c)(k_2 + k_c) - k_c^2 = 0 \quad (9.47)$$

Substitution of these eigenvalues into either of the original equations of motion, again neglecting damping and the external force, yields the eigenvectors (mode shapes). Note that the eigenvectors must be normalized to the force location (coordinate X_{1b} in this case). Selecting the top equation from Eq. 9.46, for example, gives:

$$\frac{X_{1a}}{X_{1b}} = \frac{k_c}{m_1 s^2 + (k_1 + k_c)} \quad (9.48)$$

The mass, damping, and stiffness matrices are diagonalized using the modal matrix (composed of columns of the eigenvectors), P, defined in Eq. 9.49.

$$P = \begin{bmatrix} \frac{X_{1a}}{X_{1b}}(s_1^2) & \frac{X_{1a}}{X_{1b}}(s_2^2) \\ 1 & 1 \end{bmatrix} \quad (9.49)$$

Specifically, we have that:

$$[m_q] = [P]^T[m][P] = \begin{bmatrix} m_{q1} & 0 \\ 0 & m_{q2} \end{bmatrix},$$

$$[c_q] = [P]^T[c][P] = \begin{bmatrix} c_{q1} & 0 \\ 0 & c_{q2} \end{bmatrix}, \text{ and}$$

$$[k_q] = [P]^T[k][P] = \begin{bmatrix} k_{q1} & 0 \\ 0 & k_{q2} \end{bmatrix}.$$

Based on these modal mass, damping, and stiffness values, we calculate the associated damping ratios, $\zeta_{q1,2} = \frac{c_{q1,2}}{2\sqrt{k_{q1,2}m_{q1,2}}}$. The modal solution for the direct FRF at coordinate X_{1b} of the assembled system is then expressed as shown in Eq. 9.50, where $r_{1,2} = \frac{\omega}{\omega_{n1,2}}$. As we discussed in Chap. 5, the direct FRF is the sum of the modal contributions.

$$H_{1b1b} = \frac{X_{1b}}{F_{1b}} = \frac{1}{k_{q1}} \left(\frac{(1 - r_1^2) - i(2\zeta_{q1}r_1)}{(1 - r_1^2)^2 + (2\zeta_{q1}r_1)^2} \right) + \frac{1}{k_{q2}} \left(\frac{(1 - r_2^2) - i(2\zeta_{q2}r_2)}{(1 - r_2^2)^2 + (2\zeta_{q2}r_2)^2} \right)$$

(9.50)

9.5.2   Complex Matrix Inversion

Equation 9.46 can be compactly written as $[A]\{\vec{X}\} = \{\vec{F}\}$. As shown in Sect. 5.2, complex matrix inversion is carried out using $\{\vec{X}\}\{\vec{F}\}^{-1} = [A]^{-1}$ to determine the assembly's direct and cross FRFs; it is applied when the damping may not be proportional. The inverted $[A]$ matrix for this two-degree-of-freedom example is:

$$[A]^{-1} = \frac{\begin{bmatrix} a_{22} & -a_{12} \\ -a_{21} & a_{11} \end{bmatrix}}{a_{11} \cdot a_{22} - a_{12} \cdot a_{21}}$$

$$= \frac{\begin{bmatrix} -\omega^2 m_2 + i\omega c_2 + (k_2 + k_c) & k_c \\ k_c & -\omega^2 m_1 + i\omega c_1 + (k_1 + k_c) \end{bmatrix}}{(-\omega^2 m_1 + i\omega c_1 + (k_1 + k_c))(-\omega^2 m_2 + i\omega c_2 + (k_2 + k_c)) - k_c^2},$$

where we have replaced s with $i\omega$ relative to Eq. 9.46. The individual terms in the inverted $[A]$ matrix are:

$$[A]^{-1} = \begin{bmatrix} \dfrac{X_{1a}}{F_{1a}} & \dfrac{X_{1a}}{F_{1b}} \\ \dfrac{X_{1b}}{F_{1a}} & \dfrac{X_{1b}}{F_{1b}} \end{bmatrix} = \begin{bmatrix} H_{1a1a} & H_{1a1b} \\ H_{1b1a} & H_{1b1b} \end{bmatrix}.$$

(9.51)

9.5.3   Receptance Coupling

This case is the same as the two-component flexible coupling example shown in Fig. 9.6. Replacing k with k_c in Eq. 9.22, we obtain Eq. 9.52.

Parameter	Value
m_1	3 kg
c_1	200 N-s/m
k_1	2×10^6 N/m
m_2	2 kg
c_2	100 N-s/m
k_2	1×10^6 N/m
k_c	5×10^5 N/m

Table 9.2 Mass, damping, and stiffness values for the two degree of freedom system in Fig. 9.11

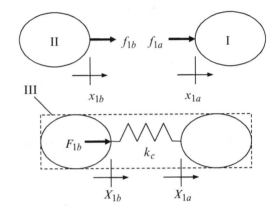

Fig. 9.12 Receptance-coupling representation of joining spring–mass–damper systems I and II to form the two degree of freedom assembly III

$$\frac{X_{1b}}{F_{1b}} = \frac{x_{1b}}{F_{1b}} = \frac{h_{1b1b}f_{1b}}{F_{1b}} = \frac{h_{1b1b}\left(1 - \left(h_{1a1a} + h_{1b1b} + \frac{1}{k_c}\right)^{-1} h_{1b1b}\right)F_{1b}}{F_{1b}}$$

$$\frac{X_{1b}}{F_{1b}} = H_{1b1b} = h_{1b1b} - h_{1b1b}\left(h_{1a1a} + h_{1b1b} + \frac{1}{k_c}\right)^{-1} h_{1b1b} \tag{9.52}$$

To compare the three methods, we apply the mass, damping, and stiffness values shown in Table 9.2 to the model displayed in Fig. 9.11. We note that proportional damping exists ($\alpha = 0$ and $\beta = 1 \times 10^{-4}$) for the selected system, so the modal approach may be applied. The code used to produce Fig. 9.13, which displays both the component receptances and the assembly receptance computed using the three methods, is provided in MATLAB® Mojo 9.1. The frequency-dependent differences between the complex matrix inversion result, which was obtained through vector manipulations by calculating the H_{1b1b} result directly:

$$H_{1b1b} = \frac{-\omega^2 m_1 + i\omega c_1 + (k_1 + k_c)}{(-\omega^2 m_1 + i\omega c_1 + (k_1 + k_c))(-\omega^2 m_2 + i\omega c_2 + (k_2 + k_c)) - k_c^2},$$

and the modal and receptance coupling method results are displayed in Fig. 9.14. It is seen that the errors introduced by the modal method (top) are approximately 4×10^{13} times greater than the errors associated with the receptance technique (bottom). The differences between the three techniques are introduced by numerical

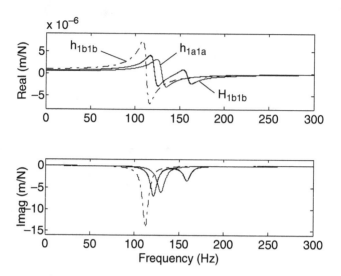

Fig. 9.13 Comparison of three methods for H_{1b1b} calculation. It is seen that the modal analysis, complex matrix inversion, and receptance coupling methods nominally agree (*superimposed solid lines*). The component receptances, h_{1a1a} and h_{1b1b}, are also shown

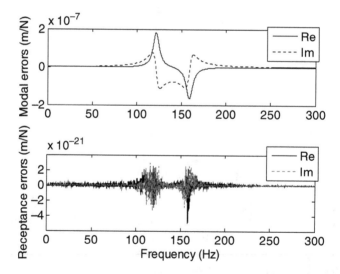

Fig. 9.14 Real and imaginary parts of the difference between complex matrix inversion and modal analysis (*top*) and real and imaginary parts of difference between complex matrix inversion and receptance coupling (*bottom*). Receptance coupling agrees more closely

round-off errors in the mathematical manipulations. However, the improved numerical accuracy obtained with receptance coupling (vector manipulations) over modal coupling (matrix manipulations) is another benefit of the receptance coupling approach.

IN A NUTSHELL   Our comparison presumes that the receptances are accurately known. If they are obtained from measurements, this may not be the case. Measured FRFs are only known with finite frequency and amplitude resolution. Large modes require more digital bits to represent them and small modes may fall below the amplitude resolution (and not appear in the measurement). For high measurement bandwidths with reduced frequency resolution, modes with very low damping (and a corresponding narrow frequency range) may be misrepresented as well.

MATLAB® MOJO 9.1

```
% matlab_mojo_9_1.m

clc
clear all
close all

% Define parameters in local coordinates
m1 = 3;              % kg
c1 = 200;            % N-s/m
k1 = 2e6;            % N/m

m2 = 2;              % kg
c2 = 100;            % N-s/m
k2 = 1e6;            % N/m

kc = 5e5;            % N/m

% Define I and II FRFs
w = (0:0.1:300)'*2*pi; % frequency, rad/s
FRF_I = 1./(-w.^2*m1 + 1i*w*c1 + k1);
FRF_II = 1./(-w.^2*m2 + 1i*w*c2 + k2);

% Receptance coupling
FRF_III_rc = FRF_II - FRF_II./(FRF_I + FRF_II + 1/kc).*FRF_II;

% Modal analysis
s_squared = roots([(m1*m2) (m1*(k2+kc)+m2*(k1+kc)) ((k1+kc)*(k2+kc)-kc^2)]);
s1_squared = s_squared(1);
s2_squared = s_squared(2);
% Order natural frequencies so that wn1 < wn2
if s1_squared < s2_squared
    temp = s1_squared;
    s1_squared = s2_squared;
    s2_squared = temp;
end
wn1 = sqrt(-s1_squared);
wn2 = sqrt(-s2_squared);

p1 = kc/(m1*s1_squared + (k1+kc));
p2 = kc/(m1*s2_squared + (k1+kc));

% Local matrices
m = [m1 0; 0 m2];
c = [c1 0; 0 c2];
k = [k1+kc -kc; -kc k2+kc];
```

```
% Modal matrices
P = [p1 p2; 1 1];
mq = P'*m*P;
cq = P'*c*P;
kq = P'*k*P;

mq1 = mq(1,1);
mq2 = mq(2,2);
cq1 = cq(1,1);
cq2 = cq(2,2);
kq1 = kq(1,1);
kq2 = kq(2,2);

zetaq1 = cq1/(2*sqrt(kq1*mq1));
zetaq2 = cq2/(2*sqrt(kq2*mq2));

r1 = w/wn1;
r2 = w/wn2;

FRF_III_modal = 1/kq1*((1-r1.^2) - 1i*(2*zetaq1*r1))./((1-r1.^2).^2 +
(2*zetaq1*r1).^2) + 1/kq2*((1-r2.^2) - 1i*(2*zetaq2*r2))./((1-r2.^2).^2 +
(2*zetaq2*r2).^2);

% Complex matrix inversion
FRF_III_inversion = (-w.^2*m1 + 1i*w*c1 + k1 + kc)./((-w.^2*m1 + 1i*w*c1 + k1
+ kc).*(-w.^2*m2 + 1i*w*c2 + k2 + kc) - kc^2);

figure(1)
subplot(211)
plot(w/2/pi, real(FRF_I), 'k:', w/2/pi, real(FRF_II), 'k-.', w/2/pi,
real(FRF_III_rc), 'k', w/2/pi, real(FRF_III_modal), 'k', w/2/pi,
real(FRF_III_inversion), 'k')
ylim([-8e-6 9e-6])
set(gca,'FontSize', 14)
ylabel('Real (m/N)')
subplot(212)
plot(w/2/pi, imag(FRF_I), 'k:', w/2/pi, imag(FRF_II), 'k-.', w/2/pi,
imag(FRF_III_rc), 'k', w/2/pi, imag(FRF_III_modal), 'k', w/2/pi,
imag(FRF_III_inversion), 'k')
ylim([-16e-6 16e-7])
set(gca,'FontSize', 14)
xlabel('Frequency (Hz)')
ylabel('Imag (m/N)')

% Calculate differences
diff_modal = FRF_III_inversion - FRF_III_modal;
diff_rc = FRF_III_inversion - FRF_III_rc;

figure(2)
subplot(211)
plot(w/2/pi, real(diff_modal), 'k', w/2/pi, imag(diff_modal), 'k:')
legend('Re', 'Im')
ylim([-2e-7 2e-7])
set(gca,'FontSize', 14)
ylabel('Modal errors (m/N)')
subplot(212)
plot(w/2/pi, real(diff_rc), 'k', w/2/pi, imag(diff_rc), 'b:')
legend('Re', 'Im')
ylim([-6e-21 4e-21])
set(gca,'FontSize', 14)
xlabel('Frequency (Hz)')
ylabel('Receptance errors (m/N)')
```

9.6 Advanced Receptance Coupling

In the previous sections we only considered transverse deflections, x_i and X_i, for the components and assembly due to internal and external forces, f_j and F_j. However, as we saw in Sect. 8.5 we must also consider *rotations* about lines perpendicular to the beam axis, θ_i and Θ_i, and *bending couples*, m_j and M_j, to completely describe the transverse dynamic behavior of beams.[2] To begin this discussion, let's consider the solid cylinder-prismatic cantilever beam assembly shown in Fig. 9.15. To determine the assembly dynamics, all four bending receptances must be included in the component descriptions (i.e., displacement-to-force, h_{ij}, displacement-to-couple, l_{ij}, rotation-to-force, n_{ij}, and rotation-to-couple, p_{ij}).

Let's now summarize the steps required to predict the Fig. 9.15 assembly receptances.

1. Define the components and coordinates for the model. In this example, we can select two components: a prismatic beam with fixed-free (or cantilever) boundary conditions and a cylinder with free-free (or unsupported) boundary conditions; see Fig. 9.16.
2. Determine the component receptances. We can use either measurements or models. For the models, an elegant choice is the closed-form receptances for flexural vibrations of uniform Euler–Bernoulli beams with free, fixed, sliding,

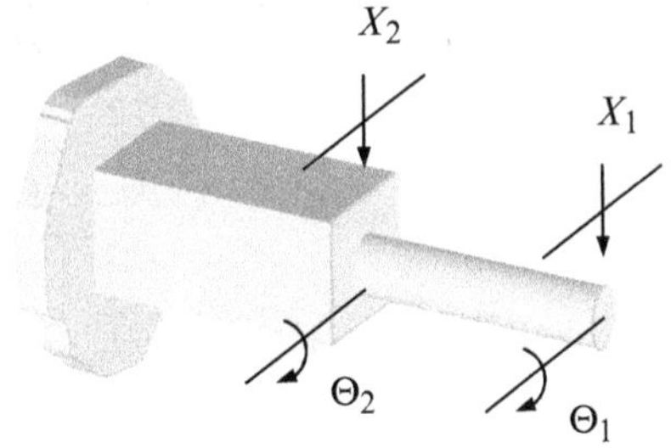

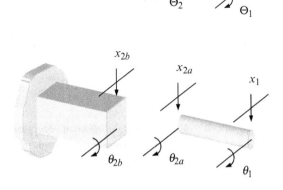

Fig. 9.15 Rigid coupling of solid cylinder and prismatic beam to form a cantilevered assembly

Fig. 9.16 Solid cylinder and prismatic beam components used to form cantilevered assembly

---

[2] We will not consider axial or torsional vibrations in this analysis.

and pinned boundary conditions that we discussed in Chap. 8 (Bishop and Johnson 1960). Of course, the Timoshenko beam model (Weaver et al. 1990) may also be applied when increased accuracy is required, but we will leave the details of this analysis to a more advanced course. For measurements, we can follow the procedures outlined in Chap. 7.

3. Based on the model from step 1, express the assembly receptances as a function of the component receptances. As demonstrated in Sects. 9.2 through 9.4, we determine the assembly receptances using the component displacements/ rotations, equilibrium conditions, and compatibility conditions.

We begin the analysis of the system shown in Figs. 9.15 and 9.16 by writing the component receptances. Note that we have placed coordinates at the prediction location (1) and coupling locations (2a and 2b) on the two components. For the cylinder, we have the following direct receptances at the coordinate 1 end:

$$h_{11} = \frac{x_1}{f_1} \quad l_{11} = \frac{x_1}{m_1} \quad n_{11} = \frac{\theta_1}{f_1} \quad p_{11} = \frac{\theta_1}{m_1}. \tag{9.53}$$

The corresponding cross receptances at the same location are:

$$h_{12a} = \frac{x_1}{f_{2a}} \quad l_{12a} = \frac{x_1}{m_{2a}} \quad n_{12a} = \frac{\theta_1}{f_{2a}} \quad p_{12a} = \frac{\theta_1}{m_{2a}}. \tag{9.54}$$

At coordinate 2a on the cylinder, the direct and cross receptances are written as shown in Eqs. 9.55 and 9.56, respectively.

$$h_{2a2a} = \frac{x_{2a}}{f_{2a}} \quad l_{2a2a} = \frac{x_{2a}}{m_{2a}} \quad n_{2a2a} = \frac{\theta_{2a}}{f_{2a}} \quad p_{2a2a} = \frac{\theta_{2a}}{m_{2a}} \tag{9.55}$$

$$h_{2a1} = \frac{x_{2a}}{f_1} \quad l_{2a1} = \frac{x_{2a}}{m_1} \quad n_{2a1} = \frac{\theta_{2a}}{f_1} \quad p_{2a1} = \frac{\theta_{2a}}{m_1} \tag{9.56}$$

Similarly, for the prismatic cantilever beam, the direct receptances at the coupling location 2b are described by Eq. 9.57.

$$h_{2b2b} = \frac{x_{2b}}{f_{2b}} \quad l_{2b2b} = \frac{x_{2b}}{m_{2b}} \quad n_{2b2b} = \frac{\theta_{2b}}{f_{2b}} \quad p_{2b2b} = \frac{\theta_{2b}}{m_{2b}} \tag{9.57}$$

To simplify notation, the component receptances can be compactly represented in matrix form as shown in Eqs. 9.58 through 9.61 for the cylinder and Eq. 9.62 for the prismatic beam:

$$\begin{Bmatrix} x_1 \\ \theta_1 \end{Bmatrix} = \begin{bmatrix} h_{11} & l_{11} \\ n_{11} & p_{11} \end{bmatrix} \begin{Bmatrix} f_1 \\ m_1 \end{Bmatrix} \text{ or } \{u_1\} = [R_{11}]\{q_1\}, \tag{9.58}$$

$$\begin{Bmatrix} x_{2a} \\ \theta_{2a} \end{Bmatrix} = \begin{bmatrix} h_{2a2a} & l_{2a2a} \\ n_{2a2a} & p_{2a2a} \end{bmatrix} \begin{Bmatrix} f_{2a} \\ m_{2a} \end{Bmatrix} \text{ or } \{u_{2a}\} = [R_{2a2a}]\{q_{2a}\}, \tag{9.59}$$

$$\begin{Bmatrix} x_1 \\ \theta_1 \end{Bmatrix} = \begin{bmatrix} h_{12a} & l_{12a} \\ n_{12a} & p_{12a} \end{bmatrix} \begin{Bmatrix} f_{2a} \\ m_{2a} \end{Bmatrix} \text{ or } \{u_1\} = [R_{12a}]\{q_{2a}\}, \tag{9.60}$$

$$\begin{Bmatrix} x_{2a} \\ \theta_{2a} \end{Bmatrix} = \begin{bmatrix} h_{2a1} & l_{2a1} \\ n_{2a1} & p_{2a1} \end{bmatrix} \begin{Bmatrix} f_1 \\ m_1 \end{Bmatrix} \text{ or } \{u_{2a}\} = [R_{2a1}]\{q_1\}, \text{ and} \tag{9.61}$$

$$\begin{Bmatrix} x_{2b} \\ \theta_{2b} \end{Bmatrix} = \begin{bmatrix} h_{2b2b} & l_{2b2b} \\ n_{2b2b} & p_{2b2b} \end{bmatrix} \begin{Bmatrix} f_{2b} \\ m_{2b} \end{Bmatrix} \text{ or } \{u_{2b}\} = [R_{2b2b}]\{q_{2b}\}, \tag{9.62}$$

where R_{ij} is the *generalized receptance matrix* that describes both translational and rotational component behavior (Burns and Schmitz 2004, 2005; Park et al. 2003) and u_i and q_j are the corresponding *generalized displacement/rotation* and *force/couple* vectors. To visualize R_{ij}, we can think of each frequency-dependent 2×2 R_{ij} matrix as a page in a book where each page represents a different frequency value. Flipping through the book from front to back scans the frequency values from low to high through the modeled, or measured, bandwidth. Naturally, all receptances in the coupling analysis must be based on the same frequency vector (resolution and range).

We write the component receptances using the new notation as $u_1 = R_{11}q_1 + R_{12a}q_{2a}$ and $u_{2a} = R_{2a1}q_1 + R_{2a2a}q_{2a}$ for the cylinder and $u_{2b} = R_{2b2b}q_{2b}$ for the prismatic beam. If we apply a rigid connection between the two components, the compatibility condition is $u_{2b} - u_{2a} = 0$. Additionally, if we again specify that the component and assembly coordinates are at the same physical locations, then we have that $u_1 = U_1$, and $u_{2a} = u_{2b} = U_2$ (due to the rigid coupling).

We can write the assembly receptances, as shown in Eq. 9.63, which again incorporates the generalized notation:

$$\begin{Bmatrix} U_1 \\ U_2 \end{Bmatrix} = \begin{bmatrix} G_{11} & G_{12} \\ G_{21} & G_{22} \end{bmatrix} \begin{Bmatrix} Q_1 \\ Q_2 \end{Bmatrix} \tag{9.63}$$

where $U_i = \begin{Bmatrix} X_i \\ \Theta_i \end{Bmatrix}$, $G_{ij} = \begin{bmatrix} H_{ij} & L_{ij} \\ N_{ij} & P_{ij} \end{bmatrix}$, and $Q_j = \begin{Bmatrix} F_j \\ M_j \end{Bmatrix}$. To determine the assembly receptance at the free end of the cylinder, G_{11}, we apply Q_1 to coordinate U_1 as shown in Fig. 9.17, where the generalized U_i and u_i vectors are shown schematically as "displacements," although we recognize that they describe both transverse deflection and rotation. The associated equilibrium conditions are $q_{2a} + q_{2b} = 0$ and $q_1 = Q_1$. By substituting the component displacements/rotations and equilibrium conditions into the compatibility condition, we obtain the expression for q_{2b} shown in Eq. 9.64. The component force q_{2a} is then determined from the equilibrium condition $q_{2a} = -q_{2b}$. The expression for G_{11} is given by Eq. 9.65.

Fig. 9.17 Receptance-
coupling model for
determining G_{11} and G_{21}.
Rigid coupling is assumed

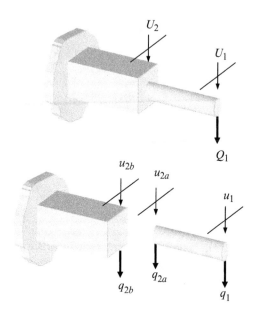

We find the corresponding cross receptance matrix, G_{21}, in a similar manner;
see Eq. 9.66. Note that G_{11} and G_{21} comprise the first column of the receptance
matrix in Eq. 9.63.

$$u_{2b} - u_{2a} = 0$$
$$R_{2b2b}q_{2b} - R_{2a1}q_1 - R_{2a2a}q_{2a} = 0$$
$$(R_{2a2a} + R_{2b2b})q_{2b} - R_{2a1}Q_1 = 0$$
$$q_{2b} = (R_{2a2a} + R_{2b2b})^{-1}R_{2a1}Q_1 \tag{9.64}$$

$$G_{11} = \frac{U_1}{Q_1} = \frac{u_1}{Q_1} = \frac{R_{11}q_1 + R_{12a}q_{2a}}{Q_1} = \frac{R_{11}Q_1 - R_{12a}(R_{2a2a} + R_{2b2b})^{-1}R_{2a1}Q_1}{Q_1}$$

$$G_{11} = R_{11} - R_{12a}(R_{2a2a} + R_{2b2b})^{-1}R_{2a1} = \begin{bmatrix} H_{11} & L_{11} \\ N_{11} & P_{11} \end{bmatrix} \tag{9.65}$$

$$G_{21} = \frac{U_2}{Q_1} = \frac{u_{2a}}{Q_1} = \frac{R_{2a1}q_1 + R_{2a2a}q_{2a}}{Q_1} = \frac{R_{2a1}Q_1 - R_{2a2a}(R_{2a2a} + R_{2b2b})^{-1}R_{2a1}Q_1}{Q_1}$$

$$G_{21} = R_{2a1} - R_{2a2a}(R_{2a2a} + R_{2b2b})^{-1}R_{2a1} = \begin{bmatrix} H_{21} & L_{21} \\ N_{21} & P_{21} \end{bmatrix}$$

$$\tag{9.66}$$

To find the receptances in the second column of Eq. 9.63, we apply Q_2 at U_2,
as shown in Fig. 9.18. The component receptances are $u_1 = R_{12a}q_{2a}$ and $u_{2a} =
R_{2a2a}q_{2a}$ for the cylinder and $u_{2b} = R_{2b2b}q_{2b}$ for the prismatic beam. For the rigid
connection, the compatibility condition is again $u_{2b} - u_{2a} = 0$. The equilibrium
condition is $q_{2a} + q_{2b} = Q_2$. By substituting the component displacements/

Fig. 9.18 Receptance-coupling model for determining G_{22} and G_{12}. Rigid coupling is assumed

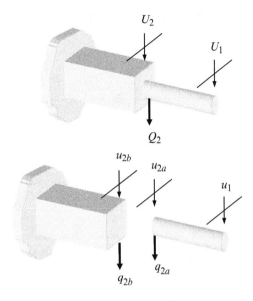

rotations and equilibrium condition into the compatibility condition, we obtain the expression for q_{2b} shown in Eq. 9.67. The component force q_{2a} is then determined from the equilibrium condition $q_{2a} = Q_2 - q_{2b}$. The expression for G_{22} is provided by Eq. 9.68. We find the corresponding cross-receptance matrix, G_{12}, in a similar manner, as shown in Eq. 9.69.

$$u_{2b} - u_{2a} = 0$$
$$R_{2b2b}q_{2b} - R_{2a2a}q_{2a} = 0$$
$$R_{2b2b}q_{2b} - R_{2a2a}Q_2 + R_{2a2a}q_{2b} = 0$$
$$(R_{2a2a} + R_{2b2b})q_{2b} - R_{2a2a}Q_2 = 0$$
$$q_{2b} = (R_{2a2a} + R_{2b2b})^{-1}R_{2a2a}Q_2 \tag{9.67}$$

$$G_{22} = \frac{U_2}{Q_2} = \frac{u_{2a}}{Q_2} = \frac{R_{2a2a}q_{2a}}{Q_2} = \frac{R_{2a2a}\left(1 - (R_{2a2a} + R_{2b2b})^{-1}R_{2a2a}\right)Q_2}{Q_2}$$
$$G_{22} = R_{2a2a} - R_{2a2a}(R_{2a2a} + R_{2b2b})^{-1}R_{2a2a} = \begin{bmatrix} H_{22} & L_{22} \\ N_{22} & P_{22} \end{bmatrix} \tag{9.68}$$

$$G_{12} = \frac{U_1}{Q_2} = \frac{u_1}{Q_2} = \frac{R_{12a}q_{2a}}{Q_2} = \frac{R_{12a}\left(1 - (R_{2a2a} + R_{2b2b})^{-1}R_{2a2a}\right)Q_2}{Q_2}$$
$$G_{12} = R_{12a} - R_{12a}(R_{2a2a} + R_{2b2b})^{-1}R_{2a2a} = \begin{bmatrix} H_{12} & L_{12} \\ N_{12} & P_{12} \end{bmatrix} \tag{9.69}$$

Table 9.3 Direct and cross receptances for generalized two component coupling. The connection type is rigid, R. The receptance type is direct, D, or cross, C. The figure and equation numbers are also included. Similarities to the corresponding entries in Table 9.1 are evident

Substructure coordinates			Receptances			
R/F	I	II	D/C		Figures	Equations
R/F	I	II	D/C			
R	u_1, u_{2a}	u_{2b}	D	$G_{11} = R_{11} - R_{12a}(R_{2a2a} + R_{2b2b})^{-1}R_{2a1}$	9.17	9.65
			C	$G_{21} = R_{2a1} - R_{2a2a}(R_{2a2a} + R_{2b2b})^{-1}R_{2a1}$		9.66
			D	$G_{22} = R_{2a2a} - R_{2a2a}(R_{2a2a} + R_{2b2b})^{-1}R_{2a2a}$	9.18	9.68
			C	$G_{12} = R_{12a} - R_{12a}(R_{2a2a} + R_{2b2b})^{-1}R_{2a2a}$		9.69

We see that the procedure to model the systems with both displacements and rotations is analogous to the examples provided in Sects. 9.2 through 9.4. Let's again summarize the receptance terms in tabular form; see Table 9.3. Due to the clear similarities to Table 9.1, we will not derive the receptances for the other two component coupling cases. The only consideration is that for non-rigid coupling, we replace the scalar stiffness term, $\frac{1}{k}$, from the displacement-to-force analyses with the matrix expression $[k]^{-1}$, where:

$$[\tilde{k}] = \begin{bmatrix} k_{xf} & k_{\theta f} \\ k_{xm} & k_{\theta m} \end{bmatrix}.$$

The subscripts for the stiffness matrix entries indicate their function. For example, $k_{\theta f}$ represents resistance to rotation due to an applied force. As shown in Eq. 9.45, these four real-valued stiffness terms are augmented by the corresponding damping expressions if viscous damping is included at the coupling location (Schmitz et al. 2007). The new complex, frequency-dependent stiffness matrix is:

$$[\tilde{k}'] = \begin{bmatrix} k_{xf} + i\omega c_{xf} & k_{\theta f} + i\omega c_{\theta f} \\ k_{xm} + i\omega c_{xm} & k_{\theta m} + i\omega c_{\theta m} \end{bmatrix}.$$

IN A NUTSHELL   More information about the dynamic behavior of a system enables improved modeling accuracy. Measurement of rotational FRFs is more complicated than the measurement of translational FRFs. Is it required? The answer, as in most of engineering analyses, depends on the application. It is more important for short, bulky components than it is for long, slender components.

9.7 Assembly Receptance Prediction

In Sect. 9.6, we provided the building blocks for assembly receptance predictions. In this section, we will detail coupling examples to demonstrate their implementation.

9.7.1 Free-Free Beam Coupled to Rigid Support

As a test of the receptance coupling procedure, let's couple a free-free beam to a rigid support (i.e., a wall) to verify that it matches the fixed-free beam response we derived in Sect. 8.3.1. As described in Sect. 9.6, we have three primary tasks to complete in order to predict the assembly response. First, we must define the components and coordinates for the model. Here we have two components: a uniform beam with free-free boundary conditions and a rigid support (which exhibits zero receptances); see Fig. 9.19. Second, we need to determine the component receptances. We will apply the Euler–Bernoulli beam receptances provided in Table 8.3. Third, based on the selected model, we express the assembly receptances as a function of the component receptances, as shown in Table 9.3.

Let's define the free-free beam to be a solid steel cylinder with a diameter of 10 mm and a length of 125 mm. The elastic modulus is 200 GPa and the density is 7,800 kg/m^3. We will select the solid damping factor to be 0.01 for plotting purposes, but in practice a value closer to 0.001 would be more realistic. The free-free cylinder's direct and cross receptance equations are provided in Table 8.3,

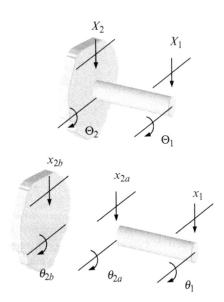

Fig. 9.19 Rigid coupling of a free-free cylinder to a wall

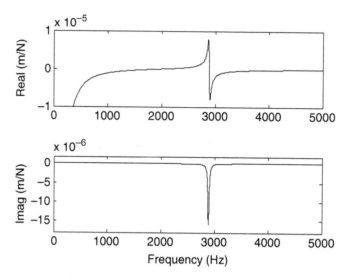

Fig. 9.20 Free-free receptance, h_{11}, for 10 mm diameter by 125 mm long steel cylinder

while the wall receptances are zero. To calculate λ, we need the frequency vector ω (rad/s), cross-sectional area, A, and second moment of area, I. We will use a frequency range of 5,000 Hz with a resolution of 0.1 Hz. The variables A and I are defined in Eqs. 9.70 and 9.71 for the cylinder, where d is the cylinder diameter. The displacement-to-force direct receptance for the free-free cylinder, h_{11}, is shown in Fig. 9.20. We see first a bending natural frequency of 2,884.9 Hz. The rigid body behavior is exhibited as the rapid decrease in the real part as the frequency approaches zero.

$$A = \frac{\pi d^2}{4} \tag{9.70}$$

$$I = \frac{\pi d^4}{64} \tag{9.71}$$

To rigidly couple the free-free cylinder to the wall, we apply Eq. 9.65:

$$G_{11} = \begin{bmatrix} H_{11} & L_{11} \\ N_{11} & P_{11} \end{bmatrix} = R_{11} - R_{12a}(R_{2a2a} + R_{2b2b})^{-1}R_{2a1},$$

where the generalized receptance matrices R_{11}, R_{12a}, R_{2a2a}, and R_{2a1} correspond to the free-free cylinder, and R_{2b2b} characterizes the wall response. The code provided in MATLAB® MOJO 9.2 is used to complete the receptance coupling procedure. The results are displayed in Figs. 9.21 and 9.22. Figure 9.21 shows the assembly-H_{11} response from the $G_{11}(1,1)$ position (solid line). The dotted line in the figure is the

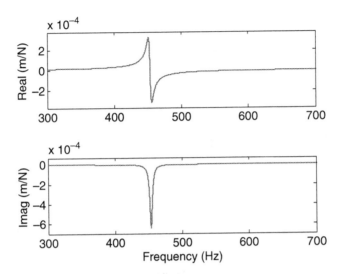

Fig. 9.21 Comparison of H_{11} receptance coupling result (*solid line*) and fixed-free response (*dotted*) for 10 mm diameter by 125 mm long steel cylinder

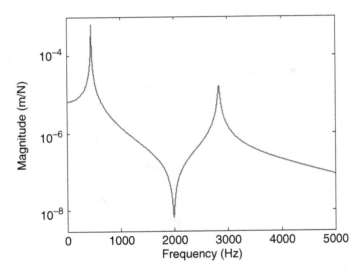

Fig. 9.22 Semi-logarithmic plot showing the first two bending modes for H_{11} tip receptances obtained from: (1) rigid coupling of free-free beam to wall (*solid line*); and (2) fixed-free response (*dotted*)

clamped-free response, $H_{11} = \frac{-c_1}{\lambda^3 c_8}$, from Table 8.3. We see that the two curves are identical and the rigid body behavior is no longer present due to the coupling conditions. A limited frequency range is displayed in Fig. 9.22 to enable closer comparison of the first bending mode. However, all bending modes are included in the Euler–Bernoulli beam receptances. The frequency range is increased in

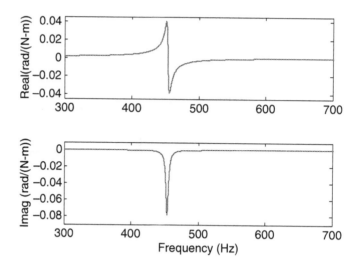

Fig. 9.23 Comparison of P_{11} receptance coupling result (*solid line*) and fixed-free response (*dotted*) for 10 mm diameter by 125 mm long steel cylinder

Fig. 9.23 to show the first two assembly bending modes. The vertical axis (response magnitude) is logarithmic in this plot because the second mode magnitude is much smaller than the first. Again, we observe exact agreement between the receptance-coupling result (solid) and clamped-free receptance (dotted). Figure 9.23 displays not only the two resonant peaks at 453.4 and 2,841.4 Hz, but also the antiresonance at 1,988.1 Hz. At this frequency, the response is very small, even for large input force magnitudes.

The rotation-to-couple free-end receptance determined from the rigid free-free beam coupling to the wall is also calculated using the code in MATLAB® MOJO 9.2. This $G_{11}(2,2)$ entry is shown in Fig. 9.23 (solid line). The fixed-free response (dotted line) from Table 8.3, $P_{11} = \frac{c_5}{\lambda c_8}$, again agrees with the receptance coupling result. We also see that the first mode natural frequency matches the H_{11} result (453.4 Hz), but the magnitude is quite different; note the new units of rad/(N-m).

We have already noted that the assembly cross receptances, G_{12} and G_{21}, and the direct receptances at the fixed end, G_{22}, are zero. We can verify this by direct application of Eqs. 9.66, 9.68, and 9.69. For the fixed end direct receptance, Eq. 9.68 simplifies as shown in Eq. 9.72.

$$G_{22} = R_{2a2a} - R_{2a2a}(R_{2a2a} + R_{2b2b})^{-1}R_{2a2a}$$

$$G_{22} = R_{2a2a} - R_{2a2a}\left(R_{2a2a} + \begin{bmatrix} 0 & 0 \\ 0 & 0 \end{bmatrix}\right)^{-1}R_{2a2a}$$

$$G_{22} = R_{2a2a} - R_{2a2a}(R_{2a2a})^{-1}R_{2a2a} = R_{2a2a} - R_{2a2a} = 0 \qquad (9.72)$$

Similar results are obtained for the cross receptances in Eqs. 9.66 and 9.69 when substituting $R_{2b2b} = \begin{bmatrix} 0 & 0 \\ 0 & 0 \end{bmatrix}$.

MATLAB® MOJO 9.2

```matlab
% matlab_mojo_9_2.m

clc
clear all
close all

% Define free-free cylinder receptances
w = (1:0.1:5000)*2*pi;      % frequency, rad/s
E = 200e9;                   % elastic modulus, N/m^2
d = 10e-3;                   % diameter, m
L = 125e-3;                  % length, m
I = pi*d^4/64;              % 2nd moment of area, m^4
rho = 7800;                  % density, kg/m^3
A = pi*d^2/4;               % cross sectional area, m^2
eta = 0.01;                  % solid damping factor
EI = E*I*(1+1i*eta);        % complex stiffness, N-m^2
lambda = (w.^2*rho*A/EI).^0.25;
c1 = cos(lambda*L).*sinh(lambda*L) - sin(lambda*L).*cosh(lambda*L);
c2 = sin(lambda*L).*sinh(lambda*L);
c3 = sin(lambda*L) - sinh(lambda*L);
c4 = cos(lambda*L) - cosh(lambda*L);
c5 = cos(lambda*L).*sinh(lambda*L) + sin(lambda*L).*cosh(lambda*L);
c6 = sin(lambda*L) + sinh(lambda*L);
c7 = EI*(cos(lambda*L).*cosh(lambda*L)-1);
c8 = EI*(cos(lambda*L).*cosh(lambda*L)+1);

h11 = -c1./(lambda.^3.*c7);
l11 = c2./(lambda.^2.*c7);
n11 = l11;
p11 = c5./(lambda.*c7);

h2a2a = -c1./(lambda.^3.*c7);
l2a2a = -c2./(lambda.^2.*c7);
n2a2a = l2a2a;
p2a2a = c5./(lambda.*c7);

h12a = c3./(lambda.^3.*c7);
l12a = -c4./(lambda.^2.*c7);
n12a = c4./(lambda.^2.*c7);
p12a = c6./(lambda.*c7);

h2a1 = h12a;
l2a1 = n12a;
n2a1 = l12a;
p2a1 = p12a;

% Define wall receptances
h2b2b = zeros(1, length(w));
l2b2b = zeros(1, length(w));
n2b2b = zeros(1, length(w));
p2b2b = zeros(1, length(w));

% Calculate assembly receptances
for cnt = 1:length(w)
```

```
    % Define generalized receptance matrices
    % Free-free cylinder
    R11 = [h11(cnt) l11(cnt); n11(cnt) p11(cnt)];
    R12a = [h12a(cnt) l12a(cnt); n12a(cnt) p12a(cnt)];
    R2a2a = [h2a2a(cnt) l2a2a(cnt); n2a2a(cnt) p2a2a(cnt)];
    R2a1 = [h2a1(cnt) l2a1(cnt); n2a1(cnt) p2a1(cnt)];

    % Rigid wall
    R2b2b = [h2b2b(cnt) l2b2b(cnt); n2b2b(cnt) p2b2b(cnt)];

    % Generalized assembly receptance matrix
    G11 = R11 - R12a/(R2a2a + R2b2b)*R2a1;

    % Individual terms in G11
    H11(cnt) = G11(1,1);
    L11(cnt) = G11(1,2);
    N11(cnt) = G11(2,1);
    P11(cnt) = G11(2,2);
end

% Define fixed-free cylinder receptances
H11cf = -c1./(lambda.^3.*c8);
L11cf = c2./(lambda.^2.*c8);
N11cf = L11cf;
P11cf = c5./(lambda.*c8);

figure(1)
subplot(211)
plot(w/2/pi, real(h11), 'k')
ylim([-1e-5 1e-5])
set(gca,'FontSize', 14)
ylabel('Real (m/N)')
subplot(212)
plot(w/2/pi, imag(h11), 'k')
ylim([-1.8e-5 1.8e-6])
set(gca,'FontSize', 14)
xlabel('Frequency (Hz)')
ylabel('Imag (m/N)')

figure(2)
subplot(211)
plot(w/2/pi, real(H11), 'k', w/2/pi, real(H11cf), 'r:')
axis([300 700 -3.75e-4 3.75e-4])
set(gca,'FontSize', 14)
ylabel('Real (m/N)')
subplot(212)
plot(w/2/pi, imag(H11), 'k', w/2/pi, imag(H11cf), 'r:')
axis([300 700 -7e-4 7e-5])
set(gca,'FontSize', 14)
xlabel('Frequency (Hz)')
ylabel('Imag (m/N)')

figure(3)
semilogy(w/2/pi, abs(H11), 'k', w/2/pi, abs(H11cf), 'r:')
ylim([3e-9 1e-3])
```

```
set(gca,'FontSize', 14)
xlabel('Frequency (Hz)')
ylabel('Magnitude (m/N)')

figure(4)
subplot(211)
plot(w/2/pi, real(P11), 'k', w/2/pi, real(P11cf), 'r:')
axis([300 700 -0.045 0.045])
set(gca,'FontSize', 14)
ylabel('Real (rad/(N-m))')
subplot(212)
plot(w/2/pi, imag(P11), 'k', w/2/pi, imag(P11cf), 'r:')
axis([300 700 -0.09 0.009])
set(gca,'FontSize', 14)
xlabel('Frequency (Hz)')
ylabel('Imag (rad/(N-m))')
```

9.7.2 Free-Free Beam Coupled to Fixed-Free Beam

Let's now consider the case depicted in Fig. 9.15. A 10 mm diameter by 100 mm long steel cylinder (free-free boundary conditions) is to be rigidly coupled to a fixed-free 50 by 50 by 200 mm long steel prismatic beam. The steel elastic modulus, density, and solid damping factor are 200 GPa, 7,800 kg/m^3, and 0.01, respectively. (Again, we selected the solid damping value to be artificially high for display purposes.) The analysis is the same as described in Sect. 9.7.1 except that the R_{2b2b} receptances are no longer zero. They are now defined, as shown in Table 8.3, for a fixed-free beam, $h_{2b2b} = \frac{-c_1}{\lambda^3 c_8}$, $l_{2b2b} = n_{2b2b} = \frac{c_2}{\lambda^2 c_8}$, and $p_{2b2b} = \frac{c_5}{\lambda c_8}$. We will again use a frequency range of 5,000 Hz with a resolution of 0.1 Hz to calculate λ. The variables A and I are defined in Eqs. 9.73 and 9.74 for the square prismatic beam, where s is the side length of 50 mm. The displacement-to-force direct receptance for the free-free cylinder, h_{11}, is shown in Fig. 9.24 (solid line). The fixed-free square beam tip receptance, h_{2b2b}, is also displayed (dotted line). We see a first bending natural frequency of 4,507.6 Hz for the free-free beam. The fixed-free beam has a first bending frequency of 1,022.5 Hz.

$$A = s^2 \tag{9.73}$$

$$I = \frac{s^4}{12} \tag{9.74}$$

The application of Eq. 9.65 to this scenario using the code provided in MATLAB® MOJO 9.3 gives Fig. 9.25, which shows H_{11} for the assembly. We see two modes within the 1,500 Hz frequency range: one at 1,045.2 Hz, near the original fixed-free response and a second more flexible mode at 680.9 Hz due to the now coupled cylinder. Because the prismatic beam is much stiffer than the cylinder, it appears to

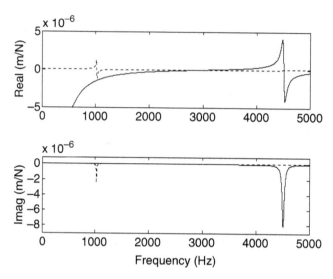

Fig. 9.24 Free-free receptance, h_{11}, for 10 mm diameter by 100 mm long steel cylinder (*solid line*) and fixed-free receptance, h_{2b2b}, for 50 mm square by 200 mm long steel prismatic beam (*dotted line*)

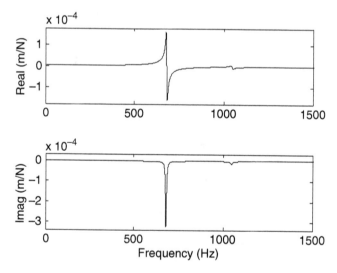

Fig. 9.25 Assembly displacement-to-force tip receptance H_{11} for the rigidly coupled cylinder and prismatic beam shown in Fig. 9.15

serve as a nearly rigid support for the cylinder. This may lead us to believe that approximating the assembly as a cylinder clamped to a wall is adequate. However, let's investigate what happens if we modify the prismatic beam to reduce its first bending frequency to a value near the fixed-free cylinder's first bending frequency.

MATLAB® MOJO 9.3
```
% matlab_mojo_9_3.m

clc
clear all
close all

% Define free-free cylinder receptances
w = (1:0.1:5000)*2*pi;        % frequency, rad/s
E = 200e9;                    % elastic modulus, N/m^2
d = 10e-3;                    % diameter, m
L = 100e-3;                   % length, m
I = pi*d^4/64;               % 2nd moment of area, m^4
rho = 7800;                   % density, kg/m^3
A = pi*d^2/4;                % cross sectional area, m^2
eta = 0.01;                   % solid damping factor
EI = E*I*(1+1i*eta);         % complex stiffness, N-m^2
lambda = (w.^2*rho*A/EI).^0.25;
c1 = cos(lambda*L).*sinh(lambda*L) - sin(lambda*L).*cosh(lambda*L);
c2 = sin(lambda*L).*sinh(lambda*L);
c3 = sin(lambda*L) - sinh(lambda*L);
c4 = cos(lambda*L) - cosh(lambda*L);
c5 = cos(lambda*L).*sinh(lambda*L) + sin(lambda*L).*cosh(lambda*L);
c6 = sin(lambda*L) + sinh(lambda*L);
c7 = EI*(cos(lambda*L).*cosh(lambda*L)-1);
c8 = EI*(cos(lambda*L).*cosh(lambda*L)+1);


h11 = -c1./(lambda.^3.*c7);
l11 = c2./(lambda.^2.*c7);
n11 = l11;
p11 = c5./(lambda.*c7);


h2a2a = -c1./(lambda.^3.*c7);
l2a2a = -c2./(lambda.^2.*c7);
n2a2a = l2a2a;
p2a2a = c5./(lambda.*c7);


h12a = c3./(lambda.^3.*c7);
l12a = -c4./(lambda.^2.*c7);
n12a = c4./(lambda.^2.*c7);
p12a = c6./(lambda.*c7);


h2a1 = h12a;
l2a1 = n12a;
n2a1 = l12a;
p2a1 = p12a;


% Define fixed-free prismatic beam receptances
E = 200e9;                    % elastic modulus, N/m^2
s = 50e-3;                    % square side, m
L = 200e-3;                   % length, m
I = s^4/12;                  % 2nd moment of area, m^4
rho = 7800;                   % density, kg/m^3
A = s^2;                     % cross sectional area, m^2
```

```
eta = 0.01;                         % solid damping factor
EI = E*I*(1+1i*eta);                % complex stiffness, N-m^2
lambda = (w.^2*rho*A/EI).^0.25;
c1 = cos(lambda*L).*sinh(lambda*L) - sin(lambda*L).*cosh(lambda*L);
c2 = sin(lambda*L).*sinh(lambda*L);
c3 = sin(lambda*L) - sinh(lambda*L);
c4 = cos(lambda*L) - cosh(lambda*L);
c5 = cos(lambda*L).*sinh(lambda*L) + sin(lambda*L).*cosh(lambda*L);
c6 = sin(lambda*L) + sinh(lambda*L);
c7 = EI*(cos(lambda*L).*cosh(lambda*L)-1);
c8 = EI*(cos(lambda*L).*cosh(lambda*L)+1);
h2b2b = -c1./(lambda.^3.*c8);
l2b2b = c2./(lambda.^2.*c8);
n2b2b = l2b2b;
p2b2b = c5./(lambda.*c8);


% Calculate assembly receptances
for cnt = 1:length(w)
    % Define generalized receptance matrices
    % Free-free cylinder
    R11 = [h11(cnt) l11(cnt); n11(cnt) p11(cnt)];
    R12a = [h12a(cnt) l12a(cnt); n12a(cnt) p12a(cnt)];
    R2a2a = [h2a2a(cnt) l2a2a(cnt); n2a2a(cnt) p2a2a(cnt)];
    R2a1 = [h2a1(cnt) l2a1(cnt); n2a1(cnt) p2a1(cnt)];

    % Prismatic beam
    R2b2b = [h2b2b(cnt) l2b2b(cnt); n2b2b(cnt) p2b2b(cnt)];

    % Generalized assembly receptance matrix
    G11 = R11 - R12a/(R2a2a + R2b2b)*R2a1;

    % Individual terms in G11
    H11(cnt) = G11(1,1);
    L11(cnt) = G11(1,2);
    N11(cnt) = G11(2,1);
    P11(cnt) = G11(2,2);
end

figure(1)
subplot(211)
plot(w/2/pi, real(h11), 'k', w/2/pi, real(h2b2b), 'k:')
ylim([-5e-6 5e-6])
set(gca,'FontSize', 14)
ylabel('Real (m/N)')
subplot(212)
plot(w/2/pi, imag(h11), 'k', w/2/pi, imag(h2b2b), 'k:')
ylim([-9e-6 9e-7])
set(gca,'FontSize', 14)
xlabel('Frequency (Hz)')
ylabel('Imag (m/N)')

figure(2)
subplot(211)
plot(w/2/pi, real(H11), 'k')
axis([0 1500 -1.8e-4 1.8e-4])
set(gca,'FontSize', 14)
ylabel('Real (m/N)')
subplot(212)
plot(w/2/pi, imag(H11), 'k')
axis([0 1500 -3.4e-4 3.4e-5])
set(gca,'FontSize', 14)
xlabel('Frequency (Hz)')
ylabel('Imag (m/N)')
```

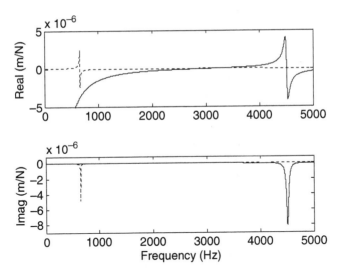

Fig. 9.26 Free-free receptance, h_{11}, for 10 mm diameter by 100 mm long steel cylinder (*solid line*) and clamped-free receptance, h_{2b2b}, for 50 mm square by 250 mm long steel prismatic beam (*dotted line*)

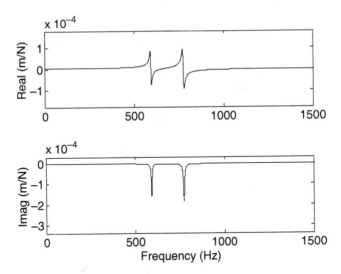

Fig. 9.27 The displacement-to-force tip receptance, H_{11}, for rigid coupling of the 10 mm diameter by 100 mm long cylinder to the 50 mm square by 250 mm long prismatic beam is displayed

Figure 9.26 displays h_{11} for the free-free cylinder (solid line), as well as h_{2b2b} for a longer (250 mm) fixed-free prismatic beam (dotted line). The cylinder's first bending natural frequency remains at 4,507.6 Hz for the free-free boundary conditions. However, the first bending frequency for the extended fixed-free beam is reduced to 654.4 Hz. Figure 9.27 shows H_{11} for the cylinder rigidly coupled to the 50 mm square by 250 mm long prismatic beam. The response is now quite different than the assembly receptance shown in Fig. 9.25 for the 200 mm long prismatic beam. Even though the cylinder is coupled to a more flexible base (i.e., a longer fixed-free beam), the assembly response has a smaller peak magnitude. The minimum imaginary value for the new assembly is -1.865×10^{-4} m/N, while the corresponding value for the shorter (and stiffer) prismatic beam assembly was -3.222×10^{-4} m/N; this represents a 42% compliance[3] reduction. The compliance reduction, or, equivalently, the stiffness increase, is due to interaction between the two beams in a manner analogous to the dynamic absorber we discussed in Sect. 5.4. When the fixed-free prismatic beam's natural frequency is near the coupled cylinder's natural frequency, some energy is able to "pass through" the cylinder and excite the stiffer base. The result is that the energy is more equally partitioned between the two modes and the assembly response appears stiffer (Duncan et al. 2005). An electrical equivalent is the impedance matching strategy used at cable connections. For example, it is common to use 50 Ω terminations at all connections to encourage signal transmission and avoid reflection.

9.7.3   Comparison Between Model and BEP Measurement

In Sect. 8.6.1, we compared an Euler–Bernoulli fixed-free beam prediction to a measurement completed on the BEP. The impact test carried out on the BEP is described in Sect. 7.4, where the 12.7 mm diameter cantilevered steel rod was extended 130 mm beyond the base (see Fig. 7.20). Now let's model the extended portion of the steel rod as a free-free beam and couple it to a rigid support (wall) to predict the fixed-free response of the assembly. We can then compare the prediction to the measurement result (see Fig. 7.21).

For the 12.7 mm diameter rod, the second moment of area is:

$$I = \frac{\pi d^4}{64} = \frac{\pi (0.0127)^4}{64} = 1.277 \times 10^{-9} \, \text{m}^4$$

and the cross-sectional area, A, is:

$$A = \frac{\pi d^2}{4} = \frac{\pi (0.0127)^2}{4} = 1.267 \times 10^{-4} \, \text{m}^2.$$

For the steel rod, let's use $\rho = 7{,}800 \, \text{kg/m}^3$, $E = 200$ GPa, and $\eta = 0.002$. The coupling proceeds as detailed in MATLAB® MOJO 9.2, but with the new rod dimensions. Figure 9.28 shows H_{11} for the rod rigidly coupled to the wall. We

---

[3] Compliance is the inverse of stiffness.

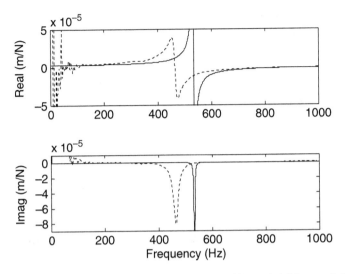

Fig. 9.28 Comparison between the free-free boundary condition rod rigidly coupled to a wall (*solid line*) and the BEP measurement (*dotted line*)

observe that the fixed-free prediction has a natural frequency that is too high and the damping is too low. As discussed in Sect. 8.6.1, the split-clamp used to secure the rod in the BEP holder does not provide an ideal fixed boundary condition. In order to incorporate the flexibility and damping in the clamping interface into the receptance coupling model, we can modify Eq. 9.65 using the complex stiffness matrix defined in Sect. 9.6:

$$[\tilde{k}'] = \begin{bmatrix} k_{xf} + i\omega c_{xf} & k_{\theta f} + i\omega c_{\theta f} \\ k_{xm} + i\omega c_{xm} & k_{\theta m} + i\omega c_{\theta m} \end{bmatrix}.$$

The modified Eq. 9.65 used to predict H_{11} is:

$$G_{11} = R_{11} - R_{12a}\left(R_{2a2a} + R_{2b2b} + [\tilde{k}']^{-1}\right)^{-1}R_{2a1} \qquad (9.75)$$

If the complex stiffness matrix is specified to be:

$$[\tilde{k}'] = \begin{bmatrix} 3 \times 10^6 + i\omega 70 & 3.5 \times 10^5 + i\omega 20 \\ 3.5 \times 10^5 + i\omega 20 & 2 \times 10^3 + i\omega 5 \end{bmatrix},$$

the result displayed in Fig. 9.29 is obtained. We see that the additional flexibility and damping at the connection between the free-free rod and wall improves the agreement between the model and measurement. The code used to carry out this receptance coupling exercise is provided in MATLAB® MOJO 9.4.

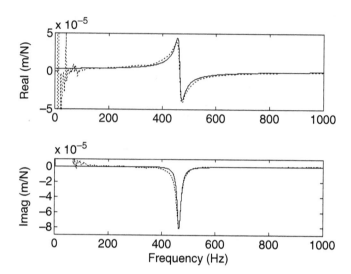

Fig. 9.29 Comparison between the free-free boundary condition rod coupled to a wall using a flexible-damped connection (*solid line*) and the BEP measurement (*dotted line*)

MATLAB® MOJO 9.4

```
% matlab_mojo_9_4.m

clc
clear all
close all

% Define free-free cylinder receptances
w = (1:0.1:1000)*2*pi;      % frequency, rad/s
E = 200e9;                  % elastic modulus, N/m^2
d = 12.7e-3;                % diameter, m
L = 130e-3;                 % length, m
I = pi*d^4/64;              % 2nd moment of area, m^4
rho = 7800;                 % density, kg/m^3
A = pi*d^2/4;               % cross sectional area, m^2
eta = 0.002;                % solid damping factor
EI = E*I*(1+1i*eta);        % complex stiffness, N-m^2
lambda = (w.^2*rho*A/EI).^0.25;
c1 = cos(lambda*L).*sinh(lambda*L) - sin(lambda*L).*cosh(lambda*L);
c2 = sin(lambda*L).*sinh(lambda*L);
c3 = sin(lambda*L) - sinh(lambda*L);
c4 = cos(lambda*L) - cosh(lambda*L);
c5 = cos(lambda*L).*sinh(lambda*L) + sin(lambda*L).*cosh(lambda*L);
c6 = sin(lambda*L) + sinh(lambda*L);
c7 = EI*(cos(lambda*L).*cosh(lambda*L)-1);
c8 = EI*(cos(lambda*L).*cosh(lambda*L)+1);

h11 = -c1./(lambda.^3.*c7);
l11 = c2./(lambda.^2.*c7);
n11 = l11;
p11 = c5./(lambda.*c7);
```

```
h2a2a = -c1./(lambda.^3.*c7);
l2a2a = -c2./(lambda.^2.*c7);
n2a2a = l2a2a;
p2a2a = c5./(lambda.*c7);

h12a = c3./(lambda.^3.*c7);
l12a = -c4./(lambda.^2.*c7);
n12a = c4./(lambda.^2.*c7);
p12a = c6./(lambda.*c7);

h2a1 = h12a;
l2a1 = n12a;
n2a1 = l12a;
p2a1 = p12a;

% Define wall receptances
h2b2b = zeros(1, length(w));
l2b2b = zeros(1, length(w));
n2b2b = zeros(1, length(w));
p2b2b = zeros(1, length(w));

% Calculate assembly receptances
for cnt = 1:length(w)
    % Define generalized receptance matrices
    % Free-free cylinder
    R11 = [h11(cnt) l11(cnt); n11(cnt) p11(cnt)];
    R12a = [h12a(cnt) l12a(cnt); n12a(cnt) p12a(cnt)];
    R2a2a = [h2a2a(cnt) l2a2a(cnt); n2a2a(cnt) p2a2a(cnt)];
    R2a1 = [h2a1(cnt) l2a1(cnt); n2a1(cnt) p2a1(cnt)];

    % Rigid wall
    R2b2b = [h2b2b(cnt) l2b2b(cnt); n2b2b(cnt) p2b2b(cnt)];

    % Complex connection stiffness
    k = [3e6 + 1i*w(cnt)*70 3.5e5 + 1i*w(cnt)*20; 3.5e5 + 1i*w(cnt)*20 2e3 +
1i*w(cnt)*5];
    invk = inv(k);

    % Generalized assembly receptance matrix
    G11 = R11 - R12a/(R2a2a + R2b2b + invk)*R2a1;

    % Individual terms in G11
    H11(cnt) = G11(1,1);
    L11(cnt) = G11(1,2);
    N11(cnt) = G11(2,1);
    P11(cnt) = G11(2,2);
end

figure(1)
subplot(211)
plot(w/2/pi, real(H11), 'k')
axis([0 1000 -5e-5 5e-5])
set(gca,'FontSize', 14)
ylabel('Real (m/N)')
subplot(212)
plot(w/2/pi, imag(H11), 'k')
axis([0 1000 -9e-5 1e-5])
set(gca,'FontSize', 14)
xlabel('Frequency (Hz)')
ylabel('Imag (m/N)')
```

IN A NUTSHELL   The ability to mathematically predict dynamic system behavior, to experimentally measure dynamic system behavior, and to combine measurements and computations is a powerful tool. The objective of this book is the description and demonstration of these techniques and illustration of their wide application range.

Chapter Summary

- The vibrating behavior of structures can be described using discrete models, continuous beam models, or measurements.
- The responses of individual components, or substructures, can be combined using receptance coupling to predict the assembly's response.
- The coupling between components can be rigid, flexible, or flexible with damping.
- The receptance coupling approach can incorporate not only transverse deflections due to forces, but also rotations and bending couples.

Acknowledgments The authors gratefully acknowledge the contributions of Dr. T. Burns, National Institute of Standards and Technology, in developing the Sect. 9.4 damping analysis.

The authors gratefully acknowledge contributions from Dr. T. Burns, National Institute of Standards and Technology, Dr. M. Davies, University of North Carolina at Charlotte, and Dr. G.S. Duncan, Valparaiso University, to the development of the Sect. 9.6 receptance coupling analysis.

Exercises

1. Determine the direct frequency response function, $\frac{X_2}{F_2}$, for the two degree of freedom system shown in Fig. P9.1 using receptance coupling. Express your final result as a function of m, c, k, and the excitation frequency, ω. You may assume a harmonic forcing function, $F_2 e^{i\omega t}$, is applied to coordinate X_2.

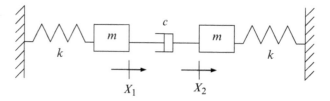

Fig. P9.1 Two degree of freedom assembly

2. Determine the direct frequency-response function, $\frac{X_1}{F_1}$, for the two degree of freedom system shown in Fig. P9.2 using receptance coupling. Express your final result as a function of m, c, k, and the excitation frequency, ω. You may assume a harmonic forcing function, $F_1 e^{i\omega t}$, is applied to coordinate X_1.

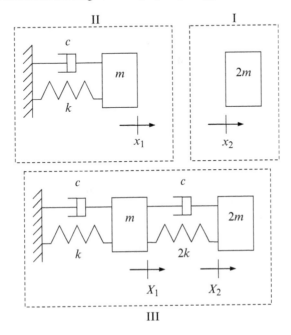

Fig. P9.2 Flexible damped coupling of mass (I) to spring–mass–damper (II) to form the two degree of freedom assembly III

3. Use receptance coupling to rigidly join two free-free beams and find the free-free assembly's displacement-to-force tip receptance. Both steel cylinders are described by the following parameters: 12.7 mm diameter, 100 mm length, 200 GPa elastic modulus, and 7,800 kg/m³ density. Assume a solid damping factor of 0.0015. Once you have determined the assembly response, verify your result against the displacement-to-force tip receptance for a 12.7 mm diameter, 200 mm long free-free steel cylinder with the same material properties. Select a frequency range that encompasses the first three bending modes and display your results as the magnitude (in m/N) vs. frequency (in Hz) using a semi-logarithmic scale.

4. Plot the displacement-to-force tip receptance for a sintered carbide cylinder with free-free boundary conditions. The beam is described by the following parameters: 19 mm diameter, 150 mm length, 550 GPa elastic modulus, and 15,000 kg/m³ density. Assume a solid damping factor of 0.002. Select a frequency range that encompasses the first three bending modes and display

your results as magnitude (m/N) versus frequency (Hz) in a semi-logarithmic format.

5. Determine the fixed-free displacement-to-force tip receptance for a sintered carbide cylinder by coupling the free-free receptances to a rigid wall (with zero receptances). The beam is described by the following parameters: 19 mm diameter, 150 mm length, 550 GPa elastic modulus, and 15,000 kg/m^3 density. Assume a solid damping factor of 0.002. Select a frequency range that encompasses the first two bending modes and display your results as magnitude (m/N) versus frequency (Hz) in a semi-logarithmic format. Verify your result by comparing it to the displacement-to-force tip receptance for a fixed-free beam with the same dimensions and material properties.

6. For a rigid coupling between two component coordinates x_{1a} and x_{1b}, the compatibility condition is _____.

7. For a flexible coupling (spring stiffness k) between two component coordinates x_{1a} and x_{1b}, the compatibility condition is _____. An external force is applied to the assembly at coordinate X_{1a}.

8. For a flexible-damped coupling (spring stiffness k and damping coefficient c) between two component coordinates x_{1a} and x_{1b}, the compatibility condition is _____. An external force is applied to the assembly at coordinate X_{1a}.

9. What are the units for the rotation-to-couple receptance, p_{ij}, used to describe the transverse vibration of beams?

10. What are the (identical) units for the displacement-to-couple, l_{ij}, and rotation-to-force, n_{ij}, receptances used to describe the transverse vibration of beams?

References

Bishop R, Johnson D (1960) The mechanics of vibration. Cambridge University Press, Cambridge

Burns T, Schmitz T (2005) A study of linear joint and tool models in spindle-holder-tool receptance coupling. In: Proceedings of 2005 American Society of Mechanical Engineers International Design Engineering Technical Conferences and Computers and Information in Engineering Conference, DETC2005-85275, Long Beach

Burns T, Schmitz T (2004) Receptance coupling study of tool-length dependent dynamic absorber effect. In: Proceedings of American Society of Mechanical Engineers International Mechanical Engineering Congress and Exposition, IMECE2004-60081, Anaheim

Duncan GS, Tummond M, Schmitz T (2005) An investigation of the dynamic absorber effect in high-speed machining. Int J Mach Tool Manu 45:497–507

Park S, Altintas Y, Movahhedy M (2003) Receptance coupling for end mills. Int J Mach Tool Manu 43:889–896

Schmitz T, Powell K, Won D, Duncan GS, Sawyer WG, Ziegert J (2007) Shrink fit tool holder connection stiffness/damping modeling for frequency response prediction in milling. Int J Mach Tool Manu 47(9):1368–1380

Weaver W Jr, Timoshenko S, Young D (1990) Vibration problems in engineering, 5th edn. Wiley, New York

Appendix A
Beam Experimental Platform

The beam experimental platform (BEP) is used throughout the text to demonstrate various concepts in mechanical vibrations. The BEP is composed of a base plate, a holder, and a rod; see Fig. A1. The base plate and holder are aluminum and can be machined using the dimensions provided in Fig. A2. The rod is a 12.7 mm diameter, 152.5 mm long high-speed steel tool blank.

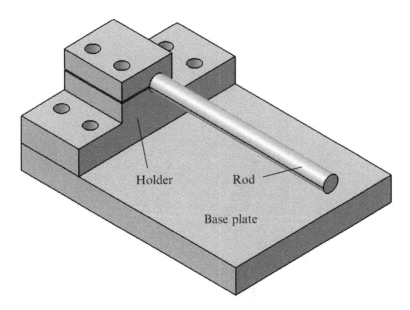

Fig. A1 BEP components

T.L. Schmitz and K.S. Smith, *Mechanical Vibrations: Modeling and Measurement*, DOI 10.1007/978-1-4614-0460-6, © Springer Science+Business Media, LLC 2012

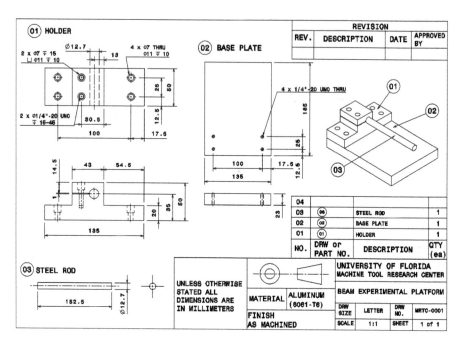

Fig. A2 BEP dimensions

Appendix B
Orthogonality of Eigenvectors

The orthogonality of eigenvectors with respect to the system mass and stiffness matrices is the basis for modal analysis. In general, we can say that two vectors are perpendicular if their scalar, or dot, product is zero. Consider the two vectors:

$$[U] = \left\{ \begin{matrix} u_{11} \\ u_{21} \end{matrix} \right\} \ and \ [V] = \left\{ \begin{matrix} v_{11} \\ v_{21} \end{matrix} \right\}. \tag{B1}$$

Their dot product is:

$$[U] \bullet [V] = [U]^T [V] = \{ u_{11} \quad u_{21} \} \left\{ \begin{matrix} v_{11} \\ v_{21} \end{matrix} \right\} = u_{11} \cdot v_{11} + u_{21} \cdot v_{21}. \tag{B2}$$

This product is zero if the vectors are perpendicular. Orthogonality can be considered a generalization of the concept of perpendicularity.

From Chap. 4, we have seen that we can write the matrix form of the system equations of motion $([m]s^2 + [k])\{\vec{X}\}e^{st} = \{0\}$ if we assume harmonic vibration. We used the characteristic equation, $\left| [m]s^2 + [k] \right| = 0$, to find the eigenvalues, s_1^2 and s_2^2. We then substituted the eigenvalues into either of the linearly dependent equations of motion to find the eigenvectors, or mode shapes. Using $s_1^2 = -\omega_{n1}^2$, we can write:

$$(-[m]\omega_{n1}^2 + [k])\{\psi_1\} = \{0\}, \tag{B3}$$

where ψ_1 is the corresponding mode shape. Eq. B3 can be expanded to:

$$- \omega_{n1}^2 [m]\{\psi_1\} + [k]\{\psi_1\} = \{0\}. \tag{B4}$$

Premultiplying Eq. B4 by the transpose of the second mode shape ψ_2, which corresponds to vibration at ω_{n2}, yields:

$$- \omega_{n1}{}^{2}\{\psi_{2}\}^{T}[m]\{\psi_{1}\} + \{\psi_{2}\}^{T}[k]\{\psi_{1}\} = 0. \tag{B5}$$

Performing the transpose operation on Eq. B5 gives:

$$- \omega_{n1}{}^{2}\{\psi_{1}\}^{T}[m]\{\psi_{2}\} + \{\psi_{1}\}^{T}[k]\{\psi_{2}\} = 0, \tag{B6}$$

where the transpose properties $([A][B])^{T} = [B]^{T}[A]^{T}$ and $\left([A]^{T}\right)^{T} = [A]$ (using matrices of appropriate dimensions) have been applied.
Completing the same operations using $s_{2}{}^{2} = -\omega_{n2}{}^{2}$ gives:

$$- \omega_{n2}{}^{2}\{\psi_{1}\}^{T}[m]\{\psi_{2}\} + \{\psi_{1}\}^{T}[k]\{\psi_{2}\} = 0. \tag{B7}$$

Taking the difference of Eqs. B6 and B7 yields:

$$\left(\omega_{n2}{}^{2} - \omega_{n1}{}^{2}\right)\{\psi_{1}\}^{T}[m]\{\psi_{2}\} = 0. \tag{B8}$$

Provided $\omega_{n2}{}^{2} \neq \omega_{n1}{}^{2}$, then $\{\psi_{1}\}^{T}[m]\{\psi_{2}\} = 0$. Substituting this result into either Eq. B6 or Eq. B7 gives $\{\psi_{1}\}^{T}[k]\{\psi_{2}\} = 0$. Collecting these results, we obtain the orthogonality conditions shown in Eqs. B9 through B12.

$$\begin{aligned} \{\psi_{1}\}^{T}[m]\{\psi_{2}\} = 0 \\ \{\psi_{2}\}^{T}[m]\{\psi_{1}\} = 0 \end{aligned} \tag{B9}$$

$$\begin{aligned} \{\psi_{1}\}^{T}[m]\{\psi_{1}\} = m_{q1} \\ \{\psi_{2}\}^{T}[m]\{\psi_{2}\} = m_{q2} \end{aligned} \tag{B10}$$

The products in Eq. B10 are not necessarily zero.

$$\begin{aligned} \{\psi_{1}\}^{T}[k]\{\psi_{2}\} = 0 \\ \{\psi_{2}\}^{T}[k]\{\psi_{1}\} = 0 \end{aligned} \tag{B11}$$

$$\begin{aligned} \{\psi_{1}\}^{T}[k]\{\psi_{1}\} = k_{q1} \\ \{\psi_{2}\}^{T}[k]\{\psi_{2}\} = k_{q2} \end{aligned} \tag{B12}$$

The products in Eq. B12 are not necessarily zero. Using the modal matrix, $[P] = [\psi_{1} \quad \psi_{2}]$, and the orthogonality conditions we obtain the diagonalized modal mass and stiffness matrices:

$$[P]^{T}[m][P] = \begin{bmatrix} \{\psi_{1}\}^{T}[m]\{\psi_{1}\} & \{\psi_{1}\}^{T}[m]\{\psi_{2}\} \\ \{\psi_{2}\}^{T}[m]\{\psi_{1}\} & \{\psi_{2}\}^{T}[m]\{\psi_{2}\} \end{bmatrix} = \begin{bmatrix} m_{q1} & 0 \\ 0 & m_{q2} \end{bmatrix} = [m_{q}] \tag{B13}$$

and

$$[P]^T[k][P] = \begin{bmatrix} \{\psi_1\}^T[k]\{\psi_1\} & \{\psi_1\}^T[k]\{\psi_2\} \\ \{\psi_2\}^T[k]\{\psi_1\} & \{\psi_2\}^T[k]\{\psi_2\} \end{bmatrix} = \begin{bmatrix} k_{q1} & 0 \\ 0 & k_{q2} \end{bmatrix} = [k_q]. \quad (B14)$$

These diagonal modal mass and stiffness matrices uncouple the equations of motion and enable the solution of independent single degree of freedom systems in modal coordinates. The individual modal contributions can then be transformed back into local (model) coordinates as discussed in Chap. 4.

Index